Library of
Davidson College

HISTORICAL STUDIES IN THE PHYSICAL SCIENCES

3

Editor
RUSSELL MCCORMMACH, *University of Pennsylvania*

Editorial Board
CLAUDE K. DEISCHER, *University of Pennsylvania*
JOHN L. HEILBRON, *University of California, Berkeley*
ARMIN HERMANN, *University of Stuttgart*
TETU HIROSIGE, *Nihon University, Tokyo*
GERALD HOLTON, *Harvard University*
MARTIN J. KLEIN, *Yale University*
HERBERT S. KLICKSTEIN, *Albert Einstein Medical Center, Philadelphia*
BORIS KUZNETSOV, *Institute for the History of Science, Moscow*
THOMAS S. KUHN, *Princeton University*
HENRY M. LEICESTER, *University of the Pacific*
JEROME R. RAVETZ, *University of Leeds*
NATHAN REINGOLD, *Smithsonian Institution*
LÉON ROSENFELD, *Nordic Institute for Theoretical Atomic Physics, Copenhagen*
ROBERT E. SCHOFIELD, *Case Western Reserve University*
ROBERT SIEGFRIED, *University of Wisconsin*
ARNOLD THACKRAY, *University of Pennsylvania*

Note on Contributors

JOAN BROMBERG was at the Niels Bohr Institute in Copenhagen from September 1969 to January 1971 while on leave from the Social Sciences Department of the Polytechnic Institute of Brooklyn. She is presently Lecturer in the Department of the History and Philosophy of Science at the Hebrew University of Jerusalem. Her current research subject is the development of theoretical physics between the world wars.

PAUL FORMAN is Assistant Professor of History at the University of Rochester. Following his collaboration on the project *Sources for History of Quantum Physics*, 1961–1964, he has devoted his researches almost exclusively to the history of physics in the first third of this century.

P. M. HEIMANN lectures in history of science at the University of Cambridge, where he specializes in eighteenth- and nineteenth-century natural philosophy. He is currently engaged in a study of conceptions of nature in British thought in this period.

KARL HUFBAUER teaches the history of science and related subjects at the University of California at Irvine. He is presently working on a book-length study of the formation of the German chemical community in the eighteenth century. His general interest is the historical sociology of science, particularly specialty-oriented communities.

ROBERT E. KOHLER, JR., is assistant director of the Burndy Library. His current research concerns the application of chemistry to biological problems in the nineteenth and twentieth centuries.

BORIS KUZNETSOV is Professor and Director of Studies in the Institute for the History of Science of the Academy of Sciences of the U.S.S.R., Moscow. His writings are devoted to problems of philosophy, theoretical physics, and the history and economic effect of science.

NOTE ON CONTRIBUTORS

J. E. McGuire is Professor of History of Science at the University of Pittsburgh. He is currently finishing a book on Newton's philosophy of nature, and is in the process of editing some of Newton's early notebooks and related materials. He is also engaged in studies of eighteenth-century intellectual history.

J. B. Morrell is lecturer in history of science at the University of Bradford, England. His special research interest is science in the Scottish Enlightenment.

R. Steven Turner is a former graduate student of the Program in History and Philosophy of Science at Princeton University, now lecturer in the history of science at the University of New Brunswick. He is currently completing a dissertation on the Prussian universities in the first half of the nineteenth century with special reference to science teaching and research.

New Pathways of Science

1. For science there is no longer any money; the observatory has gone to complete rack and ruin. 2. The astronomy professor begs the government for funds, but in vain. 3. In uttermost desperation he resolves to take up astrology.

4. As astrologer he casts horoscopes for war profiteers. 5. His prophesies are soon in much demand; he makes lots of money. 6. Now the professor has the funds to renovate himself and the observatory; the newest and best instruments are procured.

Th. Th. Heine, "Neue Wege der Wissenschaft," *Simplicissimus*, 25 (26 January 1921), 595.

Edgar F. Smith Memorial Collection, University of Pennsylvania

Historical Studies
in the
Physical Sciences

RUSSELL MCCORMMACH, *Editor*

Third Annual Volume 1971

UNIVERSITY OF PENNSYLVANIA PRESS • PHILADELPHIA

Copyright © 1971 by the UNIVERSITY OF PENNSYLVANIA PRESS, INC.
All rights reserved
Library of Congress Catalog Card Number: 77-75220

NOTICE TO CONTRIBUTORS

Historical Studies in the Physical Sciences, an annual publication issued by the University of Pennsylvania Press, is devoted to articles on the history of the physical sciences from the eighteenth century to the present. The modern period has been selected since it holds especially challenging and timely problems, problems that so far have been little explored. An effort is made to bring together articles that expose new directions and methods of research in the history of the modern physical sciences. Consideration is given to the professional communities of physical scientists, to the internal developments and interrelationships of the physical sciences, to the relations of the physical to the biological and social sciences, and to the institutional settings and the cultural and social contexts of the physical sciences. Historiographic articles, essay reviews, and survey articles on the current state of scholarship are welcome in addition to the more customary types of articles.

All manuscripts should be accompanied by an additional carbon- or photocopy. Manuscripts should be typewritten and double-spaced on $8\frac{1}{2}'' \times 11''$ bond paper; wide margins should be allowed. No limit has been set on the length of manuscripts. Articles may include illustrations; these may be either glossy prints or directly reproducible line drawings. Articles may be submitted in foreign languages; if accepted, they will be published in English translation. Footnotes are to be double-spaced, numbered sequentially, and collected at the end of the manuscript. Contributors are referred to the *MLA Style Sheet* for detailed instructions on documentation and other stylistic matters. (*Historical Studies* departs from the MLA rules in setting book and journal volume numbers in italicized Arabic rather than Roman numerals.) All correspondence concerning editorial matters should be addressed to Russell McCormmach, Department of History and Sociology of Science, University of Pennsylvania, Philadelphia, Penna. 19104.

One hundred and fifty free reprints accompany each article.

Historical Studies in the Physical Sciences incorporates *Chymia,* the history of chemistry annual.

ISBN: 0-8122-7646-9
Printed in the United States of America

Contents

Note on Contributors ... ii

Editor's Foreword ... ix

PAUL FORMAN
Weimar Culture, Causality, and Quantum Theory, 1918–1927: Adaptation by German Physicists and Mathematicians to a Hostile Intellectual Environment ... 1

BORIS KUZNETSOV
Quantum-Relativistic Retrospection and the History of Classical Physics: Classical Rationalism and Nonclassical Science ... 117

R. STEVEN TURNER
The Growth of Professorial Research in Prussia, 1818 to 1848—Causes and Context ... 137

J. B. MORRELL
Individualism and the Structure of British Science in 1830 ... 183

KARL HUFBAUER
Social Support for Chemistry in Germany during the Eighteenth Century: How and Why Did It Change? ... 205

P. M. HEIMANN and J. E. MC GUIRE
Newtonian Forces and Lockean Powers: Concepts of Matter in Eighteenth-Century Thought ... 233

JOAN BROMBERG
The Impact of the Neutron: Bohr and Heisenberg ... 307

ROBERT E. KOHLER, JR.
The Origin of G. N. Lewis's Theory of the Shared Pair Bond ... 343

Editor's Foreword

A marked feature of this third volume of *Historical Studies in the Physical Sciences* is the large scope of the themes of the contributions. The themes include the broad impact of philosophy, government, institutions, and patronage on the development of the modern physical sciences. I am glad to have this periodical associated with contributions of this nature, for they serve to identify historical problems, to exhibit the power of methods, and to establish connections between the history of science and other branches of history.

In the foreword to the second volume of this series, I discussed some directions for enlarging our historical understanding of modern physics. There I referred my remarks to an individual physicist, to Einstein. I will continue my remarks in this foreword; and in connection with the emphasis of the present volume I will refer them to the context in which an individual physicist works, to his discipline. The discipline provides a useful historical perspective, one which has been little explored by historians of physics. By tracing the growth of the physics discipline to the time of Einstein, I will show how connections between physical ideas, methods, worldviews, culture, and technology arise naturally from this perspective.

By members of a scientific discipline I mean qualified scientists who cultivate a special domain of natural knowledge. The members are concerned with goals, standards, methods, and recognition in their domain of scholarship, with facilities for conducting and communicating research, and with the training and certification of the generation of specialists who will succeed them. The discipline prescribes the scientist's problems and methods, channels his personal ambition, rewards his achievements, and conditions his attitudes, values, and behavior. The discipline touches on every aspect of the scientist's work and career; the most pervasive features of scientific thought and the collective scientific life are bound up with the discipline organization of scientific scholarship.

I will mention some advantages I think historians of science may derive from a discipline perspective. The biographical historian of science may deepen his understanding of his subject's motivations by studying his subject's discipline; the values held by the members of a discipline define the immediate psychological world of the individual scientist and constitute a major source of his identity *qua* scientist. The intellectual historian of science and the social historian of science will jointly profit by recognizing that arguments over theories, methods, and worldviews define intradisciplinary groupings; once identified the groupings may be examined for their members' extrascientific associations, enhancing our understanding of the total context and meaning of scientific arguments. The historian of science who is dissatisfied with the traditional disjunctions of his specialty—social vs. intellectual history, external vs. internal history—will find the discipline a natural unit of study for relating the scientific to the nonscientific world; the prevailing institutions and culture affect the scientist's thought and career largely through the mediation of the discipline.

I will limit my remarks to German physics. A partial justification is that the leading characteristics of the modern physics discipline became clearly delineated first in Germany, serving as a model for discipline organization elsewhere. More to the point, a comparative discussion of the physics discipline in different national cultures is scarcely feasible in the space of an editorial foreword. I will further limit my remarks to three developments in German physics. They are broad and important developments, ones which have not been amply treated before; they illustrate the scope of problems a discipline historian confronts. The developments are: first, that of a new method of physical research tending to unify the members of the physics discipline; second, that of the parallel emergence of the subspecialty of experimental physics and precision technology; and, third, that of the emergence of the subspecialty of theoretical physics and its interaction with the larger intellectual culture. After discussing the three developments, I will show how they bear on a particular event in physics—Max Planck's introduction of quanta in 1900. Planck's intensive scholarly activity in 1900 was imbedded in a discipline matrix of collegial relations, societal and institutional arrangements, and worldview debates. The matrix will emerge from the three developments I am going to discuss. I will now turn to the first of these.

EDITOR'S FOREWORD

The early organizers of the German physics discipline were often organizers of more than one branch of science, often of natural science as a whole. In large part, physics arose as an unwitting product of uncoordinated programmatic goals and entrepreneurial scholarly activities, as an intersection of groupings of scientific interests shaped by the cultural and institutional forces responsible for specialization in German scholarship in general. Physics emerged comparatively late, as the residue of natural philosophy after its other branches had organized as separate specialties.

One of the developments that brought coherence to physics was a new understanding of method. Early in the nineteenth century, German mathematical physics, rational mechanics, and applied mathematics tended to be practiced by people trained in mathematics; and experimental physics tended to be practiced by those trained in chemistry. The two groups published in different places and worked in distinct, essentially noncommunicating sciences. Many early organizers of the German physics discipline such as Johann Christian Poggendorff and Heinrich Gustav Magnus strongly condemned the infusion of mathematical theory into physics. In their mood of reaction against the influence of nature philosophy in physical science they used their intellectual and institutional authority to try to constrain German physics within a purely experimental methodology. For them experimental physics was more than a subspecialty of physics; it was a whole specialty in itself, one closely allied to chemistry and in no need of mathematical guidance.

In the 1830's and 1840's the division between mathematicians and experimentalists became less sharp, due in part to the development of a method of mathematical physics that drew attention to the common purposes of the two groups. By far the most influential contributor to this development was the Königsberg physicist Franz Neumann. I do not intend to go into the details of Neumann's method beyond pointing out that its ultimate inspiration was Fourier's analytical heat theory and its natural language was partial differential equations. I want instead to concentrate on the method's significance for the coherence of experimental and mathematical physics. The main characteristic of Neumann's method of theory construction—as he exhibited it in his published research in optics and electrodynamics—is that it proceeded from purely observable properties of matter, from precisely the properties that experimentalists measured. With it Neumann circumvented empirically unproved starting assumptions, such as hypothetical atoms and their hypothetical properties; he was not, however, dogmatic about method and was prepared to use atomistic assumptions when he needed them. Helmholtz—one of the German physicists greatly influenced by Neumann—spoke of Neumann's method as having made mathematical physics a wholly *experimental* science, stripping it of its unwanted metaphysical or a prioristic associations. He correctly identified it as a discipline-uniting innovation in the most profound sense; for now there was only one physics, a physics totally controlled by experience, instead of two mutually indifferent or even antagonistic sciences. The mathematical and the experimental physicist were concerned with the same things, and neither could make progress without the other. The ideal, naturally, was for the mathematical and the experimental physicist to be united in the same person, as they were in Neumann and Helmholtz.

The chief agency through which Neumann's innovation in method permeated German physics was the discipline-oriented university seminar. From its eighteenth-century origins as an institution for teacher training, the nineteenth-century seminar came increasingly to acquire an additional research function with heavy methodological emphasis. Together with the mathematician Gustav Jacobi, Neumann in 1834 organized the first German training and research seminar devoted expressly to mathematical physics. Through his seminar he trained an influential school of physicists, imbuing them with his method and his unified conception of mathematical and experimental physics. Over the course of the century he placed his pupils in German universities where they set up seminars on the Königsberg model. The most famous of his pupils was Gustav Kirchhoff, who began publishing original research from his seminar work. One of the first generation of trained specialists in German physics, Kirchhoff left Königsberg in the 1840's with a clear notion of the new method of mathematical physics and, more important, with a clear notion of the new specialist—the physicist who was equally at home with mathematics and experiment. Neumann's work and teaching contributed greatly to bringing German physicists around to recognizing mathematical physics as a part of their empirical specialty.

I want to emphasize that a particular conception of method had an influence on the social structuring of German science through the way it shaped German physicists' understanding of the compass of their discipline.

The second broad development I want to discuss is the rise of the subspecialty of experimental physics in its late nineteenth-century form. In accord with his belief in a unified mathematical and experimental physics, Neumann early petitioned educational officials to finance a physics laboratory in Königsberg to complement his mathematical physics seminar. His efforts failed, and in the 1840's he fell back on the expedient of frustrated German natural science professors of organizing student laboratory work largely at his own expense. Magnus in Berlin similarly opened his own laboratory to students at this time. Eventually Königsberg, Berlin, and other German universities acquired their proper laboratory facilities, and with them German physics acquired a new subspecialty with an accompanying emphasis on precision measurement.

Friedrich Kohlrausch was the prototype of the confident, new specialist in precision measurement. He was trained in the late 1850's and early 1860's, at a time when physicists were expected to command both practical and mathematical techniques. When he first tried the mathematical lectures, he believed that he had no talent for the subject. He was frightened off, and for a time he considered choosing a science other than physics, one such as chemistry that needed no mathematics. His unease lessened as his mathematics improved, and in time he discovered that he did not need a particularly strong mathematical talent. He built an immensely successful career wholly on the basis of experimental skill. It was fortunate for him that he received part of his training at Göttingen, for nowhere else was there so complete a collection of precision instruments—a collection assembled by Karl Friedrich Gauss and Wilhelm Weber, early proponents of precision measurement in Germany.

Kohlrausch made a lifelong reputation through his innovative laboratory teaching. While laboratory teaching in physics went back at least to the 1830's, it was not yet standardized when Kohlrausch began his career. In the early 1860's the only places where one could receive systematic instruction in laboratory techniques were Königsberg, Berlin, and—due to the presence of Neumann's pupil, Kirchhoff—Heidelberg. Before the end of the decade the need for an extended organization of laboratory teaching was sensed by a widening circle of German physicists, in particular by Kohlrausch's former teacher, Weber. When Kohlrausch returned to Göttingen as an instructor in 1866, Weber asked him to reorganize the physics laboratory. Kohlrausch handled his assignment well; he and several assistants guided large numbers of beginning students through the elementary manipulations and the mathematics of data reduction, while in the same laboratory a small number of advanced students did original research leading to publication. Since the existing textbooks contained little that the measuring physicist wanted to know, Kohlrausch prepared a practical manual for use in his laboratory. He published the manual in 1870, the first work of its kind; physicists elsewhere soon took it up and established courses after Kohlrausch's precepts in their own physical institutes. In the historical preface to the 1910 edition of his manual—the eleventh and last edition under his direction—Kohlrausch accurately dates the period of rapid growth of German physics from the 1860's; it was at this time that the physics staffs of German universities began to expand greatly, owing in large measure to the growth of laboratory work.

The standardization of laboratory training was important for the consolidation of the physics discipline. Aspiring physicists were uniformly indoctrinated into the methods and values of precision measurement. Kohlrausch worked a change in discipline attitudes through the influence of his textbook and teaching on laboratory organization as well as through the example of his own research. Seizing upon a shift in emphasis toward laboratory teaching, he urged upon physicists a recognition of precision measurement as a superlative research goal. No physical fact was too humble or demeaning to merit measurement; to make his point—to parody it in a way—Kohlrausch said he would be delighted to measure the velocity of water in a gutter. It was an irresistible custom with him to count steps, stairs, ways to save time, to go back to his desk and handle numbers from his laboratory notebooks over and over again; such activity was greatly satisfying, the best comfort he knew in life. Laboratory physics offered a career ideally suited to his temperament and skill, and it rewarded him well. He finished as Helmholtz' successor as president of the Physical-Technical Institute, the most prestigious position a physicist could hold in Germany.

When Kohlrausch moved to the Institute in the mid-1890's he said that physics had recently acquired a position of first rank in German scholarship. He explained that the reason it had was its connection with technology, a connection that had elicited millions of marks from the government for splendid new physical facilities throughout the country. Earlier, in 1869, Helmholtz had observed that science had transformed the entire life of modern humanity, first through steam and now through electricity. He explained that physics was the science of forces, the agents of all change in the material world. The comprehension of these forces granted its possessor both an intellectual and a material domination over nature. German physicists organized themselves into a discipline-community at the time that Germany underwent a belated, rapid industrialization, at the time that German society was with unprecedented singlemindedness bent on economic, cultural, and political aggrandizement through the intellectual and material domination of nature. The physicist, the industrialist, and the politician were jointly committed to capitalizing upon the powers of nature. Leading German physicists such as Helmholtz and Neumann were intensely nationalistic; they were sincere in their identification of German physics with German power in both its cultural and material aspects. For Helmholtz the true German physicist or scholar was a man of action, either applying or advancing his field of knowledge; in his characteristic activity the physicist was a full participant in the struggle for nationhood. When Helmholtz moved from Heidelberg to Berlin in 1871, his friends thought it was entirely right that Germany's greatest scientist should be in the same place as her greatest general and greatest statesman, a succinct comment on the association between scientific culture and national might that seemed natural to Helmholtz and his countrymen.

The importance German physicists placed on industrial technology in Helmholtz' day is easily overlooked, since the locus of technical education had shifted from the universities to special schools, and since university physicists prized the cultural mission of their science above any other. University physicists objected to the earlier tendency to regard physics as an adjunct to technology; they cooperated in largely removing technology from the university on the grounds that the university was for the pursuit and diffusion of basic knowledge and that industrial progress was causally posterior to scientific progress. This does not mean that they turned their backs on industrial progress, but that they truly believed that the basic researcher advanced industrial technology most efficiently by ignoring it.

The exact expression of the distinctness of the activities of basic physics and physical technology and, at the same time, of their symbiotic relationship was the Physical-Technical Institute in Charlottenburg, a suburb of Berlin. The initial idea and support for the Institute in the early 1870's arose from a concern over the stagnation of a branch of German industry—the manufacture of scientific precision instruments for both basic research and precision engineering. The Institute, which was completed in 1887, was divided into two distinct, roughly equal departments, one for research in basic physics, the other for research in physical technology. Helmholtz and his intimate friend Werner von Siemens jointly planned the Institute. Siemens was an industrialist with a fervent belief in the eventual technological utility of basic physical research. Siemens' commitment to physics was stimulated by the young scientists of the Berlin Physical Society—he and Helmholtz had joined the Society in its first year, 1845. Siemens became immensely wealthy through his pioneering work in the new, burgeoning electrical industry that stemmed from recent developments in basic physics. At the time of German unification, electricity was emerging as the new industrial might, and it was everywhere received as dramatic evidence of the revolutionary technological potential of physics. The great national physics laboratory in Charlottenburg embodied the scientific-technological vision that Siemens derived from his association with Helmholtz and the Berlin Physical Society and from his successful industrial exploitation of a new domain of applied physics. Siemens provided the money for the combined physical and technical laboratory, and the imperial German government financed its continuing operation, an arrangement that underscores the interlocking of basic physics, physical technology, industry, and government at the end of the nineteenth century. The economic and political conditions that favored Helmholtz and Siemens' proposal of a national physics laboratory were fostered by the intense industrial consciousness following the wars of German unification.

The Physical-Technical Institute represented not only the recognition by government and industry of the national stake in physics, but also the apotheosis of the research ethos that had been nurtured within German scientific circles from early in the century. Siemens designed the Institute's presidency expressly for Helmholtz, to free him from his onerous teaching and administrative responsibilities at Berlin University so that he could devote his energies more completely to his research. The Institute provided full-time research opportunity for a large permanent staff and for temporary guests. In terms of publications it was the single most important center of German research in both basic and applied physics before the end of the century.

The third and last development in physics I will discuss is the emergence of theoretical physics as a subspecialty. I will relate my remarks to Max Planck, one of the first of the new specialists, one whose major achievement, the quantum theory, bears closely on activities in the Physical-Technical Institute and on other matters I have discussed.

Planck devoted himself to physics because of his youthful discovery that the laws of human reasoning parallel the laws of the impressions we receive from the outside world. For him the outside world existed wholly independently of man; it was something absolute and at the same time accessible to pure reasoning. Physics was the science that dealt most directly with the basic laws of the natural world, and the work of the theoretical physicist was to grasp its laws by pure reasoning. Planck decided very early to become a theoretical physicist, to devote his life to the universal laws of nature; it was his understanding of the highest scientific calling. He studied at the University of Munich, where he went through a standard training. He learned practical techniques in the physical laboratory; and in the mathematical physics seminar, he studied with Ludwig Seidel, one of Neumann's pupils.

EDITOR'S FOREWORD

The trouble was that when Planck earned his doctorate in 1879, theoretical physics had hardly begun to be acknowledged as a separate subdiscipline. It is true that the universities had already made a beginning at a convenient division of labor. Weber had brought about a separation of the Göttingen Physical Institute into a thoretical and an experimental department in 1849; and Helmholtz had brought Kirchhoff to Berlin in 1875 expressly to give theoretical lectures. But in the creation of chairs, the theoretical subspecialty was, and would remain for a long time, markedly undynamic in its career opportunities. For years after his degree, Planck stayed on at Munich as an instructor with few prospects. In his frank autobiography, he tells of his intense ambition in his postdoctoral years to win a reputation with the established physicists. There was no contradiction between his ambition for reputation and his commitment to nature's absolutes. German university careers were structured so as to render the two drives complementary. For over a generation before Planck, ministries of education were accustomed to making decisions on university appointments and promotions primarily on the basis of published research. University careers and scientific reputations were determined by the same, overriding discipline value—original research. The university scientist could obey the research ethos and advance the state of knowledge and, at the same time, advance his discipline reputation and his financial security. As a student, Planck went to Berlin for a time to hear Helmholtz and Kirchhoff lecture. Emphatically unimpressed by their lecturing, he was overwhelmed by their presence. For the first time, he realized that while Munich had good teachers, its physics had decidedly only local significance. Helmholtz and Kirchhoff had won world reputations through original research, and they defined for Planck the measure of his life ambition. It was immaterial that Helmholtz was a boring, halting lecturer who fumbled in his notebook for material he had not troubled to prepare beforehand, or that Kirchhoff's lectures sounded like a monotonous memorized textbook. It was not on such local performances that careers were founded in the 1870's, a fact that had for Planck the force of a truism.

Planck's opportunity came when Helmholtz called him to Berlin in 1889 as Kirchhoff's replacement and as head of the new Theoretical Physics Institute. Planck saw himself as a physicist sui generis—

the only pure theorist in Berlin. As such, he was met with a good deal of suspicion. The people in the Berlin Physical Institute kept him at arm's length at first, and he had to work to cement cooperative relations across the reseparated halves of the physics discipline. He succeeded in winning the trust and friendship of Heinrich Rubens in the Physical Institute, without whom, Planck later said, the quantum theory would not have come about as it did, perhaps not even in Germany. In his autobiography, we see Planck knocking on doors, striking up friendships, initiating correspondences; channels of communication are essential to a discipline, and Planck was assiduous in assuring and extending them, not least to prevent his new subspecialty, theoretical physics, with its separate organization from being isolated from the larger discipline.

Planck saw as his primary task the consolidation of a physical worldview. I will comment on this and on the cultural function of worldviews generally, and on the fact that worldview concerns had a basis in intradisciplinary specialization. While worldview issues were not the exclusive concern of wholly theoretical physicists such as Planck, they were their specific concern. Their concern was based in part on individual temperament, on their need for global order, on the reason they were drawn to physics in the first place. It was based in part, too, on their ambition to make their mark through contributions of the highest significance—Planck confided to his young son at the time that the results of his quantum theory of blackbody radiation might stand with Newton's achievement. The theorists' concern with worldviews had a discipline basis as well; it expressed the traditional unifying imperative of German disciplinary scholarship, opposing the fragmentation endemic to specialist research. The theoretical physicist lectured over the whole domain of physics, devoting particular attention to the unity of the parts. He was a specialist of the general in his teaching, and he tended to be so in his research. He was and saw himself as worldview critic and worldview synthesizer.

The dissolution of the mechanical worldview at the end of the nineteenth century brought about a state of anomie in theoretical research. The mechanical view of nature was no longer a cohering force in the physics discipline, but a divisive one. Physicists tended to cluster around the problems appropriate to one or another worldview, mechanical or antimechanical, rather than as before to develop partial theories of physics within a secure mechanical framework. Their arguments were sometimes bitter. A worldview consensus serves the discipline by contributing to the conditions for rational argument; lacking common worldview principles, physicists descended at times to ad hominem. The worldview debates that began in the late nineteenth century and continued into the quantum and relativity developments of the early twentieth were related to philosophical discontents that occupied German intellectuals increasingly through the 1890's. The sciences with their tendencies toward mechanism, materialism, positivism, and disunity were charged with having lost their philosophical bearings. German physical scientists responded to the charges in varying ways and degrees; certain of them advocated overthrowing the mechanical worldview that had guided physical research for over two centuries. It is on the level of worldviews that the connections between scientific and philosophical thought are most operative. The most persuasive evidence for the connections is the broad coherence of thought within the sciences and German learning generally. In philosophy there were calls for antimaterialist, neoidealist, monistic syntheses, for life-philosophy, for spiritualism. In biology there was a revival of antimaterialist, neovitalist theories; in the social and psychological sciences there were similar antimaterialist trends. And in physics two well-articulated worldviews—the energetic and the electromagnetic—were proposed as alternatives to the mechanical worldview.

The common objective of the alternative worldviews was to bring unity to natural knowledge by denying the reality of mass, the basic concept of mechanics. When Wilhelm Ostwald suggested that matter is nothing but the spatial distribution of energy, he was reinterpreting physics to express philosophical trends he favored. So was Emil Wiechert when he suggested that matter is nothing but a condition of an immaterial electromagnetic world-ether, a substance visible to spiritual eyes alone. Defending the mechanical worldview, Boltzmann admitted that it was not yet complete, but he correctly observed that the challenger, in this case the energetic worldview, was far less complete. The satisfaction that certain scientists derived from a physics founded on immaterial energetic or electromagnetic concepts at the end of the nineteenth century was rooted in the philosophical climate as well as in the narrower circle of physical problems.

I will, finally, recall the circumstances surrounding Max Planck's introduction of quanta at the end of the century. My intention is not to explain in depth the genesis of the quantum theory, but to provide a concrete historical illustration of the complex interdependence of the physicist's work and his discipline. I want to suggest that our understanding of a specific achievement in modern physics—even something as technical and seemingly internal to physics as the introduction of quanta—will remain incomplete until we have a history of the larger physics discipline.

Wilhelm Wien, working in the Physical-Technical Institute, proposed a theoretical formula for the energy distribution of blackbody radiation in 1896. Otto Lummer and Ernst Pringsheim, also working in the Institute, found that Wien's law failed for long wavelengths of blackbody radiation; they reported their findings to the German Physical Society, meeting in the Berlin Physical Institute in February 1900. Later that year, Friedrich Paschen of the Hannover Technical High School, Heinrich Rubens of the Berlin Physical Institute, and Ferdinand Kurlbaum of the Physical-Technical Institute produced further experimental evidence of the limited validity of Wien's law. A few days before the October meeting of the Physical Society, Rubens and Kurlbaum told Planck of their findings, giving him time to draw out their consequences. At the meeting Planck offered a formula that fit the new data; the formula was partly empirical and partly guided by theoretical considerations. Rubens, who attended the meeting, checked Planck's new formula against his and Kurlbaum's latest measurements that same night; the next morning he called on Planck, telling him that he had found satisfactory agreement. With this encouragement Planck spent the next two months in the most strenuous intellectual activity of his life, searching for a full theoretical basis of his formula. The outcome was his quantum theory of blackbody radiation, which he read to the Physical Society in its December meeting, and which was quickly published in the Society's proceedings and shortly after in an expanded version in the *Annalen der Physik*. This communication proved to be the starting point for an entire reorientation of the physics community about a new worldview.

These were the events of 1900. I will now suggest some of the ways in which they are related to worldviews, methods, and precision measurements. In 1900 Planck did not know he had raised a worldview challenge, nor had that been his purpose. His quantum theory was, all the same, deeply influenced by the strident and unyielding worldview controversies that flared in Germany in the 1890's, controversies he had taken vigorous part in. Planck's original motivation to become a physicist—his dedication to nature's absolutes—led him to thermodynamics, for him the branch of physics that dealt directly with nature's absolutes. His whole career up to 1900 was devoted to clarifying and applying the first and, especially, the second laws of thermodynamics. For five or six years immediately prior to his announcement of the quantum theory, he had been thinking closely about the problem of blackbody radiation. For him the laws of the spectral distribution of blackbody radiation, like the two laws of thermodynamics themselves, belonged to the category of the absolute. His long-range program was to explain the irreversibility of the second law by conservative forces and, in the process, to derive the blackbody radiation law. The conservative forces he looked to were electromagnetic; he hoped to attach thermodynamics to an incipient electromagnetic worldview. By 1899 Planck had progressed toward his goal to the point where he could say that if Wien's blackbody law had limited validity, then so had the second law of thermodynamics. When the new measurements in 1900 pointed to a limitation of Wien's law, Planck was immensely interested. Nothing less than the absolute character of the basic laws of thermodynamics was at stake.

It was Planck's special faith in thermodynamics' absoluteness that determined his position in the worldview debates. He rejected the energetic worldview because its proponents misunderstood both laws of thermodynamics. He rejected the mechanical worldview because its proponents attributed only statistical, not absolute, validity to the second law and because he, following in Neumann's tradition, preferred whenever possible to avoid atomistic hypotheses, hypotheses that were closely bound up with the mechanical worldview. Planck was working completely outside the mechanical worldview by 1900;

one evidence of this is his indifference to the equipartition theorem of statistical mechanics, a theorem central to the theoretical difficulties of Wien's blackbody law in the context of the mechanical worldview. In 1900 Planck tried at first to build a theory of blackbody radiation solely on the basis of the two laws of thermodynamics and electromagnetic theory. Failing, he turned in what he later called an act of desperation to Boltzmann's statistical-mechanical interpretation of the second law. His quantum theory thus rested upon principles drawn eclectically from three worldviews—mechanical, electromagnetic, and thermodynamic-energetic. It was in part because his theory fell outside the program of any particular worldview that it was so long ignored. The experimentalists used his formula, but the theorists did little with it, not at least before Einstein in 1905. It is significant that it should be Einstein who first fully grasped the profound worldview implications of Planck's work. Einstein was as committed as Planck to the worldview responsibility of the pure theorist, and he was at the same time deeply critical of each of the existing worldview positions.

Both the immediate theoretical and experimental starting points of Planck's 1900 quantum theory issued from the Physical-Technical Institute under Kohlrausch's presidency. The Institute's official publications included both Wien's 1896 derivation of the spectral distribution of blackbody radiation and the precision measurements of Lummer and his colleagues. The Institute's basic commitment complemented Planck's perfectly. The most important work of the physical department dealt with precision measurements of universal constants and functions—e.g., the blackbody law and the constants it contained. The unparalleled facilities of the Institute supplied the technical opportunity for pursuing the absolute, for vindicating the accepted view that precision measurements could be the starting point of new understanding—even, as it turned out, of new worldviews.

Planck recalled that when he took up the blackbody problem, a host of outstanding physicists were working on it, most of whom he was in close communication with. They were mainly German, in fact mainly Berlin physicists who were associated with August Kundt's school of radiation investigation. The main exception was Paschen in Hannover, but he too was a product of Kundt's school.

The group of blackbody researchers had had the same training in the Berlin Physical Institute; they moved between the University and Charlottenburg; they were in constant personal touch; and they often collaborated in researches and coauthored publications.

Like Kohlrausch, Planck began his career at a time when the members of the discipline were ready to make room for one of his special interest and ability, and, again like Kohlrausch, he left a strong imprint on the discipline through his pioneering of a new, successful learned role. Planck demonstrated the potentiality of the pure theorist's contribution, and not only for his own subdiscipline—Warburg, Kohlrausch's successor at the Physical-Technical Institute, liked to speak of the immense impulse Planck's quantum theory gave to the purely experimental subspecialty. Planck was supreme among the first generation of theorists in establishing theoretical physics as a highly prestigious subspecialty.

I have suggested some of the important connections that emerge from a discipline approach to the history of physics. I have discussed ideas in their social context, emphasizing the ways in which physical theories, methods, and worldviews relate to discipline identity and differentiation. I have referred to the motivations and self-images of physicists in the context of the research ethos of the discipline. I have suggested how the physicist's individuality—Kohlrausch's and Planck's—responds to and, at the same time, reshapes the discipline's needs at a given time. I have illustrated the school phenomenon by the group of blackbody researchers, locating them within the network of physical institutions the discipline's organizers promoted in the course of the century. And I have indicated some of the connections between the nineteenth-century physics discipline and government, industry, and culture. The historical development of the physics discipline is an enormous subject. Needless to say I have sketched only a small part of it here.

Weimar Culture, Causality, and Quantum Theory, 1918–1927: Adaptation by German Physicists and Mathematicians to a Hostile Intellectual Environment

BY PAUL FORMAN[*]

> "It is interesting to observe that even physics, a discipline rigorously bound to the results of experiment, is led into paths which run perfectly parallel to the paths of the intellectual movements in other areas [of modern life]." Gustav Mie, inaugural lecture as Professor of Physics, University of Freiburg i.B., 26 January 1925.

I. Weimar Culture as a Hostile Intellectual Environment
  1. As Perceived by the Physicists and Mathematicians 8
  2. As Confirmed by Other Observers 15
  3. Intellectual Allies: Vienna Circle and Bauhaus 19
  4. Educational Ideals and Reforms 23
  5. The Crisis of *Wissenschaft* 26
  6. Spengler's *Decline of the West* 30

II. Adaptation of Ideology to the Intellectual Environment
  1. Introduction 38
  2. From Positivism to *Lebensphilosophie* 40
  3. Capitulation to Spenglerism 48
  4. A Craving for Crises 58

[*] Department of History, University of Rochester, Rochester, New York 14627.

III. "Dispensing with Causality": Adaptation of Knowledge to the Intellectual Environment
1. Introduction: The Concept of Causality 63
2. The First Intimations of an Issue, 1919–1920 70
3. Conversions to Acausality, 1919–1925
   a. The Earliest Converts: Exner and Weyl 74
   b. 1921, Summer and Fall: von Mises, Schottky, Nernst, *et al.* 80
   c. Later Notable Conversions: Schrödinger and Reichenbach 87
4. Unregenerates against the Tide, 1922–1923 91
5. The Situation circa 1924 96
6. Causality's Last Stand, 1925–1926 100
7. Conclusion 108

In perhaps the most original and suggestive section of his book on *The Conceptual Development of Quantum Mechanics* Max Jammer contended "that certain philosophical ideas of the late nineteenth century not only prepared the intellectual climate for, but contributed decisively to, the formation of the new conceptions of the modern quantum theory"[1]; specifically, "contingentism, existentialism, pragmatism, and logical empiricism, rose in reaction to traditional rationalism and conventional metaphysics. . . . Their affirmation of a concrete conception of life and their rejection of an abstract intellectualism culminated in their doctrine of free will, their denial of mechanical determinism or of metaphysical causality. United in rejecting causality though on different grounds, these currents of thought prepared, so to speak, the philosophical background for modern quantum mechanics. They contributed with suggestions to the formative stage of the new conceptual scheme and subsequently promoted its acceptance."[2]

These are far-reaching propositions. Properly construed they are,

---

1. M. Jammer, *The Conceptual Development of Quantum Mechanics* (New York: McGraw-Hill, 1966), section 4.2, "The Philosophical Background of Nonclassical Interpretations"; on pp. 166-167.
2. *Ibid.*, p. 180. The search for philosophic precedents and influences has otherwise focused almost exclusively upon Bohr's doctrine of complementarity. This issue, which I am not directly concerned with here, has been recently examined once again and the literature reviewed by Gerald Holton, "The Roots of Complementarity," *Daedalus, 99* (Fall, 1970), 1015-1055.

I think, essentially correct. But it must be said that Jammer did not go very far toward demonstrating them. He displayed such anticausal sentiments among a variety of late nineteenth-century philosophers—French, Danish, and American—but adduced scarcely any evidence to bridge the wide gaps of a quarter century of time, a cultural tradition, and the disciplines of philosophy and physics, which separated their philosophical theses from the development of quantum mechanics by German-speaking Central-European physicists circa 1925. It is not my aim to fill in these gaps, but rather to examine closely the lay of the land on the far side of them. The result is, on the one hand, overwhelming evidence that in the years after the end of the First World War but before the development of an acausal quantum mechanics, under the influence of "currents of thought," large numbers of German physicists, for reasons only incidentally related to developments in their own discipline, distanced themselves from, or explicitly repudiated, causality in physics.

Thus the most important of Jammer's theses—that extrinsic influences led physicists to ardently hope for, actively search for, and willingly embrace an acausal quantum mechanics—is here demonstrated for, but only for, the German cultural sphere. This cultural qualification is essential; it forms the basis of my attempt to provide, on the other hand, an answer to the question—in its general form crucial to all intellectual history—why and how these "currents of thought," evidently of negligible effect upon physicists at the turn of the century, came to exert so strong an influence upon German physicists after 1918. For it seems to me that the historian cannot rest content with vague and equivocal expressions like "prepared the intellectual climate for," or "prepared, so to speak, the philosophical background for," but must insist upon a causal analysis, showing the circumstances under which, and the interactions through which, scientific men are swept up by intellectual currents.

Such an analysis may be either "psychological" or "sociological." That is, it may either consider the mental makeup of the individual scientists concerned, stressing previous intellectual environments and conditioning experiences as determinative of present attitudes, or, on the contrary, it may ignore these factors, treating present mental posture as socially determined response to the immediate intellectual environment and current experiences. I have chosen the latter course, and sought a model in which certain "field variables" and their derivatives at a given place and time are regarded as evoking corresponding attitudes. Though it may seem harsh to stress the social pressure and ignore the emotional pain, though it may seem unsatisfactory to break off our explanatory endeavors at the level of the individual decision, nonetheless I do think the "sociological" the more general and fruitful approach.

The inquiry must begin, then, by characterizing the intellectual milieu in which the German physicists were working and quantum mechanics was developed. This is a formidable problem, above all on account of methodologic difficulties. And the task is especially unattractive to the historian of science, for it obliges him to deal with the "expressions" of nonscientists as well as those of scientists, thus forcing the abandonment of the demarcation criterion by which he seeks to identify and delimit his subject. Nevertheless, with aid and guidance from previous studies by general intellectual historians, especially the work of Fritz K. Ringer, I have addressed this problem in Part I. I show that in the aftermath of Germany's defeat the dominant intellectual tendency in the Weimar academic world was a neo-romantic, existentialist "philosophy of life," reveling in crises and characterized by antagonism toward analytical rationality generally and toward the exact sciences and their technical applications particularly. Implicitly or explicitly, the scientist was the whipping boy of the incessant exhortations to spiritual renewal, while the concept—or the mere word—"causality" symbolized all that was odious in the scientific enterprise.

Now if, as is largely the case even at this late date, the interest of the historian of science is held exclusively by the substantive scientific achievements, he will immediately be struck by a remarkable paradox: this place and period of deep hostility to physics and mathematics was also one of the most creative in the entire history of these enterprises. Faced with this paradox many of us would be tempted to rub our hands with satisfaction, to regard it a welcome refutation of any attempt to impugn the autonomy of these sciences and the sufficiency of intellectualist-internalist history of them. But such an inference would be too hasty. Presupposing the hostility of the intellectual environment, the crucial question is the *nature* of the response of the exact scientists to this circumstance. I had myself previously assumed that in the face of antiscientific currents the predominant response in these highly professionalized sciences would be retrenchment, withdrawal into the science and the community of its practitioners, reaffirmation of the discipline's traditional ideology—i.e., its notion of the value, function, motive, goal, and future of scientific activity.[3] *Were* that the case, then, a fortiori, any attempt to attribute a strong and direct influence of that same intellectual environment upon the scientific discourse and dispositions of these same men would appear implausible.

Yet the historian who takes even the most casual notice of the valuations of physical science in contemporary American society, on the one hand, and the present ideological tendencies in these sciences, on the other hand, could scarcely maintain that the predominant response to a hostile intellectual environment is retrenchment. On the contrary, as sentiments of resentment and antagonism toward the scientific enterprise—coupled with a revival of existentialist *Lebensphilosophie*—have become prominent in the last few years, so also have the expressions of and concessions to these same sentiments within the sciences themselves. We are indeed witnessing in America today a widespread and far-reaching accommodation of scientific ideology to a hostile intellectual milieu. As the distinguished physical chemist Franklin A. Long recently stated in both explanation and advocacy of this development: "Faculty, and especially students, are sensitive to social problems, are eager to work on them, and are often prepared to change their previous ways of life to do so. The pressures of discipline orientation and the tradition of individual scholarship are strong among faculty members, but not strong enough to counter the pressures of social concern." And in all of this "responsiveness" there is an astonishing sincerity, a striking absence of cynical, calculated image projection, testifying to a surprising participation of the physical scientists themselves in those fundamentally, often manifestly, antiscientific sentiments.[4]

3. P. Forman, *The Environment and Practice of Atomic Physics in Weimar Germany* (Ph.D. dissertation, Berkeley, 1967; Ann Arbor: University Microfilms, 1968), pp. 11-24.

4. F. A. Long, "Interdisciplinary Problem-Oriented Research in the University [editorial]," *Science, 171* (12 March 1971), 961. Marvin L. Goldberger, "Physics and Environment: How Physicists Can Contribute," *Physics Today* (December 1970), 26-30, and the reply by John Boardman, *ibid.* (February 1971), 9. The new mood, especially the neo-Spenglerianism, in the scientific community is discussed by Bentley Glass in his presidential address to the AAAS, 28 December 1970, "Science: Endless Horizons or Golden Age?" *Science, 171* (8 January 1971), 23-29.

But our contemporary experience does not merely lead us to anticipate an ideological accommodation by the Weimar physicists and mathematicians; it also suggests a simple model for the circumstances under which such accommodation is likely to occur. We may suppose that when scientists and their enterprise are enjoying high prestige in their immediate (or otherwise most important) social environment, they are also relatively free to ignore the specific doctrines, sympathies, and antipathies which constitute the corresponding intellectual milieu. With approbation assured, they are free of external pressure, free to follow the internal pressure of the discipline—which usually means free to hold fast to traditional ideology and conceptual predispositions. When, however, scientists and their enterprise are experiencing a loss of prestige, they are impelled to take measures to counter that decline. Drawing upon Karl Hufbauer's factorization of prestige into image and values, one sees that such countermeasures will in general be attempts to alter the public image of science so as to bring that image back into consonance with the public's altered values. But if this is not mere image projection, then such alterations of the image of the scientist and his activity will also involve an alteration of the values and ideology of the science, and may even affect the doctrinal foundations of the discipline—as Theodore Brown has shown of the beleaguered College of Physicians in the latter seventeenth century.[5]

In Parts II and III, I apply this model to the German-speaking exact scientists working in academic environments in the Weimar period. Bearing in mind the radically rearranged scale of values ascendant in the aftermath of Germany's defeat, I explore in Part II the response of these scientific men at the ideological level. This response I have sought primarily in addresses by exact scientists to academically educated general audiences, and especially in their addresses to their assembled universities. The historian is fortunate that the institutions of German academic life provided frequent occasions for addresses before university convocations, and doubly fortunate that it was customary to publish such *Reden*. Conversely,

5. K. Hufbauer, "Social Support for Chemistry in Germany During the Eighteenth Century: How and Why Did It Change?" in this volume, pp. 205-231; T. M. Brown, "The College of Physicians and the Acceptance of Iatromechanism in England, 1665-1695," *Bulletin of the History of Medicine*, 44 (1970), 12-30.

the existence of these institutions is both an index and an instrument of the extraordinarily heavy social pressure which the German academic environment could and did exert upon the individual scholar or scientist placed within it. As I illustrate in Part II, there was in fact a strong tendency among German physicists and mathematicians to reshape their own ideology toward congruence with the values and mood of that environment—a repudiation of positivist conceptions of the nature of science, of utilitarian justifications of the pursuit of science, and, in some cases, of the very possibility and value of the scientific enterprise.

Was the tendency toward accommodation, which predominated in the response of this highly professionalized scientific community to its hostile intellectual environment, confined to the ideological level, or did it extend beyond it into the substantive doctrinal content of the science itself? Specifically, are there indications that German physicists and mathematicians were anxious to, and deliberately tried to, alter the character of their disciplines as cognitive enterprises and to alter specific concepts employed within them in order to bring their sciences in closer conformity with the values of the Weimar intellectual milieu? I strongly suspect that the intuitionist movement in mathematics, which won so many adherents and created so much furor in Germany in this period, was primarily an expression of just such inclinations and aims. I am convinced, and in Part III endeavor to demonstrate, that the movement to dispense with causality in physics, which sprang up so suddenly and blossomed so luxuriantly in Germany after 1918, was primarily an effort by German physicists to adapt the content of their science to the values of their intellectual environment.

The explanation of the creativity of this place and period must therefore be sought, in part at least, in the very hostility of the Weimar intellectual milieu. The readiness, the anxiousness of the German physicists to reconstruct the foundations of their science is thus to be construed as a reaction to their negative prestige. Moreover the nature of that reconstruction was itself virtually dictated by the general intellectual environment: if the physicist were to improve his public image he had first and foremost to dispense with causality, with rigorous determinism, that most universally abhorred feature of the physical world picture. And this, of course, turned out to be precisely what was required for the solution of those problems in atomic physics which were then at the focus of the physicists' interest.

ACKNOWLEDGMENTS: I am indebted to Stephen G. Brush, Stanley Goldberg, John L. Heilbron, Karl Hufbauer, Hans Kangro, Fritz K. Ringer, Donald E. Strebel, and the editor of this journal, Russell McCormmach, for their close, critical readings of the typescript and their numerous queries, suggestions, objections, and corrections. I also sincerely thank Ann Schertz and the other staff of the Interlibrary Loan Department of Rush Rhees Library, without whose constant aid it would not have been possible to prosecute this inquiry in Rochester, New York.

I. WEIMAR CULTURE AS A HOSTILE INTELLECTUAL ENVIRONMENT

I.1. As Perceived by the Physicists and Mathematicians

Through the summer of 1918 the German physical scientists, like the rest of the German public, continued to look forward with confidence and satisfaction to a victorious conclusion of the war in which they had been engaged four years. They, perhaps more than any other segment of the German academic world, also felt *self*-confidence and *self*-satisfaction due to their contributions to Germany's military success and to their anticipation of a postwar political and intellectual environment highly favorable to the prosperity and progress of their disciplines. The botanist looking about his institute, bleak and vacant, had to conclude that "probably it will also remain so after the war, for youth will turn to technology and leave so 'unpractical' a discipline as botany lying by the wayside."[6] The chemist, the physicist, the mathematician, however, emphasizing the great practical importance of their subjects during the war and the desirability and inevitability of still closer collaboration with technology in the future, looked forward to yet more, larger, and better stocked institutes and to substantially increased public

---

6. Karl v. Goebel to Th. Herzog, Munich 19 July 1917, in Goebel, *Ein deutsches Forscherleben in Briefen aus sechs Jahrzehnten, 1870-1932*, ed. Ernst Bergdolt, 2nd ed. (Berlin, 1940), p. 170.

esteem and academic prestige. "The closer we appear to approach the victorious conclusion of the war," Felix Klein observed in June 1918 before an audience including leaders of German industry and the Prussian government, "the more our thoughts are dominated by the question what, after peace is successfully won, ought then to come." Klein's desiderata ranged from a mathematical institute for himself and his university, through a general reorientation of academic research in the exact sciences to achieve a "preestablished harmony" with the requirements of industry and the military, to a corresponding reorientation of German education at all levels.[7] And at least the first of these desiderata seemed assured as the Prussian Minister of Education, Friedrich Schmidt, came forward to announce a grant of 300,000 Mark. Who, participating in these festivities, could have foreseen that the Göttingen mathematical institute would not be built for another ten years, and then only with American money?[8]

When that "victorious end" which seemed imminent in the summer of 1918 turned suddenly to utter defeat in the fall, the exact scientists found themselves confronting a dramatically transformed scale of public values and thus a drastically altered valuation of their field. That, certainly, was their perception of the situation. Had we no explicit testimony to this effect, we could nonetheless infer it from the defensive tone of the talks given by exact scientists before the assembled faculties and students at academic convocations. While during the latter years of the war such speeches convey self-

---

7. F. Klein, "Festrede zum 20. Stiftungstage [22 June 1918] der Göttinger Vereinigung zur Förderung der Angewandten Physik und Mathematik," *Jahresbericht der Deutschen Mathematiker-Vereinigung*, 27 (1918), Part I, pp. 217-228; on pp. 217, 219. As the philologists noted with some bitterness, during the war the scientists and mathematicians had raised substantially their demands upon secondary school curricula: Robert Neumann, "Politik und Schulreform," *Monatsschrift für höhere Schulen*, 18 (1919), 93-106, "Vortrag, gehalten im Berliner Philologen-Verein Februar 1918"; Friedrich Poske and R. von Hanstein, *Der naturwissenschaftliche Unterricht an den höheren Schulen*, Schriften des Deutschen Ausschusses für den mathematischen und naturwissenschaftlichen Unterricht, II. Folge, Heft 5 (Leipzig-Berlin, 1918).

8. *Jahresbericht der D. M.-V.*, 27 (1918), Part 2, p. 47. In 1926 the International Education Board of the Rockefeller Foundation appropriated $275,000 for a mathematical institute. (Geo. W. Gray, *Education on an International Scale, A History of the International Education Board, 1923-1938* [New York, 1941], p. 30; Otto Neugebauer, "Über die Einrichtung des Mathematischen Institutes der Universität Göttingen," *Minerva-Zeitschrift*, 4 [1928], 107-111.)

assurance, confidence in the esteem and good will of the audience, in the Weimar period that is seldom the case. And while it is difficult to display this *tone,* one can at least point to passages alluding more or less explicitly to reproaches against exact science which the speaker clearly supposes to be in his audience's mind. Thus in November 1925 Wilhelm Wien described the great scientific discoveries of the early modern period, especially Newton's derivation of the motion of the planets from the laws of mechanics, as "the first convincing demonstration of the causality [n.b.] of natural processes which revealed to man for the first time the possibility of comprehending nature by the logical force of his intellect." But he then immediately conceded that this program, which the natural scientist finds so grand, has its limitations, and he proceeded to quote Schiller: "Without feeling even for its creator's honor/ Like the dead stroke of the pendulum clock/ Nature devoid of God follows knavishly the law of gravity."[9] The quotation is clearly in response to popular demand, as the astrophysicist, Hans Rosenberg, makes still clearer in his academic address on 18 January 1930: " 'Your subject is, to be sure, the most sublime in space/ But, friend, the sublime does not reside in space,' I hear Schiller-Goethe call out to us."[10]

It is, of course, their audience which Wien, Rosenberg, *et al.* hear calling out these sentiments, and they seek to escape half the reproach by showing that they are themselves at least familiar with the classical literary expressions of German idealism. When, however, the physicist or mathematician was in the audience he had to listen to far sharper reproaches. In March 1921, Friedrich Poske came away from the funeral of the poet Carl Hauptmann smarting at the accusations against the exact natural sciences which he encountered there,[11] accusations apparently much like those which poor Max

9. W. Wien, *Universalität und Einzelforschung. Rektorats-Antrittsrede, gehalten am 28. November 1925,* Münchener Universitätsreden, Heft 5 (Munich, 1926), 19 pp., on 14.

10. H. Rosenberg, *Die Entwicklung des räumlichen Weltbildes der Astronomie. Rede zur Reichsgründungsfeier . . . am 18. Januar 1930* (Kiel, 1930), 27 pp., on 26. The same lines are quoted—with a still deeper "bow before the secret which the other side hides from us"—by Hans Kienle, "Vom Wesen astronomischer Forschung. Rede, gehalten bei der Verfassungsfeier der Universität Göttingen am 29. Juli 1932," *Bremer Beiträge zur Naturwissenschaft, 1* (1933), 113-125, on 125.

11. F. Poske at the Hauptversammlung of the Deutscher Verein zur Förderung des mathematischen und naturwissenschaftlichen Unterrichtes, 31 March 1921. (*Unterrichtsblätter für Mathematik und Naturwissenschaft, 27* [1921], 34.)

Born had to listen to daily from his wife, a would-be poet and playwright. Hedwig Born derived a masochistic pleasure from "the feeling of being cast upon an icy lunar landscape" which the company of "objective" natural scientists aroused in her.[12] Nor did she hesitate to let her husband's colleagues know that "it is always like a revelation to me whenever behind the *physicist* I suddenly discover the human being; there are, I mean, also inhuman physicists."[13] Certainly there is no reason to think that Einstein's explanation— "what you call 'Max's materialism' is simply the causal [n.b.] mode of considering things"—alleviated Mrs. Born's disquiet.[14]

Painful as it may have been for the theoretical physicist to have to live with such attitudes, the accusation of *Entseelung*, of destruction of the soul, of the world was not the worst he encountered. As Max von Laue saw it in the summer of 1922, the school of Rudolf Steiner "raises the most serious charges against today's natural science. It is represented as bearing the guilt for the world crisis [n.b.] in which we stand at present, and the whole of the intellectual and material misery bound up with that crisis is charged to natural science's account."[15] The counterattack which Laue published was read "with much pleasure" by his mentor and colleague Max Planck, who thought it "will certainly achieve good effects in wider circles."[16]

---

12. H. Born, "Albert Einstein ganz privat," *Helle Zeit—dunkle Zeit. In memoriam Albert Einstein*, ed. C. Seelig (Zurich, 1956), pp. 35-39, on 36.

13. H. Born to H. A. Kramers, 29 September 1925: "Offengestanden hatte ich früher fast etwas Angst vor Ihnen! Aber die ist ganz verschwunden, seit ich hier die Wärme, den Ernst und die ungekünstelte Kraft Ihres Wesens kennen lernen durfte. Es ist mir immer wie eine Offenbarung, wenn ich neben dem *Physiker* plötzlich den Menschen finde; es gibt nämlich auch unmenschliche Physiker!" (Archive for History of Quantum Physics, Sources for History of Quantum Physics Microfilm 8, Section 2; for descriptions and locations of this archive, see Thomas S. Kuhn, et al., *Sources for History of Quantum Physics. An Inventory and Report*, Memoirs of the American Philosophical Society, Vol. 68 [Philadelphia, 1967].)

14. Einstein to H. Born, 1 September 1919, in Albert Einstein, Hedwig and Max Born, *Briefwechsel, 1916-1955*, edited and annotated by M. Born (Munich, 1969), p. 32. An English translation is being published.

15. M. v. Laue, "Steiner und die Naturwissenschaft," *Deutsche Revue*, 47 (1922), 41-49; reprinted in Laue's *Aufsätze und Vorträge = Gesammelte Schriften und Vorträge*, Band III (Braunschweig, 1962), pp. 48-56, on 48.

16. Planck to Laue, 8 July 1922: "Ihren Aufsatz über R. Steiner habe ich mit vielem Vergnügen gelesen. Er . . . wird gewiss in weiteren Kreisen gute Wirkung erzielen." (Handschriftensammlung, Bibliothek, Deutsches Museum, Munich.)

Clearly Planck saw Rudolf Steiner as merely providing the occasion and the ostensible target for rebutting a set of attitudes which he and Laue felt to be widespread among the German educated public. Planck himself adverted to these attitudes and to their danger for science in an address in the Prussian Academy of Sciences a few weeks later.[17] Early in the following year he complained bitterly in a public lecture that "precisely in our age, which plumes itself so highly on its progressiveness, the belief in miracles in the most various forms—occultism, spiritualism, theosophy, and all the numerous shadings, however they may be called—penetrates wide circles of the public, educated and uneducated, more mischievously than ever, despite the stubborn defensive efforts directed against it from the scientific side." Compared to this movement, the agitation of Planck's former *bête noir*, the Monist League, has had, he now allows, "only very meagre success."[18]

It is thus not surprising that the remnants of this largely defunct positivist-monist movement thoroughly agreed with Planck that the Weimar intellectual environment was fundamentally and explicitly antagonistic to science. Drawing upon the universally accepted analogy between contemporary Germany and the period following its defeat by Napoleon, Wilhelm Ostwald thought it evident that "In Germany today we suffer again from a rampant mysticism,

17. M. Planck, "Ansprache des vorsitzenden Sekretärs, gehalten in der öffentlichen Sitzung zur Feier des Leibnizischen Jahrestages, 29. Juni 1922," *Preuss. Akad. d. Wiss., Sitzungsber.* (1922), pp. lxxv-lxxvii, reprinted in *Max Planck in seinen Akademie-Ansprachen; Erinnerungsschrift der Deutschen Akademie der Wissenschaften zu Berlin* (Berlin, 1948), pp. 41-48. A similar characterization of the intellectual environment had been given in the fall of 1920 by Artur Schoenflies, *Über allgemeine Gesetzmässigkeiten des Geschehens* [Rektoratsantrittsrede], Frankfurter Universitätsreden XI (Frankfurt, 1920), 16 pp., on 4: "In increasing measure in recent [letzten] years there has developed a conscious hostility to the natural-scientific mode of thought. . . . The fact is that the new mode of thought with force and bluster has fought its way through to success in all fields—in *Wissenschaft* and art, literature and politics, in writing and speaking."

18. M. Planck, *Kausalgesetz und Willensfreiheit. Öffentlicher Vortrag gehalten in der Preuss. Akad. d. Wiss. am 17. Februar 1923* (Berlin, 1923), 52 pp.; reprinted in Planck, *Vorträge und Erinnerungen* (Stuttgart, 1949), pp. 139-168; on 162-163. And again, eight years later, "It is astonishing how many people, particularly from educated circles . . . fall under the sway of these new religions, iridescing with every hue from the most confused mysticism on out to the crassest superstition." ("Wissenschaft und Glaube. Weihnachtsartikel vom Jahre 1930," *ibid.*, pp. 246-249; also quoted at length in Hans Hartmann, *Max Planck als Mensch und Denker* [1953; reprinted Frankfurt, 1964], pp. 52-55, on 52-53.)

which, as at that time, turns against science and reason as its most dangerous enemies."[19] And even where, as with the theory of relativity, there was great public interest in particular results of physical research, that interest was never, to my knowledge, construed by the physicists as evidencing appreciation and approbation of their enterprise. Rather, it struck Einstein as "peculiarly ironical that many people believe that in the theory of relativity one may find support for the anti-rationalistic tendency of our days."[20]

Arnold Sommerfeld was thus clearly speaking for most of his colleagues when, responding to a request from the most prestigious of the South German monthlies for a contribution to a special number on astrology, he asked:

> Doesn't it strike one as a monstrous anachronism that in the twentieth century a respected periodical sees itself compelled to solicit a discussion about astrology? That wide circles of the educated or half-educated public are attracted more by astrology than by astronomy? That in Munich probably more people get their living from astrology than are active in astronomy? Certainly in Germany this anachronism is based in part upon the misery of the present. The belief in a rational [vernünftig] world order was shaken by the way the war ended and the peace dictated; consequently one seeks salvation in an irrational [unvernünftig] world order. But the reason must lie deeper, for astrology, spiritualism, and Christian Science are flourishing among our enemies also. We are thus evidently confronted once again with a wave of irrationality and romanticism like that which a hundred years ago spread over Europe as a reaction against the rationalism of the eighteenth century and its tendency to make the solution of the riddle of the universe a little too easy. Even though I [wir] have no illusions about being able to hold back this wave by means of arguments based upon reason, nonetheless I [wir] want to throw myself decisively against it.[21]

19. W. Ostwald, *Lebenslinien. Eine Selbstbiographie* (Berlin, 1926-1927), *3,* 442. And again, *ibid., 2,* 309, "It is at present considered modern to speak all conceivable evil of the intellect."

20. A. Einstein, *Vossische Zeitung,* 10 July 1921, as quoted by Siegfried Grundmann, "Der Deutsche Imperialismus, Einstein und die Relativitätstheorie (1914-1933)," *Relativitätstheorie und Weltanschauung* (Berlin, 1967), pp. 155-285, on 194.

21. A. Sommerfeld, "Über kosmische Strahlung." *Südd. Monatshefte, 24* (1927), 195-198; reprinted in Sommerfeld's *Gesammelte Schriften* (Braunschweig, 1968), *4,* 580-583. Cf. Lewis M. Branscomb, Director of the U.S. National Bureau of Standards, *Science, 171* (12 March 1971), 972: "Astrology is booming; there are three professional astrologers in this country for every astronomer."

Although the German physical scientists, regardless of their special discipline, agreed that irrationalism and mysticism were characteristic of the postwar mood, altogether it was the mathematicians and the theoretical physicists who, more than the experimental physicists or the chemists, felt themselves to be the particular objects of odium, both public and private. One cannot withhold a certain sympathy for the Nazi Theodor Vahlen as he confesses in 1923 before the assembled members of his university how "a friendly attitude toward mathematics is so rare that, if we run across it, it really strikes us as especially remarkable."[22] This feeling of facing an antagonistic environment, inside and outside the university, was so generally shared among mathematicians that Gerhard Hessenberg could appeal to it in trying to persuade the theoretical physicist Arnold Sommerfeld to take a course of action which would antagonize an experimental physicist (Friedrich Paschen) to whom Sommerfeld looked for much of his raw material: "But we poor scapegoats of mathematicians have gotten to hear so much evil about ourselves these days—behind our backs as well as to our faces—what difference does a little bit more or less make. . . ."[23] Indeed, these "antimathematical currents," "this onslaught against mathematics" which sprang forth after the war seemed so strong and threatening that in 1920 the German mathematicians joined together in a defense organization, the Mathematischer Reichsverband, whose special task was to protect the position of mathematics in the schools.[24]

22. Th. Vahlen, *Wert und Wesen der Mathematik. Festrede . . . am 15. V. 1923*, Greifswalder Universitätsreden 9 (Greifswald, 1923), 32 pp., on 1. And in this, if in nothing else, Konrad Knopp agreed with Vahlen: "We mathematicians . . . have not been able to obtain, or even merely to retain, the position in public life which mathematics merits." ("Mathematik und Kultur, Ein Vortrag," *Preussische Jahrbücher, 211* [1928], 283-300, on 283.)

23. G. Hessenberg to A. Sommerfeld, 16 June 1922: "Wir armen Sündenböcke von Mathematikern aber haben in diesen Tagen so viel schlechtes, hintenherum, wie auch vorneherum, über uns zu hören bekommen, dass es uns auf ein bischen mehr oder weniger nicht ankommt; der Gerechte hat nun einmal viel zu leiden." (Sources for History of Quantum Physics Microfilm 33, Section 1.)

24. Georg Hamel, as president, at the first general assembly of the Mathematischer Reichsverband, Jena, 23 September 1921, *Jahresbericht der Deutschen Mathematiker-Vereinigung, 31* (1922), Part 2, p. 118. And again at the second general assembly, Leipzig, 22 September 1922, the *Arbeitsausschuss* stressed in its report that "With respect to its place and prestige [*Geltung*] in the schools, mathematics finds itself in a defensive position. The contemporary intellectual currents, directed against intellectualism and rationalism, are decidedly unfavorable to mathematics." (*Ibid., 32* [1923], Part 2, pp. 11-12.)

The result, then, of this first approach to the problem of establishing the tenor of the intellectual environment within which the Weimar physical scientists worked so productively is unambiguous: the environment was *perceived* by the physical scientists to be markedly hostile. Is it therefore necessary to carry our inquiry any further? One might, after all, argue that it is vain to ask whether these perceptions corresponded to "reality" and that moreover the answer would be of no consequence for the behavior of the physical scientists. Nonetheless the accuracy or inaccuracy of these perceptions is certainly an important datum about these men, a datum which is essential for any attempt to infer their perceptions, and the effects upon their science, of a given intellectual environment. For the purposes of this paper, moreover, it is important to go farther afield in exploring the attitudes toward physical science in Weimar Germany; we need a more detailed specification of those attitudes if we are to determine how far and in what sense the ideology and ideas of the physical scientists may be regarded as responses to their intellectual environment.

I.2. As Confirmed by Other Observers

Unequipped and disinclined to undertake an extensive independent exploration and reconstruction of the Weimar intellectual environment, I have turned to other observers—first intellectual historians, then contemporary observers—seeking their conclusions and their guidance.

For our period and theme there are studies by Georg Lukács,[25] Kurt Sontheimer,[26] Peter Gay,[27] and—most recent, detailed, and relevant—by Fritz Ringer.[28] While these intellectual historians are not specifically concerned with the attitudes toward exact science,

---

25. G. Lukács, *Die Zerstörung der Vernunft. Der Weg des Irrationalismus von Schelling zu Hitler* (Berlin, 1954).
26. K. Sontheimer, *Antidemokratisches Denken in der Weimarer Republik* (Munich, 1962).
27. P. Gay, *Weimar Culture: The Outsider as Insider* (New York, 1968). An only slightly abridged version, omitting however the extensive bibliography, appeared under the same title in Donald Fleming and Bernard Bailyn, eds., *The Intellectual Migration* (Cambridge, Mass., 1969), pp. 11-93.
28. F. K. Ringer, *The Decline of the German Mandarins. The German Academic Community, 1890-1933* (Cambridge, Mass.: Harvard University Press, 1969).

their characterizations of the intellectual milieu do in fact bear directly upon this question. Especially Ringer's examination of academic ideology places before us many of the attitudes toward science which pervaded the Weimar academic world and directs us to many important sources. And despite the diversity of the personal-professional backgrounds, research methods, and ethical-political motivations of these intellectual historians, by and large they give us the same general picture of the Weimar intellectual milieu: rejection of reason as an epistemological instrument because inseparable from positivism-mechanism-materialism, and because, as fundamentally disintegrative, incapable of satisfying the "hunger for wholeness";[29] glorification of "life," intuition, unmediated and unanalyzed experience, with the immediate apprehension of values, and not the dissection of causal nexus, as the proper object of scholarly or scientific activity. This "life-philosophy," of which existentialism was but a variety, Lukács sees as "the dominant ideology of the entire imperialistic period in Germany. . . . In the postwar period virtually all of the widely read bourgeois *Weltanschauungsliteratur* is *lebensphilosophisch.*"[30]

With these studies by intellectual historians giving us some confidence that we are not going seriously astray, let us look a little more closely at certain of the programmatic slogans of this life philosophy as epitomized by contemporary observers of Weimar intellectual life. Such characterizations of the intellectual environment will, I think, not merely suggest, irresistibly, a valuation of the physical scientist, but will also force us to recognize the crucial role of the concept of causality.

Within a year of the end of the war these intellectual currents, now monopolizing the movement for educational reform, were flowing everywhere. Discussing "the social-pedagogic demand of the present" in 1920, Alfred Vierkandt could see clearly that "We are generally experiencing today a full rejection of positivism; we are experiencing a new need for unity, a synthetic tendency in all the world of learning [Wissenschaft]—a type of thinking [Eindenken] which primarily emphasizes the organic rather than the mechanical, the living instead of the dead, the concepts of value, purpose, and goal,

---

29. The phrase is Gay's, *op. cit.* (note 27).
30. Lukács, *op. cit.* (note 25), p. 318.

instead of causality."[31] A sharper and more penetrating analysis of this call for a "revolution in science [Wissenschaft]" in the early Weimar period is that which Ernst Troeltsch published in 1921.[32] Here causality appears over and over again as the pejorative term epitomizing the tendency in *Wissenschaft* which the new movement rejects: "The methods of these specialized scientific disciplines are those of causal explanation, of natural causality, of psychophysical, psychological, and sociological causality. It is the ultimate intellectualization of our attitude toward the world, the disenchantment of the world, and the path toward an unlimited approximation to a totally causal system [Gesamtkausalsystem] of things."[33]

Troeltsch, in common with many other observers, cites Henri Bergson as perhaps the most important—and the only nonGerman—source of the movement against "all suffocating determinism." "A general sigh of relief follows almost audibly the ever stronger establishment of this system."[34]

> If now we draw all that together, the freedom from positivistic causalism and determinism, the overcoming of neo-Kantian formalism, . . . the orientation toward immediate experience of unanalyzable but understandable cultural tendencies, . . . a new phenomenological platonism which through visions beholds and justifies norms and essences, then one has all elements of the *wissenschaftlichen* revolution in one's hands. . . . It is a neoromanticism as formerly in the *Sturm und Drang*.[35]

---

31. A. Vierkandt, *Die sozialpaedagogische Forderung der Gegenwart* (Berlin, 1920), p. 20, as quoted by F. K. Ringer, "The German Universities and the Crisis of Learning, 1918-1932" (Ph.D. diss., Harvard University, 1960), p. 145.

32. E. Troeltsch, "Die Revolution in der Wissenschaft," *Schmoller's Jahrbuch (Jahrbuch für Gesetzgebung, Verwaltung . . .), 45* (1921), 1001-1030. Reprinted in Troeltsch's *Gesammelte Schriften,* Vol. 4: *Aufsätze zur Geistesgeschichte und Religionssoziologie,* ed. Hans Baron (Tübingen, 1925; reprinted 1961), pp. 653-677.

33. *Ibid.,* p. 1020. Cf. Max Weber, "Science as a Vocation [1919]," *From Max Weber: Essays in Sociology,* trans. and ed. H. H. Gerth and C. Wright Mills (1946; reprinted New York, 1958), p. 142: "And today? 'Science as the way to nature' would sound like blasphemy to youth. Today, youth proclaims the opposite: redemption from the intellectualism of science in order to return to one's own nature and therewith to nature in general."

34. *Ibid.,* p. 1005.

35. *Ibid.,* p. 1007. Or again, "The peculiarity of German thought, in the form in which it is nowadays so much emphasized, both outside and inside Germany, is primarily derived from the Romantic Movement . . . a revolution, above all, against the whole of the mathematico-mechanical spirit of science in Western Europe . . ." (Troeltsch, "The Idea of Natural Law and Humanity in World

These intellectual currents, whose sources lay in the prewar period, but which welled up immediately following Germany's defeat, continued to dominate the intellectual milieu in the mid-1920's as in the first years of the Weimar Republic. In 1927, Theodor Litt, reviewing contemporary philosophy and its influence upon the ideal of liberal education [Bildung], found *Lebensphilosophie* to be the strongest intellectual current. It was not a system, not a school, but a general tendency which is only to be defined by what it opposes: "On the one hand . . . the mechanism and determinism of a causal explanation which calculates everything in advance, makes everything comparable, dissolves everything into elements—on the other hand . . . the rationalism and formalism of a logical systematization which deduces everything, classifies everything, subjects everything to concepts."[36]

Litt went on to point out again the often noticed parallel between the rise of *Lebensphilosophie* and the "victorious breakthrough of 'wholistic' convictions" in biology (neovitalism) and psychology (Gestaltism, etc.).[37] It is therefore of some interest to ask what impression a biologist-philosopher of these convictions received of the Weimar intellectual milieu. Eloquent in this connection is Hans

---

Politics [1922]," in Otto Gierke, *Natural Law and the Theory of Society, 1500-1800,* trans. E. Barker [Cambridge, 1934; reprinted Boston, 1957], pp. 201-222, on 210.) Troeltsch emphasized (*op. cit.* [note 32], pp. 1003-1004, 1028-1029), that this "Revolution in der Wissenschaft" was confined to the *Geisteswissenschaften,* that the revolutionary innovations in natural science had no clear *weltanschaulich* significance, and he insisted that the close connection of the natural sciences with technology would prevent their sloughing off rigorous methods, or backsliding into "Naturphilosophie" and dilettantism. But, one would ask Troeltsch, what if, under the influence of these same intellectual currents, the exact scientists should repudiate their connection with technology—as indeed they did. Could we then expect some parallel to the romantic physics of the early nineteenth century?

36. Th. Litt, *Die Philosophie der Gegenwart und ihr Einfluss auf das Bildungsideal,* 2nd ed. (Leipzig, 1927), pp. 32-33. Cf. Friedrich Meinecke, "Über Spengler's Geschichtsbetrachtung," *Wissen und Leben, 16* (1923), 549-561, as reprinted in Meinecke's *Werke, 4* (Stuttgart, 1959), 181-195, characterizing the mood of the times: "One is also tired of having only interconnections of cause and effect [Ursache und Wirkung] demonstrated over and over again according to rational methods of cognition, and tired of performing such demonstrations oneself; one is of the opinion that there is a great deal more in life and humanity than an apparatus of mechanical causality [Kausalitäten]. One has become tired of knowing and thirsty for living. . . ."

37. Litt, *loc. cit.*

Driesch's introduction to *Man and the Universe* (1928); for despite his vitalism, wholism, and idealism he too felt the milieu to be hostile to science and reason. Recognizing that it is "unfashionable" to take account of the results of natural science and that he will be put down as betraying his origin as a scientist, he nonetheless accepts the characterization of his method by the opprobrious epithet "rational" and holds that "the modern contempt for [natural] science is due to the fact that its champions take the concept in too narrow a sense, namely, as denoting a mechanistic view of the world."[38]

The historian of science may feel impelled to object that it is a most serious misconception to regard physics after 1900 as "mechanistic," and that it is a complete misunderstanding of positivism to equate it with mechanism, materialism, or even rationalism. Indeed it is difficult to understand how contemporary observers generally failed to recognize in Mach, Ostwald, and their cohorts a quasi-romantic movement parallel in several respects to *Lebensphilosophie*.[39] But all such objections are, of course, entirely beside the point. The relevant question is only what image the educated public held of the physical scientist and his world view. The image of the mechanistic, rationalistic causalist led inevitably to a negative valuation.

I.3. Intellectual Allies: Vienna Circle and Bauhaus

But can this picture of the physical scientist's intellectual milieu be accurate? When we say "Weimar culture" do we not immediately think also of the Vienna Circle and logical positivism, of the Bauhaus and functionalism, as its typical expressions? And were not *these* movements inherently congenial to rational analysis and the achievements of modern physical science and technology?

Assuredly the Vienna Circle, with its goal of a "wissenschaftliche

---

38. H. Driesch, *Der Mensch und die Welt* (Leipzig, 1928), trans. by W. J. Johnson as *Man and the Universe* (London, 1929), pp. 5-8. Cf. Karl Jaspers, *Die geistige Situation der Gegenwart* (Berlin-Leipzig, 1932), trans. by E. and C. Paul as *Man in the Modern Age* (New York, 1933), p. 159: "Anti-science stalks abroad today amid all parties and sects and manifests its influence among persons of the most diversified outlooks, pulverizing the very substance of rational human existence."

39. Stephen G. Brush, "Thermodynamics and History," *The Graduate Journal*, 7 (1967), 477-565, on 530.

Weltauffassung" based upon empiricism and logical atomistic analysis of conceptual structures, had a very "positive" attitude toward the physical sciences and mathematics. But how characteristic was their brand of philosophy? We are sometimes led to believe that logical positivism, which in fact emerged as a coherent program only in 1929/30, was the dominant current in German philosophy throughout the 1920's. Thus H. Stuart Hughes has the movement in full flower by the early 1920's and represents Ludwig Wittgenstein's *Tractatus Logico-Philosophicus,* of which the German edition (1921) lay virtually unread in the final number of Ostwald's defunct *Annalen der Naturphilosophie,* as "the most influential philosophical work of the post-war years. . . . [T]he neo-positivists . . . were able to rehabilitate the scientific method in philosophy . . . and for another two decades Europe was to be without a philosophy that could speak to the ordinary citizen. . . ."[40]

One need, however, only glance at the manifestos of the Vienna Circle in order to recognize that Hughes has utterly misrepresented the case. In *Wissenschaftliche Weltauffassung: Der Wiener Kreis,* the brochure with which in 1929 the circle first came before the public, "their tone," as Ringer rightly points out, "was that of exasperated outsiders."[41] Indeed the opening lines tell us: "Many assert that metaphysical and theologizing thinking, not only in everyday life, but also in science and scholarship [in der Wissenschaft], are today again increasing. . . . The assertion itself is easily confirmed by a glance at the themes of lectures at the universities

---

40. H. S. Hughes, *Consciousness and Society. The Reorientation of European Social Thought, 1890-1930* (New York, 1958; reprinted New York, n.d.), pp. 399-401. Before we accept as fact that the "ordinary citizen" was deserted by philosophy we should hear what Heinrich Rickert had to say on this score in 1920: "The concept which today dominates the general intellectual atmosphere [die Durchschnittsmeinungen] in an especially high degree seems to us to be best designated by the expression *life*. For some time now it has become ever more frequently used, and plays a great role not only among the popular writers, but also among academic philosophers. 'Erlebnis' and 'lebendig' are favorite words, and there is no opinion which is counted so modern as that it is the task of philosophy to give a doctrine of life, which, shaping itself vitally and genuinely out of experience, is capable of being used by the living human being." *(Die Philosophie des Lebens. Darstellung und Kritik der philosophischen Modeströmungen unserer Zeit.* [Tübingen, 1920], p. 4.)

41. Ringer, *op. cit.* (note 28), p. 308.

and at the titles of philosophical publications."⁴² Writing in 1931 for their *Schriften zur Wissenschaftlichen Weltauffassung,* Philipp Frank, the one professional physicist in the group, repeatedly cited and quoted the "Ganzheitsphilosophie" of Othmar Spann's *Kategorienlehre* (1924) as characteristic of the negative valuation of natural science and mathematics in the prevalent "school philosophy."

> To the discipline which depicts things by means of merely external (quantitative) features the essence of things remains eternally foreign. This is the key to why mathematical-causal natural science is not a comprehending, mentally creative discipline as the *Geisteswissenschaften* are. . . . The quantifying, so-called exact, investigation is on the contrary merely measurement and, since it ignores the essence of things and must decompose them into magnitudes in order to inventory them, it does not deserve the name *Wissenschaft* in the same high sense as the *Geisteswissenschaften*. . . . The question of utility and achieved goals is one thing, the worth [Würde] of genuine *Wissenschaften* concerned with totality and essence is another. Such worth modern mathematical natural science does not possess today.⁴³

Far from dominating German philosophy in the 1920's, the Vienna Circle and the corresponding group in Berlin—the Gesellschaft für empirische Philosophie around Hans Reichenbach and Richard von Mises—with their high positive valuation of mathematical natural science represented a rather late and distinctly marginal movement. The impression which in 1929 Sidney Hook brought back to the United States from a year of philosophical study in Germany was that almost all the contemporary schools "are amazingly indifferent to the methods and results of modern physical science." Worse, "The attitude of the German Philosopher to science is not always one of indifference. It is often an attitude of open hostility." The writings of Hans Reichenbach would, Hook thought, be of great interest to the American reader, but in Germany Reichenbach is "ignored by academic philosophers as are all of his

---

42. Verein Ernst Mach, *Wissenschaftliche Weltauffassung. Der Wiener Kreis* (Vienna, 1929), 63 pp., on 9.
43. O. Spann, *Kategorienlehre* (1924), as quoted by P. Frank, *Das Kausalgesetz und seine Grenzen,* Schriften zur wissenschaftlichen Weltauffassung, Band 6 (Vienna, 1932), pp. 54-55.

kind. . . ."[44] Two decades later, sketching the history of the Vienna Circle, Victor Kraft described the great resonance the movement found in western Europe and America, adding ruefully, "It was only in Germany that the Vienna Circle's approach was not taken up at all."[45]

With the Bauhaus the case is somewhat different, for the movement of which Walter Gropius was the leading representative was indeed to a degree characteristic of Weimar culture.[46] Thus in this case we must ask, rather, if the new architecture and the associated movement in design were the expression of an impulse inherently congenial to the methods of the exact sciences or the achievements of modern technology. When one looks at the manifestos of this movement, however, one cannot but be struck by their ambivalence. In the first place, the initial conception and artistic direction of the Bauhaus was largely within the William Morris tradition of a return to handcrafts as a *reaction against* modern technology. When Gropius, in good measure out of simple financial necessity, began to reorient the institution toward industrial design, he had to face tenacious internal resistance. "With absolute conviction I reject the slogan 'Art and Technology—A New Unity,'" Lyonel Feininger wrote in a private letter in August 1923, "this misinterpretation of art is, however, a symptom of our times. And the demand for linking it with technology is absurd from every point of view." An antagonism toward science cum technology was even more explicit in the manifesto which Oskar Schlemmer drafted for the publicity pamphlet of the first Bauhaus exhibition in the summer of 1923.

---

44. S. Hook, "A Personal Impression of Contemporary German Philosophy," *Journal of Philosophy*, 27 (1930), 141-160, on 147, 159. The same view is stated less vigorously by Kurt Grelling, "Philosophy of the Exact Sciences [in Germany]," in *Philosophy Today*, ed. E. L. Schaub (Chicago, 1928), pp. 393-415.

45. V. Kraft, *Der Wiener Kreis. Der Ursprung des Neopositivismus. Ein Kapitel der jüngsten Philosophiegeschichte,* 2nd ed. (Vienna, 1968), p. 8. The first edition was published in 1950.

46. On the architectural side, this is well shown by Barbara Miller Lane, *Architecture and Politics in Germany, 1918-1945* (Cambridge, Mass., 1968). Cf. Gropius' speech before the Thuringian Landtag in Weimar on 9 July 1920: "Based on indisputable facts, I am now going to show convincingly that what the Bauhaus has accomplished is an uninterrupted and logical development that must take place, and already is taking place everywhere in the country." (Hans M. Wingler, ed., *The Bauhaus. Weimar, Dessau, Berlin, Chicago,* trans. W. Jabs and B. Gilbert [Cambridge, Mass., 1969], p. 42.)

"Reason and science, 'man's greatest powers,' are the regents, and the engineer is the sedate executor of unlimited possibilities. Mathematics, structure, and mechanization are the elements, and power and money are the dictators of these modern phenomena of steel, concrete, glass, and electricity . . . calculation seizes the transcendant world: art becomes a logarithm."[47]

Even Gropius himself, moreover, was thoroughly ambivalent on this question. "My primary aim" in planning the curriculum of the Bauhaus was "training the individual's natural capacities to grasp life as a whole, a single cosmic entity. . . . Our guiding principle was that artistic design is neither an intellectual nor a material affair, but simply an integral part of the stuff of life."[48] And so we return once again to *Lebensphilosophie*.

I.4. Educational Ideals and Reforms

Having gotten a clearer picture of the attitudes toward physical science and analytical rationality prevalent among the educated middle classes, and especially strong in the academic world in the Weimar period, we can now better appreciate the great apprehension with which the mathematicians and physicists viewed the movement for educational reform which followed in the wake of the revolution. And it is worthwhile to examine briefly the educational ideals announced by those in the Prussian Ministry of Education with the power to enact and administer such reforms, for one sees thereby both the resonance which these attitudes found across the political spectrum as well as the imminence of the threat to the physical sciences which these attitudes constituted.

The antirationalist theme was sounded at the opening of the Weimar period by Staatssekretär Carl Heinrich Becker. A distinguished Islamicist, Becker had entered the Prussian Education Ministry during the war and, qua Democrat, was elevated to the top civil service position after the Social Democrats threw out the *Kultus-*

---

47. Wingler, pp. 65-66, 69. The pamphlet was suppressed by the Bauhaus after it had been printed—not because of anything Schlemmer said about science, rationality, or technology, but because he had allowed a favorite Bauhaus slogan, "building the Cathedral of Socialism," to slip into the manifesto.

48. W. Gropius, *The New Architecture and the Bauhaus*, trans. by P. M. Shand (London, 1935), pp. 52, 89.

minister Friedrich Schmidt in November 1918.[49] "The basic evil," Becker asserted in 1919 in a widely read essay on university reform, "is the overvaluing of the purely intellectual in our cultural activity, the exclusive predominance of the rationalistic mode of thought, which had to lead, and has led, to egoism and materialism of the crassest form." And again, in another pamphlet written at this same time, Becker maintained that "our entire educational system is too exclusively oriented toward the intellect. We must acquire again reverence for the irrational."[50] This is all perhaps not too surprising in an academic *Geisteswissenschaftler*. It must give one pause, however, when one finds Becker's superior, the Social Democratic *Kultusminister* Konrad Haenisch propagating the same slogans (recall the rationalist-materialist traditions of his party!): "But if . . . the German people, having suffered for decades from the plight of mechanism and materialism, . . . if in our spiritual life not only the intellectual but also the irrational is to receive its due, then the barriers will have to be broken down which presently separate the universities and the people. . . ."[51]

One thus sees that whatever considerations may have led government officials to support and advance academic research in the physical sciences, the attitude of these "progressive" politicians and

49. Erich Wende, *C. H. Becker, Mensch und Politiker* (Stuttgart, 1959). Cf. Friedrich Schmidt-Ott, *Erlebtes und Erstrebtes, 1860-1950* (Wiesbaden, 1952).

50. C. H. Becker, *Gedanken zur Hochschulreform* (Leipzig, 1919), p. ix; *Kulturpolitische Aufgaben des Reiches* (Leipzig, 1919), p. 55: "Wir müssen wieder Ehrfurcht bekommen vor dem *Irrationalen*." For further examples see Adolf Grimme, ed., *Kulturverwaltung der zwanziger Jahre: Alte Dokumente und neue Beiträge* (Stuttgart, 1961), pp. 78-79; Wende, *op. cit.* (note 49), p. 305. Cf. the remarks of William D. McElroy, Director of the U.S. National Science Foundation, at Indiana University, 12 October 1970: "In my view, the science community generally should consider more carefully . . . 'the new romanticism,' emphasizing man as an emotional and feeling creature as well as a reasoning one. A healthy dose of this view may counterbalance some of the extreme emphasis upon rational thinking I suspect is endemic within the science community." (*Science, 170* [1970], 517.)

51. K. Haenisch, *Staat und Hochschule* (Berlin, 1920), pp. 110-111, as quoted by Ringer, *op. cit.* (note 28), p. 282. Haenisch's general readiness to adopt the political and social ideologies of the German academics has been stressed by Hans Peter Bleuel, *Deutschlands Bekenner; Professoren zwischen Kaiserreich und Diktatur* (Berne, 1968), pp. 128-129. Ten years later the Social Democrats once again claimed the Prussian *Kultusministerium*. Their man, Adolph Grimme, was soon writing to Martin Heidegger, 14 May 1930, "as admirer and in a modest sense as pupil. . . . I don't have to tell you how very anxious I am [to get Heidegger to Berlin]. With you here a particular type of philosophy, above all metaphysics, could break through in Berlin." (Grimme, *Briefe*, ed. D. Sauberzweig [Heidelberg, 1967], pp. 36-37.)

bureaucrats toward the "hard" sciences, and particularly toward the intellectual style they associated with these disciplines, was certainly not unambiguously affirmative.[52] "The mood of decisive circles," Wilhelm Hillers warned the Mathematischer Reichsverband in 1921, "is unfavorable to the natural sciences." And when the Prussian education ministry's plan for the reform of the secondary school curricula finally appeared in the spring of 1924, it proved even worse than had been feared. Taking it for granted that "the economic-political, technical, and positivistic age . . . now lies behind us," the ministry refused to justify any part of the curricula on utilitarian grounds. The claims of mathematics and natural science derived from, and only from, the fact that "not only in Kant but also in Goethe [these types of thinking] have co-determined to the very depths the vital features of German idealism"—which, however, was insufficient merit to save these subjects from substantial reductions in the amount of time allotted them. To the older generation of mathematicians and physicists—to Friedrich Poske, Georg Hamel, Felix Klein—who had been struggling since the nineties to make a generous place for their disciplines in the German secondary schools, it seemed that all had been in vain: "This school reform," Klein remarked bitterly, "signifies for our educational system the end of the century of science."[53]

52. A principal consideration underlying the relatively high level of financial support for academic research in the physical sciences was, once again, prestige—in particular the image of science as a *substitute* for political and economic power. See: Brigitte Schröder-Gudehus, *Deutsche Wissenschaft und internationale Zusammenarbeit* (diss. Geneva, 1966), pp. 181-189, 199; P. Forman, "Scientific Internationalism and the Weimar Physicists," *Science, War, and Internationalism, 1900-1939*, ed. Roger Hahn (in press).
53. W. Hillers, *Jahresbericht der Deutschen Mathematiker-Vereinigung, 31* (1922), Part 2, pp. 120-121. *Die Neuordnung des preussischen höheren Schulwesens: Denkschrift des Preussischen Ministeriums für Wissenschaft, Kunst und Volksbildung* (Berlin, 1924), reprinted in Hans Richert, ed., *Richtlinien für die Lehrpläne der höheren Schulen Preussens*, 7th ed. (Berlin, 1927), *1*, 17-77, on 68-70. F. Klein quoted indirectly by F. Poske, "Der naturwissenschaftliche Unterricht und die Neuordnung des preussischen höheren Schulwesens," *Naturwiss., 13* (1925), 73-75. Klein himself, in a letter to G. Hamel, remarked on the "remarkable circumstance that the development of the German school system has taken an entirely different direction" than the one he had foreseen in his address of June 1918. (*Unterrichtsblätter für Mathematik und Naturwissenschaft, 30* [1924], 44-45). Hamel, speaking for the Mathematischer Reichsverband (*Jahresbericht der Deutschen Mathematiker-Vereinigung, 33* [1924], Part 2, p. 63), quite agreed: "In fact the new school reform signifies the complete repudiation of [Abkehr von] the previous development . . . throws us way back before the time of the first school reform of 1892."

I.5. The Crisis of *Wissenschaft*

In the preceding sections I have explored from several directions the attitudes toward science and reason permeating the intellectual environment of the Weimar physicists and mathematicians. But the intellectual milieu is not fully characterized by a specification of such *substantive* constituents, not even when the catalog of valuations is supplemented by a measure of the intensity with which each of these attitudes is held. To fully characterize an intellectual atmosphere one must specify not merely the likes and dislikes, the sympathies and antipathies, but also the mood, the morale, the accepted view of the contemporary cultural situation, and the common notions of what that situation demanded, or where it must lead.

Turning back once again to the general intellectual historians and to the contemporary observers of the intellectual scene, we find as before remarkable unanimity about this essential dimension of the intellectual milieu: widespread among the educated middle classes, but especially oppressive in academia, was a generalized sense of crisis. Included therein was the permanent political and economic crisis, but, far from being limited to this, the fundamental phenomenon was felt to be a moral and intellectual crisis, a crisis of culture, a crisis of science and scholarship. Fritz Ringer, who has given the closest attention to German academic ideology, and, in particular, to this "crisis of learning," found that:

> Throughout the Weimar period, it was often said in academic circles that a crisis was in progress. No one felt the need to define the exact nature of this crisis, to ask where it came from or what it involved. "Sometimes [the educator Aloys Fischer wrote in 1924], the present situation is represented as a crisis of the . . . economic system only, sometimes as one of politics and of the idea of the state, or as a crisis of the social order. At other times it is conceived more deeply and inclusively as a crisis of the entire intellectual and spiritual culture. . . ." In any case, the crisis existed, if only by virtue of the fact that almost every educated German believed in its reality.[54]

---

54. Ringer, *op. cit.* (note 28), p. 245. It would be more accurate, perhaps, to say that although many academics did indeed feel a need to define the exact nature of this crisis, the diagnoses were often diametrically opposed. For example, Arthur Liebert, *Die geistige Krisis der Gegenwart*, 2nd ed. (Berlin, 1923),

This notion of a crisis in or of learning, although it had roots running back into the previous century, emerged as a universally cogent cliché only in the aftermath of Germany's defeat. "The phrase 'Krisis der Wissenschaft' has already become a popular slogan in everyone's mouth," the political economist Arthur Salz noted in 1921.[55] So it continued throughout the Weimar period: "The idea of such a crisis of culture [Kulturkrise]," Pierre Viénot observed a decade later, "belongs today to the solid stock of the common habit of thought in Germany. It is a part of the German mentality."[56] And at a quarter century's distance it remained clear to Werner Richter, Becker's lieutenant in charge of the University Section of

---

pp. 7-9, expressed this need very forcefully: the purpose of his essay "is not to establish and portray any one arbitrarily selected crisis from contemporary life, no matter how staggering a force it may possess. The intent is much rather to expose *the* crisis of our time and of simply the entire contemporary world view and life mood, i.e. the concept and meaning of all the individual crises and the common spiritual and metaphysical wellspring by which they are all conditioned and from which they are all fed." This *he* found in "the disastrous historical scepticism and relativism nourished by historicism."

55. A. Salz, *Für die Wissenschaft. Gegen die Gebildeten unter ihren Verächtern* (Munich, 1921), p. 10. Typical for this period, and perhaps for this sort of phenomenon, is the circumstance that even the "opponents" shared in large measure the attitudes they were attacking. Thus Troeltsch, *op. cit.* (note 32), p. 1026, found Salz's essay "very instructive and symptomatic, above all in its almost fatalistic surrender to the anti-scientific, and in that sense revolutionary, currents." Another, more pertinent, example of this circumstance is Adolf von Harnack, historian-theologian but President of the Kaiser-Wilhelm-Gesellschaft. Responding to Karl Barth's challenge, hurling his "Fünfzehn Fragen an die Verächter der wissenschaftlichen Theologie unter den Theologen [1923]," Harnack warned against efforts "to revile, indeed eliminate reason [Vernunft]. . . . Does not gnostic occultism even now raise itself upon the ruins?" Yet on another occasion ("Stufen wissenschaftlicher Erkenntnis [1930]") we find this spokesman for the interests of the natural sciences declaring that "our intellect [Verstand] is the born mathematical physicist; like the mathematical physicist it abstracts, it calculates, it weighs." But "this abstracting method which corresponds to mechanism" is incapable of grasping the "life," "forms," "wholes" which surround us. Moreover natural science is only the second step in the cognitive hierarchy; above it stands knowledge of life, followed by knowledge of man, while the fifth, last, and highest step is occupied by philosophy. (Harnack, *Ausgewählte Reden und Aufsätze,* ed. A. von Zahn-Harnack and Axel von Harnack [Berlin, 1951], pp. 132-134, 177-180.)

56. P. Viénot, *Ungewisses Deutschland. Zur Krise seiner bürgerlichen Kultur,* trans. Eva Mertens (Frankfurt a.M., 1931), pp. 24-25. Typical is the observation of the former Prussian Minister of Education Otto Boelitz, in *Grundsätzliches zur Kulturlage der Gegenwart,* Flugschriften der deutschen Volkspartei, Folge 77 (Berlin, 1931), p. 5: "so gehe ich aus von der Kennzeichnung der Kultur der Gegenwart als einer Kulturkrisis."

the Prussian Ministry of Education, that "The self-image [Selbstverständnis] of that period was decisively influenced by the consciousness of a crisis of culture."[57] I will return in Part II to consider how this generalized sense of crisis may have affected the rhetoric and the *Selbstverständnis* of the Weimar physicists and mathematicians; here I would only emphasize that implicit in this sense of crisis was a negative valuation of the traditional scientific disciplines, methods, and practitioners. If the educated public were convinced that "today's *Wissenschaft*, together with its methods, has run up a blind alley,"[58] then, inevitably, the stature of those who had cultivated these sciences by these methods would be considerably diminished. Conversely, if the scholar or scientist was to maintain his prestige in his academic environment and beyond, he too would have to acknowledge and affirm the crisis, would have to repudiate the traditional methods and doctrines of his discipline.[59]

It was, of course, especially the radical *Lebensphilosophen* who pressed this interpretation of the crisis of learning. As a crisis of "causal monism," of positivistic methods in *Wissenschaft*, the crisis of learning must be followed by a revolution which liquidates this barren and intolerable mechanism in favor of a "new *Wissenschaft*" of values, intuition, feeling, of the living, the organic.[60] But the

---

57. W. Richter, *Wissenschaft und Geist in der Weimarer Republik*, Arbeitsgemeinschaft für Forschung des Landes Nordrhein-Westfalen, Geisteswissenschaften, Nr. 80 (Cologne, 1958), 31 pp., on 11.

58. A. Salz, *op. cit.* (note 55), p. 10. Cf. Hermann Weyl, "Felix Kleins Stellung in der mathematischen Gegenwart," *Naturwiss.*, 18 (1930), 4-11, on 6: "der Typus des Gelehrten und die Wissenschaft in ihrer Geltung und ihrem Wert während der letzten Jahrzehnte in Frage gestellt waren, im Zeichen der Krisis standen." (Reprinted in Weyl's *Gesammelte Abhandlungen* [Berlin, 1968], *3*, 292-299.)

59. The Weimar academics were indeed much agitated by a variety of perceived threats to their social prestige, their intellectual leadership, and their economic situation. Yet, by and large, they saw democracy and republican institutions as the cause of their falling esteem, and rather looked to *Lebensphilosophie* to restore their power and prestige. Cf. note 40, above.

60. E. Troeltsch, *op. cit.* (note 32), p. 1023. To take but one of innumerable examples: Ernst Barthel, Privatdozent U. Köln, in an essay on "Mechanischer und organischer Naturbegriff," *Annalen der Philosophie, 5* (1925), 57-76, argued that the "three principles" of space, time, and causality have been sufficiently analyzed and recognized as the "fundaments of rational thought. . . . It is, however, also generally conceded that the usual rational thinking through of the world totality assuming these principles leads to essential antinomies and incomprehensibilities, which bring thought to the limits of its competence. The organic conception of nature would like to ask itself now, whether it could

conviction of the reality of this crisis of learning—including the image of science in a cul de sac—was even more widely spread than formal *Lebensphilosophie* itself. Thus "the collapse of science" was also trumpeted vigorously by Hugo Dingler, a prolific and widely read philosopher of physics, whose orientation was strongly rationalistic: "The state where nothing is any longer really certain, everything is possible and at the same time every possible position is also maintained, where there is no longer any basis and any guidelines, nothing, nothing which may be considered certain—in a word, chaos, collapse. In that state we stand; right in the middle of it. The public does not suspect it, and the *Gelehrten,* often frantically, shut their eyes."[61] And in Dingler's view "this new collapse of science, in whose midst we stand . . . consists in the collapse of the belief in the certainty of the experimental principle," i.e., of the possibility of establishing the truth of a theory upon its agreement with experiment.[62]

Here the "crisis of learning" begins to touch the Weimar physicists very nearly. And when we find the aging and conservative professor of Experimental Physics at Privatdozent Dingler's institution taking advantage of his term as rector to contradict and deplore the notion of "the collapse of science," at least in respect of his own science,[63] then I think it fair to infer that the physicists who could not or would not join the revolution saw such doctrines as diminishing their claims and threatening their prestige both in the academic world and among the public at large.

---

possibly be that the one and only way of thinking is to make three abstract principles the basis for explanation of a world full of concrete living contents, or whether, on the contrary, also the opposite way . . ." (p. 71). The exploration of this "opposite way" leads to the conclusion of the article (pp. 75-76) that "the quality of the phenomena and their mutual connection lies in a region of noncausal [nichtkausal] harmony, which is to be grasped only by the intuition" and that, in general, one may distinguish between "mechanical" and "organic" scientific research in that the aim of the former is "practically utilizable, hypothetical causal abstractions, of the latter an intuitive cognition of the immanent essential connections."

61. H. Dingler, *Der Zusammenbruch der Wissenschaft und der Primat der Philosophie,* 2nd ed. (Munich, 1931), p. 10. The second edition differs from the first, 1926, only in the addition of supplementary notes.
62. *Loc. cit.*
63. W. Wien, *Vergangenheit, Gegenwart und Zukunft der Physik. Rede, gehalten beim Stiftungsfest der Universität München am 19. Juni 1926,* Münchener Universitätsreden, Heft 7 (Munich, 1926), 18 pp., on 18.

I.6. Spengler's *Decline of the West*

The crisis of culture, the revolution in *Wissenschaft*, radical *Lebensphilosophie*, all proclaimed and epitomized by a sweeping theory of world history in which—"das ist das Neue"[64]—physics and mathematics are treated alongside art, music, and religion as wholly culturally conditioned. The first volume of Oswald Spengler's *Der Untergang des Abendlandes*,[65] in which the theory is presented and to which the extensive discussions of science are largely confined, appeared in 1918. Although at certain points it bore the stamp of the wartime mind—as, for example, in its positive valuation of technology—on the whole its fatalistic-relativistic pessimism was precisely the right tone for a defeated Germany. In five years the first volume went through thirty printings, and by 1926 the revised edition published in 1923 had gone through a further thirty printings —altogether 100,000 copies in a country with scarcely three times that number of college graduates. Almost universally read in academic circles—by the physicists, too, as we shall see—the typical professorial reaction was: "Of my discipline Spengler understands, of course, not the first thing, but aside from that the book is brilliant."[66]

Ernst Troeltsch saw the first volume of the *Untergang* as the paradigm of the revolution in science: "It is the first decisive public

---

64. E. Troeltsch's review of the first volume of *Der Untergang* in the *Historische Zeitschrift* (1919), reprinted in Troeltsch's *Ges. Schr.*, 4 (*op. cit.*, note 32), 682. Troeltsch found it, inter alia, "ein bedeutsames Kulturdokument aus der Zeit einer geistigen Krisis der deutschen Wissenschaft."

65. O. Spengler, *Der Untergang des Abendlandes. Umrisse einer Morphologie der Weltgeschichte*. Vol. 1: *Gestalt und Wirklichkeit* (Munich, 1918). The first thirty-two editions are unaltered, and of these the third through thirty-second (1920-1922) have essentially the same pagination; I refer to these latter editions as the "orig. ed." Editions 33 through 47, published in 1923, are the revised edition; they all have the same pagination and are referred to as the "rev. ed." The English translation of the revised edition by C. F. Atkinson, *The Decline of the West*, Vol. 1: *Form and Actuality* (New York: Knopf, 1926) is referred to as the "Eng. ed."

66. Or so it appeared to Gerhard Hessenberg, *Vom Sinn der Zahlen. Akademische Antrittsrede, gehalten an der Universität Tübingen am 8. Dezember 1921* (Leipzig, 1922), 56 pp., on 31. So also had it appeared to Hessenberg's friend Leonard Nelson, *Spuk; Einweihung in das Geheimnis der Wahrsagerkunst Oswald Spenglers* . . . (Leipzig, 1921). And in 1923 Friedrich Meinecke noted (*op. cit.*, note 36) that "When the first volume of the *Decline of the West* appeared one frequently heard from the circle of professional scholars [Fachgelehrten] the judgment: 'What he says about my field is, indeed, complete nonsense. But all the rest is very ingenious [geistreich].' "

revelation of the new *Wissenschaft,* and thereupon rests a great part of its captivating effect." For Lukács, Spengler is the characteristic representative of the *Lebensphilosophie* of the postwar years.[67] And for us his book is all the more valuable as an index of the attitudes toward science and reason in the intellectual environment of the early Weimar period because, on the one hand, it gives a prominent place to physics and mathematics, and because, on the other hand, it was the one expression of those attitudes to which the Weimar physicists and mathematicians uniformly exposed themselves.

The Spenglerian account of world history is based on the proposition that the principal cultures are autonomous organisms, each wholly unique apart from a common life cycle. Every cultural manifestation—art, science, or whatever—is simply and solely an expression of the soul of that particular culture and as such is neither "valid" nor even comprehensible outside that culture, i.e., at any other time or place: "Each culture has its own new possibilities of self-expression which arise, ripen, decay, and never return. There is not *one* sculpture, *one* painting, *one* mathematics, *one* physics, but many, each in its deepest essence different from the other, each limited in duration and self-contained."[68]

After sketching his program in the "Introduction," Spengler aims in his first chapter, "The Meaning of Numbers," to establish this thesis of mutual nonintelligibility once and for all by proving (by iteration) that "There is not and cannot be number as such. There are several number-worlds because there are several cultures."[69] Likewise, "There is no mathematic, but only mathematics."[70] In the following chapters this radical relativism is extended to natural science, above all, to physics.

And in fact, in the historian's view there is only a *history of physics.* All its systems now appear to him as neither correct nor incorrect, but

---

67. Troeltsch, *op. cit.* (note 32), p. 1014. Lukács, *Zerstörung der Vernunft (op. cit.,* note 25), pp. 364-378. The same general view of Spengler's role is taken by Helmut Kuhn, "Das geistige Gesicht der Weimarer Zeit," *Zeitschr. f. Politik, 8* (1961), 1-10.
68. Orig. ed., p. 29; rev. ed., p. 29; Eng. ed., p. 21.
69. Orig. ed., p. 85; rev. ed., p. 81; Eng. ed., p. 59.
70. "Es gibt keine Mathematik, es gibt nur Mathematiken." (Orig. ed., p. 88; rev. ed., p. 83; Eng. ed., p. 60.)

as historically, psychologically conditioned by the character of the epoch, and representing that character more or less completely.[71]

... a firstrate scientist of the time of Archimedes would have declared himself, after a thorough study of our modern theoretical physics, quite unable to comprehend how anyone could assert such arbitrary, grotesque, and involved notions to be science, still less how they could be claimed as necessary consequences from actual facts.[72]

Writing off all criteria for the truth of a scientific theory as themselves culture-bound illusions, and dismissing with a wave of his hand the argument from the fact that "the machine works," as Boltzmann put it, Spengler maintained: "There simply are no conceptions other than anthropomorphic conceptions ... so is it certainly with every physical theory, no matter how well founded it is supposed to be. Every such is itself a myth, and in all its features anthropomorphically preformed. There is no pure natural science, there is not even a natural science which could be designated as common to all men [als allgemein menschlich]."[73] And although Spengler was a bit less categorical in the second edition, altogether his extension of extreme cultural relativism to physics and mathematics was meant and was received as a direct challenge to the ideology of the exact scientists. At first they might refuse to hear anything of it, but, asked repeatedly for their reactions, within a year or two they all had to confront it.

Still more important for our present inquiry than Spengler's notorious theses regarding the nonobjectivity of the exact sciences are Spengler's specific interpretations of post-Renaissance physics, its content and its future. The content of Western physics and mathematics is, of course, but an expression of the soul of Western culture —of the "Faustian" culture, as Spengler calls it. And the essential, determinative characteristic of "Faustian" science is, we are no longer surprised to learn, "the *Kausalitätsprinzip*—the logical form of the Faustian world-feeling:"[74] "We see then that the causality-principle, in the form in which it is selfevidently necessary for us —the agreed basis of truth for our mathematics, physics and philos-

71. Orig. ed., p. 167; deleted in rev. ed.
72. Orig. ed., p. 530; rev. ed., p. 491; Eng. ed., p. 380.
73. Orig. ed., p. 533; deleted in rev. ed.
74. Orig. ed., p. 551; deleted in rev. ed.

ophy—is a Western and, more strictly speaking, Baroque phenomenon. . . ."[75] But although causality has reigned supreme in modern exact science, it is nonetheless—here comes the *Lebensphilosophie*—an artificial construction erected as a defense against the more fundamental, and fundamentally irrational, notion of destiny, *Schicksal*. This, indeed, is the "key" to the problem of world history:

> I mean the opposition of the *destiny-idea* and the *causality-principle*, an opposition which, in its deep world-shaping necessity, has never hitherto been recognized as such. . . . Destiny is the word for an indescribable inner certainty. One makes the essence of the causal clear by means of a physical or epistemological system, by means of numbers, by means of conceptual analyses. . . . The one requires us to dismember, the other to create, and therein lies the relation of destiny to life and causality to death.[76]

Thus we have the fundamental *lebensphilosophisch* theme, with which we are already all too familiar, inflated to cosmic proportions. Over and over again Spengler equates causality, conceptual analysis, and physics, and flays them across the stage of world history.

> For the principle of causality is late, unusual, and only for the energetic intellect of higher cultures a secure, somewhat artificial possession. Out of it speaks fear of the world. Into it the intellect banishes the demonical in the form of a continually valid necessity, which rigid [starr] and soul-destroying is spread over the physical world-picture. Causality is coextensive with the concept of law. There are only causal-laws.[77]

> The abstract savant, the natural scientist, the thinker in systems, whose entire mental existence is founded upon the principle of causality, is a "late" manifestation of the hatred of the powers of destiny, of the incomprehensible.[78]

> The words "time" and "destiny," for anyone who uses them instinctively, touch life itself in its deepest depths—life as a whole, which is not to be separated from lived experience. On the other hand, physics,

---

75. Orig. ed., p. 549; rev. ed., pp. 507-508; Eng. ed., p. 392.
76. Orig. ed., pp. 164-165; slightly, but insignificantly, altered in the rev. ed., pp. 154-155; Eng. ed., pp. 117-118.
77. Orig. ed., p. 165; first sentence modified and second and third sentences omitted in rev. ed., p. 155; Eng. ed., p. 118.
78. Orig. ed., pp. 168-169; rev. ed., p. 158; Eng. ed., p. 120.

reason, *must* separate them. The livingly-experienced in itself, detached from the living act of the observer and become an object, dead, inorganic, rigid [starr]—that is now Nature as mechanism, i.e., as something to be exhausted mathematically. . . . This is the eternal embarrassment of all physics as the expression of a soul. All physics is treatment of the problem of motion, in which lies the problem of life itself, not as if it could one day be solved, but even though it is unsolvable.[79]

Striking in this last passage is Spengler's elaboration of a notional complex advanced by Bergson and soon to be codified as existentialism in such works as Heidegger's *Being and Time*.[80] Time, Spengler assures us, is "something intensely personal," indeed, "we ourselves, insofar as we live, are time." It follows, therefore, that physics really "has nothing whatever to do with time," knows no direction of time, eliminates time in favor of a "web of cause and effect . . . of timeless duration."[81] Likewise to be noted for later reference is Spengler's favorite epithet for causality—*starr,* i.e., stiff, rigid; it is intended to evoke and reinforce an antithesis between causality and life, an association of causality with death (cf. *die Totenstarre,* rigor mortis).[82]

Spengler's indictment of physics=causality is all the weightier because he pretends to be a connoisseur of the physical sciences and modern technology, for whom "the depths and refinement of mathematical and physical theories are a joy," who "would sooner have the splendidly clear, highly intellectual forms of a fast steamer, of a steel structure, of a precision lathe, the subtlety and elegance of certain chemical and optical processes, than all the pickings and stealings of present day applied art, architecture and painting included."[83] *He* is not to be dismissed as an aesthete, a romantic; *he* is a hard-headed realist who, with a full appreciation of modern physics, "our ripest and strictest science,"[84] "the masterpiece of the Faustian spirit,"[85]

---

79. Orig. ed., pp. 542-543; rev. ed., pp. 501-502, where the sentence "This is the eternal embarrassment . . ." is deleted; Eng. ed., pp. 388-389. Cf. Harnack (*op. cit.,* note 55): "our intellect is the born mathematical physicist."
80. Published in Edmund Husserl's *Jahrbuch für Philosophie und phänomenologische Forschung, 8* (1927), 1-438.
81. Orig. ed., pp. 170-172; rev. ed., pp. 158-160; Eng. ed., pp. 120-122.
82. Thus orig. ed., pp. 69, 165, 167, 574, and p. 156 of the rev. ed. where an additional *starr* is added: "die starre Weltmaske der Kausalität."
83. Orig. ed., p. 60; rev. ed., p. 60; Eng. ed., pp. 43-44.
84. Orig. ed., p. 215; rev. ed., p. 205; Eng. ed., p. 156.
85. Orig. ed., p. 608; deleted in rev. ed.

tells us what sort of cultural manifestation it really is, and what, according to the ineluctable cycle of cultural development, its fate must be.

> Before us there stands a last spiritual crisis that will involve all Europe and America. What its course will be late Hellenism tells us. The tyranny of reason—of which we are not conscious, for the present generation is its apex—is in every culture an epoch between man and old-man, and no more. Its most distinct expression is the cult of the exact sciences, of dialectics, of demonstration, of causality.[86]

> Now, the history of the higher cultures shows that "science" is a transitory spectacle, belonging only to the autumn and winter of their life-courses, and that . . . a few centuries suffice for the complete exhaustion of its possibilities. Classical science faded out between the battle of Cannae [216 B.C.] and that of Actium [31 B.C.] and made way for the world outlook of the "second religiousness." And from this it is possible to calculate in advance the end of western natural science.[87]

> It remains now to sketch the last stage of western science. From our standpoint of today the already declining route is clearly visible. . . . In this very century, I prophesy, in the age of scientific-critical Alexandrianism, resignation will overcome the will to victory of science. European science is advancing toward self-destruction through refinement of the intellect. . . . But from skepsis a path leads to the "second religiousness." . . . No one yet believes in the exhaustion of the spirit even though we already feel it acutely in all our limbs. But two hundred years of civilization and orgies of scientificness—then one is fed up. Not the individual, but the soul of the culture itself has had enough, and expresses this by choosing to put into the historical field of the day ever smaller, narrower, and more unfruitful researchers . . . in physics as in chemistry, in biology as in mathematics, the great masters are dead, and we are experiencing today the decrescendo of the stragglers, who arrange, collect, and conclude, like the Alexandrians of the Roman period.[88]

---

86. Orig. ed., pp. 607-608; rev. ed., p. 551; Eng. ed., p. 424.
87. Orig. ed., p. 532; rev. ed., p. 492, where the phrase "and made way for the world-outlook of the 'second religiousness' " was added and the final sentence softened to: "And from this it is possible to foresee a date at which our Western scientific thought shall have reached the limit of its evolution." (Eng. ed., p. 381.)
88. Orig. ed., pp. 607-609; rev. ed., pp. 551-553, where altered slightly but insignificantly; Eng. ed., pp. 424-425.

I would happily let this stand—and it was meant to stand—as the measure of Spengler's world-historical vision, for a more erroneous description and valuation of early twentieth-century physics could scarcely be devised. Yet as perverse and denigrating as this image is, it must still be recognized for what is was—an integral part of an analysis of Western culture, its present state, and its future prospects, which expressed and shaped the notions and inclinations of the educated middle classes in postwar Germany.

In another respect, however, Spengler's analysis of contemporary physics, confused and contradictory like all else in his treatise, shows a flash of prescience. For physics in his generation is not merely plodding forward in a beaten track, tying up loose ends, it is also, according to Spengler, disintegrating *and* metamorphizing, undergoing a transformation of the goals and principles of scientific explanation paralleling the *Zeitgeist,* the "second religiousness."

> Western European physics—let no one deceive himself—has reached the limit of its possibilities. . . . This is the origin of the sudden and annihilating doubt that has arisen about things that even yesterday were the unchallenged foundation of physical theory, about the meaning of the energy principle, the concepts of mass, space, absolute time, and causal natural laws generally [n.b.] . . . this doubt extends to the very possibility of a natural science. What deep and utterly unconscious skepsis lies, for example, in the rapidly increasing use of enumerative and statistical methods, which aim only at the probability of the results, and forgo in advance the absolute exactitude of the laws of nature, as one understood it in hopeful earlier generations.[89]

It is not, of course, to the quantum theory which Spengler refers here in the original edition. In speaking of the concepts of mass, space, time, energy he evidently has above all the theory of relativity in mind, and it is primarily to the atomistic foundation of the second law of thermodynamics that the remarks about statistics and probability refer. But the talk of doubt about the concept of causal natural laws points beyond these theories; I do not myself know just what Spengler, the seer, has in mind,[90] but his images and associa-

---

89. Orig. ed., pp. 596-597; rev. ed., pp. 541-542, where again modified insignificantly; Eng. ed., pp. 417-418.

90. Uncited, like virtually all of Spengler's sources, but almost certainly important was Wilhelm Wien's "Ziele und Methoden der theoretischen Physik.

tions are certainly suggestive of those which, as I show in Part III, soon after begin to appear in the writings of German theoretical physicists.

> Statistics belong, like chronology, to the domain of the organic, to fluctuating life, to destiny and incident and not to the world of exact laws and timeless eternal mechanics. . . .
>
> The more dynamics exhausts its inner possibilities as it nears the goal . . . the more insistently the organic necessity of destiny asserts itself side by side with the inorganic necessity of causality. . . . The course of this process is marked by the appearance of a whole series of daring hypotheses, all of like sort. . . . Above all, this is manifested in the bizarre hypotheses of atomic disintegration. . . . This destiny strikes only a few individuals in an aggregate of radioactive atoms, the neighbors being entirely unaffected.[91]

Here, then, is the fate and the salvation of physics—a reunification of thought and feeling, a self-discovery of physics as a fundamentally religious-anthropomorphic expression:

> The goal reached, the vast and ever more meaningless and threadbare fabric woven by natural science falls apart. It was, after all, nothing but the inner structure of the mind. . . . But what appears under the fabric is once again the earliest and deepest, the myth, immediate becoming, life itself. . . . Out of the religious soulfulness of the gothic there grew up the urban intellect, the alter ego of irreligious natural science, overshadowing the original world feeling. But today, in the sunset of the scientific epoch, in the stage of victorious skepsis, the clouds dissolve and the quiet landscape of the morning reappears in all distinctness . . . weary after its striving, the Western science returns to its spiritual home.[92]

Festrede . . . Universität zu Würzburg. Gehalten am 11. Mai 1914," printed in several places at that time and reprinted in Wien, *Aus der Welt der Wissenschaft. Vorträge und Aufsätze* (Leipzig, 1921), pp. 150-171. If this is the case, the doubts about causality which Spengler thinks he finds among the physicists, like so much of what Spengler reads into and out of his scientific sources, is simply a confusion. If, however, Spengler is drawing upon Max Planck's *Festrede* of 4 August 1914 (see note 158 below) he has evidently understood his author very well.

91. Orig. ed., pp. 603,. 605-606; rev. ed., pp. 547 (where "timeless eternal mechanics" becomes "timeless causality"), 549-550; Eng. ed., pp. 421, 423.

92. Orig. ed., pp. 614-615; rev. ed., pp. 556-557; Eng. ed., pp. 427-428.

II. ADAPTATION OF IDEOLOGY TO THE INTELLECTUAL ENVIRONMENT

II.1. Introduction

Spengler epitomizes for us a set of attitudes, widely diffused among educated Germans, explicitly hostile to the ideology of the exact sciences and to particular concepts employed within them. In the remainder of this paper I explore some aspects of the response of the representatives of these sciences in German-speaking Central Europe—in the first instance of the response at the level of ideology; i.e., I explore the effect of this intellectual environment upon the professed justifications of scientific activity, upon the epistemological stance of the exact scientists, and upon their elan, their esprit, their confidence in the future of their discipline.

I do not, however, undertake to construct here a comprehensive typology of ideological responses to the Weimar intellectual milieu, but limit myself to illustrating a few of the more striking ideological adaptations. The resulting picture, emphasizing examples of accommodation, but not examples of resistance, is necessarily one-sided. Nonetheless, the imbalance is not as great as one might expect, both because the instances of a physicist or mathematician forcefully advancing ideals antithetic to those of his milieu are rare indeed, especially before the last years of the Weimar period,[93,94] and

---

93. Apart from the literature discussed below (Part III) in connection with the dispute over the law of causality, examples from the early Weimar period are almost exclusively in the form of, and limited to, rejections and rebuttals of Spengler's book and theses. Such are the tracts by Leonard Nelson and Gerhard Hessenberg cited in note 66; P. Riebesell, "Die Mathematik und die Naturwissenschaften in Spengler's 'Untergang des Abendlandes,'" *Naturwiss.*, 8 (1920), 507-509; and the preface to the second edition of Franz Exner's *Vorlesungen über die physikalischen Grundlagen der Naturwissenschaften* (Leipzig-Vienna, 1922), pp. vi-xiii. Other affirmations of the "scientific approach" are at the very least ambivalent: so, for example, the "Introduction" to Max Born's *Die Relativitätstheorie Einsteins* (1920), trans. H. L. Brose as *Einstein's Theory of Relativity* (London, 1924), pp. 1-6. In reprinting the "Einleitung" as the first selection in Born's *Physik im Wandel meiner Zeit*, 4th enlarged ed. (Braunschweig, 1966) the highly characteristic epigraph from Goethe was omitted: "The most perfect pleasure of the thinking human being is to have successfully researched that which is researchable, and to calmly revere that which is unresearchable."

94. Only two outspoken, unambivalent affirmations of an antithetic ideal before an academic audience are known to me; both are from the *late* Weimar period when, as Sontheimer has pointed out *(op. cit.,* note 26, pp. 43 ff.), there

because, as we shall see, it is often the case that the same scientist who in one context offers resistance to the antiscientific currents of his milieu, in another context can be found flirting with propositions intimately associated with those same currents. Moreover, this one side of the scientist's response is particularly interesting and instructive as largely contradicting the usual assumption of the intellectual autonomy of modern professionalized scientific disciplines. It thus provides essential motivation and support for my contention in Part III that the German physicists' predisposition toward acausal laws of nature likewise arose as a form of accommodation to their intellectual environment.

---

was a general stiffening of resistance among intellectuals to irrationalism: Richard von Mises, *Über das naturwissenschaftliche Weltbild der Gegenwart. Rede bei der Feier . . . der Berliner Universität . . . am 27. Juli 1930* (Berlin, 1930), 29 pp., reprinted in *Naturwiss., 18* (1930), 885-899; Konrad Knopp, "Der Einfluss der Naturwissenschaft auf das moderne Bildungsideal," in H. Gerber, ed., *Die Universität. Ihre Geschichte, Aufgabe und Bedeutung in der Gegenwart. Öffentliche Vorträge der Universität Tübingen, Wintersemester 1932-33* (Stuttgart, 1933), pp. 189-217. Mention might here also be made of Wilhelm Blaschke, *Leonardo und die Naturwissenschaften. Rede, gehalten am 10. November 1927, zum Antritt des Rektoramts an der Universität Hamburg,* Hamburger mathematische Einzelschriften 4 (Leipzig, 1928), 15 pp.; of Walther Kossel's *Die Einheit der Naturwissenschaft. Rede beim Antritt des Rektorats der . . . Universität [Kiel] am 5. März 1929* (Kiel, 1929), 22 pp.; and certainly also of the closing lines of David Hilbert's public address at the Naturforscherversammlung in Königsberg in September 1930, "Naturerkennen und Logik," *Naturwiss., 18* (1930), 959-963, reprinted in Hilbert's *Gesammelte Abhandlungen, 3* (Berlin, 1935; reprinted New York, 1965), 378-387: "he who is sensible of the truth of the liberal way of thinking and world view which shines forth from these words of Jacobi, he does not succumb to reactionary and unfruitful scepticism; he will not believe those who today with philosophic mien and superior tone prophesy the decline of western culture [den Kulturuntergang] and take pleasure in declaring 'ignorabimus.' For the mathematician there is no 'ignorabimus,' and equally for natural science, in my opinion, there is none whatsoever. . . . The true reason that Comte was not able to find an insoluble problem is, in my opinion, that there simply does not exist an insoluble problem. In place of this foolish 'ignorabimus' let our solution on the contrary be: We must know,/ We will know." Hermann Weyl, "Zu David Hilberts siebzigstem Geburtstag," *Naturwiss., 20* (22 January 1932), 57-58, reprinted in Weyl's *Ges. Abhl., 3,* 346-347, quoted this last sentence, adding that "our contemporaries are not glad to hear that sort of thing; they see in it shallow-minded rationalism or human presumption and with a torrent of confused words appeal to 'life itself' or the deeper 'existential truth' or the 'creatureliness' of man as justifying their repudiation of reason [Ratio]. And granted; a sentence here and there in Hilbert's address sounds suspiciously like the words with which Gottfried Keller ridiculed his natural scientist. . . . Nonetheless, one does Hilbert an injustice if one tosses his rationalism into the same pot with, say, that of a Haeckel."

As today with the "ecology" fad, so also in the Weimar period it was the biologist who could most easily adapt his ideology and values to those of his intellectual milieu. Life, that central symbol, was his own subject. Paraphrasing a spokesman for the discipline,[95] of all natural sciences biology merits an especially ample place in the school curriculum for it is least deserving of the reproach of aiming at "knowledge for power"; its mission is to counter the alienation from nature in our technical age; it provides the link between the *Naturwissenschaften* and the *Geisteswissenschaften* because it works in part with the concept of scientific law, but also with the techniques of understanding and imparting of meaning; it brings us to the edge of the irrational and teaches us to respect that which is beyond rational investigation. Nor are these arguments merely for public and governmental consumption; the eminent embryologist Hans Spemann, recommending to his eldest son a recent work on the philosophy of education, praises the author, Eduard Spranger, because he "above all treats the living and the spiritual with reverence and the love of the artist; he lets it live and doesn't pluck it apart into dead little pieces."[96] But although the materials are richer in some respects, and the evidence of the influence of the intellectual milieu more flagrant, biology is not my subject.

II.2. From Positivism to *Lebensphilosophie*

From his term as rector of the University of Würzburg in 1914 until his death in 1928 Wilhelm Wien was—alongside Max Planck— the most prominent spokesman for physics in Germany. His semi-popular essays and addresses fall especially densely in the period 1918–1926, and so offer us a most striking example of the very quick change of tune which followed Germany's defeat in the First World War. On 1 May 1918, speaking in Dorpat on "Physics and the

---

95. Philipp Depdolla, "Biologie," *Unterrichtsblätter für Mathematik und Naturwissenschaft, 37* (1931), 183-190.

96. Letter of 12 October 1928 quoted in H. Spemann, *Forschung und Leben,* ed. Friedrich Wilhelm Spemann (Stuttgart, 1943), p. 229. One cannot but feel some sympathy for Spemann who had to bear through the early twentieth century the cross of "Entwicklungs*mechanik*" which the late nineteenth century had fastened upon his field.

Theory of Knowledge," Wien emphasized the independence and autonomy of physics—from philosophy, especially; for aid, physics calls only upon mathematics, chemistry, and technology. Helmholtz, he stated flatly and unapologetically, was a "pure empiricist in particular opposition to the German idealist philosophy, above all to Hegel." More important, the bulk of the lecture was given over to a discussion of Ernst Mach's views, which Wien accepted in large measure with some conventionalist and some realist modifications.[97] In his Machianism and, more to the point, in his readiness to advocate it before a general academic audience, Wien was at this moment by no means unusual. That same summer, in his address as rector of the University of Göttingen, Hermann Th. Simon, applied physicist, delivered himself of a thoroughly positivist account of how we obtain knowledge of the world through adaptation of our ideas to our sensations, basing himself upon Mach and Avenarius.[98]

Compare, now, Willy Wien as he appears in September 1919 in an article commemorating the twenty-fifth anniversary of Helmholtz' death.[99] In an apologetic tone Wien explained that while it is true that Helmholtz became an empiricist "through opposition to the Hegelian school"—n.b., not to idealism—and that he was never able to give up this position, nonetheless Helmholtz was always concerned with the "totality of the sciences," and always had "ideal," not "material," goals in view, always aimed at the "dominion of the spirit." Wien thus implicitly concedes the series of equations made repeatedly by the antagonists of modern science—empiricism= positivism = narrow specialization = utilitarianism = materialism— attempting only to make an exception of Helmholtz, who, we are assured, were he alive today would look to "German idealism" to put us on our feet again. In this essay there is, of course, not the faintest whiff of positivism; no mention of Mach at all. By February 1920, Wien's own "idealism" had matured so far that in addressing

---

97. W. Wien, "Physik und Erkenntnistheorie. Vortrag, gehalten in Dorpat am 1. Mai 1918," *Aus der Welt der Wissenschaft. Vorträge und Aufsätze* (Leipzig, 1921), pp. 209-234.
98. H. Th. Simon, *Leben und Wissenschaft, Wissenschaft und Leben. Rektoratsrede zur Jahresfeier der Georgia Augusta [Universität Göttingen] am 26. Juni 1918* (Leipzig, 1918), 32 pp.
99. W. Wien, "Hermann von Helmholtz," reprinted from *Naturwiss.*, 7 (5 September 1919), 645-648, in *Aus der Welt der Wissenschaft*, pp. 86-94.

a public session of the Prussian Academy of Sciences on "The Connections Between Physics and Other Disciplines" he represented the "postulate of the cognizability of nature" as "in the final analysis not so very far from the fundamental idea of the Hegelian philosophy of identity,"[100] and in November 1925 he could be found shedding tears before his assembled university over the abandonment, some twenty years earlier, of the requirement that philosophy, the "unifying discipline," be one of the subordinate subjects in every Ph.D. examination at a German university.[101]

Wien may well be the only German physicist to have expressed regret for that requirement, but he is certainly far from unusual in announcing a recent reversal of an earlier, deplorable trend toward fragmentation and isolation of physics from other disciplines,[102] or in thoroughly suppressing his earlier positivism. I know of only one instance during the entire Weimar period of a German physicist venturing, in a general academic address, to mention Mach's name with clear approbation and to associate himself with Mach's epistemological doctrines. Nor was it mere coincidence that in taking this courageous stand at the *end* of the Weimar period Richard von Mises refused to associate himself with the demand for synthesis, "counting it"—as did Mach—"the highest philosophy to tolerate an incomplete world view."[103]

The renunciation of positivism was intimately connected—for

100. W. Wien, "Über die Beziehungen der Physik zu andern Wissenschaften. Öffentlicher Vortrag, gehalten in der Preussischen Akademie der Wissenschaften in Berlin am 27. Februar 1920," *Aus der Welt der Wiss.*, pp. 16-40, on 28. And again, eighteen months later, we can find Walther Nernst pursuing the question in his *Rektorats-Antrittsrede* "ob nicht, wie fast stets bei derartigen starken geistigen Strömungen, auch in der Identitätsphilosophie ein gesunder Kern steckt." (*Naturwiss.*, 10 [1922], 489-495, on 490.)

101. W. Wien, *Universalität und Einzelforschung. Rektorats-Antrittsrede, gehalten am 28. November 1925*, Münchener Universitätsreden, Heft 5 (Munich, 1926), 19 pp., on 14-15. All the German universities except Berlin had abandoned this requirement at the beginning of the century.

102. The rebutting of the charge of fragmentation, of disintegration of the world picture by specialized research, is surely the single most common theme of general academic lectures by physicists. So Hermann Weyl, *op. cit.* (note 58): "To be sure one hears over and over complaints about the extent of specialization in the sciences. I believe, however, that on the whole in recent decades the situation has gotten better rather than worse." Likewise, Walther Kossel, *Die Einheit der Naturwissenschaft* (*op. cit.*, note 94).

103. R. von Mises, *Über das naturwissenschaftliche Weltbild* (*op. cit.*, note 94), p. 27. W. Ostwald, *Lebenslinien* (*op. cit.*, note 19), 2, 312.

Wien as for all his fellow physicists—with a renunciation of "knowledge for power," of the harnessing of nature, of utility as the object, motive, or justification for scientific research. In June 1914, reviewing as rector of the University of Würzburg the development of the German universities in the preceding century, Wien had given only one measure of achievement for the fields of physics and chemistry, viz. that they "have created the solid foundations upon which the pillars of our industry are erected," and he had reproached the universities for failing to incorporate the *Technische Hochschulen* as technical faculties.[104] In May 1918 Wien's lecture on "Physics and Epistemology," rejecting any liaison with philosophy, was followed a few days later by one on "Physics and Technology," whose basic, endlessly exemplified theme was the "support and stimulation" which *these* two fields have received from one another and should continue to receive in the future.[105] In the following years Wien did indeed play a key role in the creation and operation of the Helmholtz-Gesellschaft, which for the first time channeled substantial financial support from German industry into the physical institutes of the German universities. Yet in his academic addresses of the Weimar period he never let slip even a single word about this compromising connection.[106]

The shift in that element of the ideology defining "the significance of physical research" was announced by Wien in his address before the Prussian Academy of Sciences in February 1920. This question, we are told, can be judged from two very different points of view. The first sees physical research as aiming at "human domination over the recalcitrant forces of nature." Wien implicitly recognizes

104. W. Wien, "Die neuere Entwicklung unserer Universitäten und ihre Stellung im deutschen Geistesleben. Rede für den Festakt in der neuen Universität am 29. Juni 1914," *Aus der Welt der Wissenschaft,* pp. 1-15, on 14.
105. W. Wien, "Physik und Technik, Vortrag, gehalten in Reval am 6. Mai 1918," *Aus der Welt der Wiss.,* pp. 235-263.
106. W. Wien, "Die Helmholtz-Gesellschaft und ihre Bedeutung für die deutsche Physik . . .," *Die Helmholtz-Gesellschaft zur Förderung der physikalisch-technischen Forschung in sieben Jahren ihres Wirkens* [privately printed], 1928), pp. 7-11. This renunciation of utility in academia was not wholly uncompensated. The slogan "knowledge for power" was exchanged for "knowledge as a substitute for power," "Wissenschaft als Macht-Ersatz." See B. Schröder, *op. cit.* (note 52). It must be said, moreover, that the academic chemists, unlike the physicists, do not appear to have become at all bashful about discussing technical applications and justifying their science through them.

that his audience is very hostile to any such conception, and after pointing out that it need not necessarily proceed from a "purely materialistic mode of thought," passes on to the second and proper point of view, one "free of all striving toward a goal [Zielstrebigen]." Physical research is, in truth, nothing but the expression of "the pure human instinct for inquiry"; "it arises solely from an inner need of the human spirit"—note the implicit *Lebensphilosophie*—"a craving," Wien explains, "to grasp the causality of events."[107] To this new line Wien held fast in all further academic addresses. In November 1925, as rector of the University of Munich, his discussion of "Universality and Specialized Research" touches on technological applications of physical science in only one paragraph, and there only in order to stress that scientific ideas which appear at first to refer to but a narrow field may turn out to have enormous practical consequences. He is very careful, however, not to appear to be praising science on that account. Throughout the lecture Wien totally abstains from any attempt to justify science by utility. On the contrary, the goal of science is *culture*: "The significance of a scientific achievement can ultimately only be measured by the effect which it has upon the intellectual life"; "the results of research are worthless if they are not taken up into the culture."[108]

As Wien's reference to causality suggests, he had very strong views on which results of research ought to be taken up into the culture and which ought not. And although, as will appear in Part III, many of Wien's colleagues took a different view of the concept of causality, they were in complete agreement with him on the motive, goal, and justification for physical research. The "common driving force" of

107. W. Wien, *op. cit.* (note 100), p. 28. Cf. the mineralogist Gottlob Linck, *Über Wesen und Wert der Universität. Rede, gehalten . . . am 19. Juni 1920 . . . zu Jena vom Rektor der Universität* (Jena, 1920), p. 4: "Wie der Hirsch schreiet nach frischem Wasser, so schreiet unsere Seele nach Erkenntnis."

108. W. Wien, *op. cit.* (note 101), pp. 13, 7, and 19, respectively. Cf. the theoretical physicist Erwin Madelung, *Die Bedeutung der Wissenschaft im Rahmen unserer Kultur. Rede anlässlich der Übernahme des Rektorates* [1931], Frankfurter Universitätsreden 39 (Frankfurt a. M., 1932), 16 pp., who explains (pp. 2-4) that by "Kultur" he understands "everything which broadens and enriches our inner life, which is, I believe, in sufficient accord with the customary usage. . . . We want, thus, to put completely aside here the consideration of practical utilizability and of the dead piling up of information, however important it may be. We want only to ask ourselves what spiritual needs [geistige Bedürfnisse] we have and to what extent these are satisfied by science." Naturally, "our needs arise out of the dark spring of our living existence."

all the research in the university—as the members of the University of Greifswald learned from their new rector, the *applied physicist* Friedrich Krüger—is "the innate human urge for ever new knowledge. . . . All external driving forces originating in considerations of utility and necessity for our physical existence and its advancement are completely without influence." And, after this introduction, Krüger devoted his *Rektoratsrede* to the question of the heat death of the universe![109]

This ideology tinged with *Lebensphilosophie*—that the value of physics lies in and derives from the fact that it is the expression of an unanalyzable and irreducible human drive (for cognition of nature, in this case)—attained by 1929 the status of orthodox physical doctrine, occupying the opening pages of the volume on "General Foundations of Physics" in the new edition of the *Handbuch der Physik*. Firmly rejecting technological application as a measure of the value of physical knowledge, and, less firmly, rejecting the foundation of a *Weltanschauung* as the goal of physical research, Hans Reichenbach explained that "the most important thing that one can say about it [i.e., doing physics] is that it is a need, that it grows up out of the human being just like the wish to live, or to play, or to form a community with others."[110]

At first sight it seems most surprising that even a Reichenbach, i.e., even a representative of "rigorous" logical empiricism, should have taken over the standards of value and the ideology for the physical enterprise from an intellectual milieu specifically hostile to his philosophical position. This circumstance becomes less surprising

---

109. F. Krüger, *Materie und Energie im Welt-Geschehen. Rektoratsrede,* Greifswalder Universitätsreden 15 (Greifswald, 1928), 29 pp. "Alle äusseren Triebkräfte des Nützlichen und für die äussere Existenz und ihre Förderung Notwendigen kommen nicht in Frage" (p. 3). Only at the end of his address, apropos of artificial disintegration of atoms, did Krüger make any reference at all to the application of scientific knowledge, namely to "extracting the energy of the atom as one of the greatest technical problems worthy of the most strenuous efforts . . . consequently we see at present a mighty contest in the laboratories of the civilized nations . . . to find the methods for extracting this energy" (pp. 28-29).

110. H. Reichenbach, "Ziele und Wege der physikalischen Erkenntnis," *Handbuch der Physik,* Band 4: *Allgemeine Grundlagen der Physik,* ed. H. Thirring (Berlin, 1929), pp. 1-80, on 1-2. As for the new discipline of "technical physics," Reichenbach dismisses it (p. 11) with the remark that even though it has now attained a place in German universities "sie ist doch ihrem Wesen nach eine Technik und keine Wissenschaft," for a *Wissenschaft* aims solely at "Erkenntnis."

if one recalls that the positivist tradition itself contained a substantial element of *Lebensphilosophie* and that, moreover, there was a solid Machian precedent for regarding natural science as the outgrowth of a basic human drive. How horrified Hedwig Born would have been had she known that her favorite Einsteinian apothegm—"I feel such solidarity with everything living, that it is all one to me where the individual begins and ends," which she heard at his bedside as he lay critically ill in 1917/18 and which she found so beautiful that she quoted it over and over again to Einstein himself[111]—was also in fact a most genuinely Machian-positivist sentiment.

Yet with such sentiments we have by no means reached the limits of the community of values and attitudes between the physical scientist and his *lebensphilosophisch* intellectual milieu. The "unbroken life force" of mathematics, the "concrete nearness to life" of applied mathematics, the *"lebensvollste* interlacing" of mathematics with natural science and technology, the training of students by "living interaction to spontaneous scientific work," the maintaining of mathematics "in contact with the concrete stuff of life," and avoiding the danger that it will "rigidify as a pure distant-from-life form"—these examples of "life" rhetoric are all drawn from a single address, a bare half dozen pages, by Richard Courant.[112] To be sure Courant's rhetoric of "life," "organism," "spontaneity," "ecstacy," "phantasy," "instinct," "intuition" is considerably more exuberant than that of most of his colleagues. It is indicative, nonetheless, of a substantial participation by the physicist and mathematician in the values of his general cultural milieu.

Another striking example of this participation is the assent of the exact scientists to the proposition that feeling and intellect are antithetic, incapable of coexisting in an exalted state in a single individual, and that feeling is the higher quality.[113] Einstein

111. Born-Einstein, *Briefwechsel (op. cit.,* note 14), p. 113.
112. R. Courant, "Über die allgemeine Bedeutung des mathematischen Denkens," *Naturwiss., 16* (1928), 89-94. "Vortrag, Tagung Deutscher Philologen und Schulmänner, Göttingen, September 1927." Under the cover of this rhetoric, however, Courant basically stands firm upon the traditional intellectualist conception and cognitive claims of mathematics.
113. The "classical" exponent of this thesis was Ludwig Klages, who is still counted a great seer in Germany today. Indeed the press of the Deutsche Physikalische Gesellschaft recently published a collection of essays—*Physik, Gleichung*

expressed this very well in a letter to H. A. Lorentz: "In your case Nature had the rare impulse to unite with a sharp mind warm feeling. If only this were often so. . . ."[114] Given this ranking, it is not surprising that the physicists and mathematicians responded so ineffectually to the charge of "intellectualism" leveled by Becker *et al.* Nor is it surprising that they themselves sought the motive, goal, and justification for their own scientific activity in feeling and instinct, in the wish for, not in the fact of, cognition. As Wilhelm Ostwald, now in his old age, so succinctly put it, "the reason, and science along with it, is only a servant of the feeling."[115]

Here, moreover, we must add that most curious simile in Reichenbach's explanation of why one does physics—it is "like the wish to form a community with others," the characteristic emotional need of the Weimar period. In the famous "Introduction" to his book on *Einstein's Theory of Relativity,* written in 1920, Max Born gave expression to this same longing for "community," for participation in some whole which transcends the individual, maintaining that "all religions, philosophies, sciences are procedures designed for the purpose of expanding the 'I' to the 'we'." What distinguishes the natural scientist is his resolve—"often shuddering"—to achieve this goal by sacrificing the absolute for the sake of objectivity. And it is in *this* way that, for the physicist, "the pain of spiritual loneliness disappears, the bridge to kindred spirits is formed."[116] And with

---

und Gleichnis. Vorträge und Aufsätze über Physik (Mosbach i. B., 1967)—by Eberhard Buchwald, a mediocre theorist who in the 1920's was struggling to make a career in applied physics. There one can read, pp. 68-80, Buchwald's contribution to *Ludwig Klages, Erforscher und Künder des Lebens* (Linz, 1947), and learn what Klages, "enthroned" alongside Heraclitus, Goethe, and Nietzsche, can mean to a "dankerfüllten Fachphysiker."

114. Einstein to H. A. Lorentz, 3 April 1917. Microfilm of Lorentz papers in the Algemeen Rijksarchief, The Hague, deposited in the Archive for History of Quantum Physics (see note 13); on microfilm nr. 6. For the higher status of feeling: Einstein to Hedwig Born, 1 Sept. 1919, *Briefwechsel (op. cit.,* note 14), pp. 32-33, and Einstein to Jacob Laub [1909?], as quoted by Carl Seelig, *Albert Einstein: Eine dokumentarische Biographie* (Zurich, 1954), p. 117.

115. W. Ostwald, "Von der Formel zur Form," *B.Z. am Mittag,* 3 Nov. 1926, as quoted in Grete Ostwald, *Wilhelm Ostwald, mein Vater* (Stuttgart, 1953), pp. 229-230.

116. M. Born, "Introduction" to *Einstein's Theory of Relativity (op. cit.,* note 93). Born's inclination toward what one might call "futurist *Lebensphilosophie*"—such as one finds expressed by Walther Rathenau, for example—is revealed in his recommendation of Richard N. Coudenhove-Kalergi's *Apologie*

this concession that science is primarily a means for satisfying certain emotional needs, we have arrived again at the very axiom of the "revolution in science."

II.3. Capitulation to Spenglerism

A remarkable readiness of the Weimar physicists to adapt their ideology to the values of their milieu has, I think, been shown in the preceding section. Yet the adaptations displayed there, while they involve redefinitions of the motivation and justification for doing physics, do not explicitly alter the fundamental conceptions of scientific method, renounce the cognitive claims of the exact sciences, or resign confidence in their future development—as demanded and predicted by Spengler. But if we shift our focus slightly, from the physicists to the theoretical physicists and mathematicians, there appears a distinct tendency to carry the ideological adaptation into these vital regions and to adopt specific propositions which at the time were attributed to, or especially closely associated with, the *Untergang des Abendlandes*.[117]

A most interesting and most suggestive case of a clearly discernible Spenglerian influence is to be found in Richard von Mises' inaugural

---

der Technik (Leipzig, 1922), "dessen Inhalt mir sehr eingeleuchtet hat." (Born to Einstein, 7 April 1923, *Briefwechsel* [*op. cit.*, note 14], p. 110.) On science as primarily a means to satisfy emotional needs, compare E. Madelung (*op. cit.*, note 108), p. 14, replying to those who place a higher value upon the *Geisteswissenschaften* than upon the *Naturwissenschaften*: "As for the value judgement, I want to emphasize once again that it obviously doesn't matter so very much what one does and what methods one employs, but far more how one views one's own activity, whether one feels richer and freer through it, and more secure vis-à-vis life's changing forms."

117. The shift in focus away from the experimental physicists is in good part simply a consequence of their relative inarticulateness. But not entirely. If in general they did not go so far in their ideological accommodation as the theoretical physicists and mathematicians, that was also in part because they had less to recant. In the Weimar period the theoretical physicists seem to have drawn closer to the mathematicians with whom they were lumped for opprobrium in the public's mind, and with whom their relations could in some respects be less constrained than those which they maintained with the experimental physicists. Some twenty notable theoretical physicists held membership in the Deutsche Mathematiker-Vereinigung in 1924, half of whom had joined in or after 1918 (*Jahresbericht, 34* [1925], Part 2, pp. 49-92), and that despite the fact that the annual meetings of the Deutsche Physikalische Gesellschaft and the Deutsche Mathematiker-Vereinigung were ordinarily held at the same time and place, thus largely obviating the need for formal membership in both organizations.

(and farewell) lecture as Professor of Mechanics at the *Technische Hochschule* Dresden—delivered in February 1920, after von Mises had accepted a chair of applied mathematics at the University of Berlin. Considerable stir was created by von Mises' contention—or rather concession to the intellectual milieu—that the "age of technology," to which the *Technische Hochschulen* owed their rise, was on its way out. His advice to these institutions was that they do their best to get onto the wave of the future by entering the field which was destined to replace technology in the "culture consciousness," namely speculative natural science, particularly relativity and atomic physics. In these subjects, he asserted, we have had for the past two decades a period like that of Copernicus, Galileo, and Kepler. "It is not a question of new facts of any sort, nor of new theoretical propositions, nor even of new methods of research, but, if I may say it—taking this word in its philosophical sense—of new intuitions [Anschauungen] of the world." Atomic physics has taken up again "the question of the old alchemists"; "numerical harmonies, even numerical mysteries play a role, reminding one no less of the ideas of the pythagoreans than of some of the cabbalists."[118]

Astonishing as these remarks are from the mouth of a convinced positivist, impossible as it may be to find their like two years earlier, much as they remind us of Spengler's prediction that a new mysticism was the fate and salvation of natural science, still *prima facie* evidence of a connection with the *Decline of the West* is wanting. In fact, the immediate precedent and probable inspiration for von Mises' reference to numerical harmonies and mysteries is an article on "A Number Mystery in the Theory of the Zeeman Effect" by Arnold Sommerfeld which appeared in *Die Naturwissenschaften* a few weeks earlier, as well as the preface to Sommerfeld's *Atomic Structure and Spectral Lines,* which had appeared late in 1919.[119]

118. R. v. Mises, *Naturwissenschaft und Technik der Gegenwart. Eine akademische Rede mit Zusätzen,* Abhandlungen und Vorträge aus dem Gebiete der Mathematik, Naturwissenschaft und Technik, Heft 8 (Leipzig, 1922), 32 pp., on 2, 5, 16, respectively. The initial publication without the *Zusätze* had been in the *Zeitschr. des Vereins deutscher Ingenieure, 64* (1920), 687-690, 717-719.

119. A. Sommerfeld, "Ein Zahlenmysterium in der Theorie des Zeemaneffektes," *Naturwiss., 8* (23 Jan. 1920), 61-64, on which see my "Alfred Landé and the Anomalous Zeeman Effect, 1919-1921," *Historical Studies in the Physical Sciences,* 2 (1970), 153-261; Sommerfeld, *Atombau und Spektrallinien,* 1st ed. (Braunschweig, 1919), p. viii. Sommerfeld was sufficiently pleased with this *Sphärenmusik* passage to reprint it in the 2nd, 3rd, and 4th editions, 1920-1924.

There Sommerfeld had spoken of "the mysterious organ upon which nature plays the spectral music" of the atomic spheres. In the future Sommerfeld was to go considerably further in this direction. A ceremonial address at a public session of the Bavarian Academy of Sciences in July 1925 offered Sommerfeld the opportunity to stress that "hand in hand with this turn toward the arithmetical goes a certain inclination of modern physics toward pythagorean number mysticism. Precisely the most successful researchers in the field of theoretical spectral analysis—Balmer, Rydberg, Ritz—were pronounced number mystics. . . . If only Kepler could have experienced today's quantum theory! He would have seen the most daring dreams of his youth realized. . . ."[120]

It is true that, having indulged himself in such rhetoric at some length, Sommerfeld concluded with the hope that he will "not be suspected of speaking in favor of mysticism in the ordinary sense, as it comes out in the astrological, metaphysical, and spiritualistic impulses of our time." Nothing is farther from his intent, he insisted; he was not speaking of human things, but only of laws of nature, and he meant rather to be attacking "conventionalism," "positivism," and "Machian philosophy."[121] Yet it is perfectly clear that, despite the disclaimers, Sommerfeld was indeed catering to the antirational as well as the antipositivist inclinations of his audience, that he was trying to project an image of physics that would find favor with his audi-

---

120. A. Sommerfeld, *Die Bedeutung der Röntgenstrahlen für die heutige Physik. Festrede, gehalten in der öffentlichen Sitzung der B. Akademie der Wissenschaften . . . am 15. Juli 1925* (Munich, 1925), 17 pp., reprinted in Sommerfeld's *Ges. Schr. (op. cit.,* note 21), *4,* 564-579, on 573-574.

121. *Ibid.,* pp. 575-576. Cf. Georg Hamel, President of the Mathematischer Reichsverband, whom we have also previously seen throwing himself against the wave of irrationalism and anti-intellectualism, summarizing his *Rektoratsrede* delivered 30 June 1928 at the Technische Hochschule Berlin: "Mathematics customarily appears as the rational science per se; to the layman the mathematician is a calculator. In opposition thereto I maintain the thesis that mathematics is an art, and that, in the last analysis, it is conditioned not logically but transcendentally. . . . The mathematician is a poet. Like the dramatist he creates a form. . . . The problem of the irrational numbers leads mathematics into metaphysics. . . . The genuine foundation for all of mathematics I see in Kant's pure *Anschauung.* . . . In conclusion I take issue with the misconception that my remarks represent a repudiation of intellectualism. Although the irrational basis of mathematics has been clearly recognized, that does not alter in the least the mathematician's obligation to proceed purely logically within his science, with the greatest care, with precise modes of inference." (Hamel, "Ueber die philosophische Stellung der Mathematik," *Forschungen und Fortschritte, 4* [1928], 267.)

tors and raise the prestige of the discipline in their eyes. And one cannot help but be struck at the close correspondence between this image and that which Spengler sketched in the final pages of the *Decline of the West*.

But let us return to von Mises—upon whom a direct influence of Spengler can in fact be established by September 1921. When at this time von Mises added an appendix to the republication of his lecture of February 1920, his tone had changed entirely, his optimism and enthusiasm had disappeared. Von Mises had largely, and *explicitly*, adopted Spengler's perspective and assumptions. It is "at least highly probable that the towering structure, under construction for the past five centuries, of a Western culture oriented entirely toward cognition and performance will collapse in the following centuries. From this standpoint one must count the theory of relativity and modern atomic physics as among the last building stones destined to crown the structure." Accepting Spengler's doctrine that cultures, as "living organisms," are fundamentally incommensurable, von Mises declared it "entirely out of the question" that the culture which succeeds ours will "continue the exact sciences in our sense." Nor can such views be dismissed as pessimism—"as if the man, who conscious of his old age and the inevitability of his death, is a pessimist because he faces the fact and acts accordingly."[122] What, one wonders, would it mean for a physicist or mathematician to "act accordingly"? Could it possibly mean that he strives to alter the content of his science and the very nature of the scientific enterprise, in order to fulfill Spengler's prophesies?

What is perhaps most striking and appalling about the von Mises of September 1921 is the failure of nerve, the complete loss—just as Spengler predicted—of the esprit, the self-confidence which we expect from the mathematical physicist. And in this von Mises was by no means unique. One can find, on the contrary, many examples—most

---

122. R. v. Mises, *op. cit.* (note 118), p. 32. Cf. note 103 and text thereto. In general the shifts in von Mises' mental posture correspond quite closely to those which Georg Steinhausen, *Deutsche Geistes- und Kulturgeschichte von 1870 bis zur Gegenwart* (Halle a. d. Saale, 1931), p. 4, found to be typical for the Weimar period. Namely, initially, despite the political and military collapse, a certain euphoria over the wholly new epoch thus begun; this mood was, however, very soon displaced by disillusionment and an "Untergangsstimmung," which, again, had entirely disappeared by the late Weimar period when there was a strong tendency "to fall back into the old mental grooves."

often in addresses before general academic audiences—of theoretical physicists and applied mathematicians denigrating the capacity of their discipline to attain true, or even valuable, knowledge. The earliest such is, perhaps, the passage which in the spring of 1918 Hermann Weyl placed as a conclusion to the first edition of *Space-Time-Matter*.[123] Theoretical physics is, Weyl maintained, entirely analogous to formal logic. "True" propositions must conform to logic, but logic is incapable of judging the "truth" of the propositions it manipulates; so also reality conforms to the laws of physics, but physics is incapable of informing us about the reality which its laws govern. Is, perhaps, this reality—these "darker depths" than the mathematician can grasp with his methods[124,125]—Spengler's "immediate becoming, life itself"? Indeed, as we shall see in Part III, it is.

A still more striking example of these same "annihilating doubts" is offered us by Gustav Doetsch in his inaugural lecture as Privatdozent for applied mathematics at the University of Halle, 27 January 1922. There, in conclusion, pointing back to his exposition of the "Meaning of Applied Mathematics," Doetsch burst forth:

> Such *rationalistic dogmatism* is the characteristic expression of *that* intellectual epoch which is at this moment perishing [im Untergehen]. It is the spirit, one could say, of the *age of natural science,* which, essentially, coincided with the 19th century, and which in our days is sinking with violent convulsions into its grave in order to make room

123. H. Weyl, *Raum-Zeit-Materie. Vorlesungen über allgemeine Relativitätstheorie,* 1st ed. (Berlin, 1918), pp. 226-227; again, somewhat more fully, in the 3rd ed. (Berlin, 1919), pp. 262-263. In the fall of 1920, when preparing the fourth edition (Berlin, 1921), Weyl struck this conclusion, replacing it by an attack on causality.

124. *Ibid.,* 1st ed., p. 9; 3rd ed., p. 9; 4th ed., p. 9.

125. Slightly later, and perhaps owing something to Weyl, is Paul Gruner, *Die Neuorientierung der Physik. Rektoratsrede, gehalten an . . . der Universität Bern den 26. November 1921* (Bern, 1922), 23 pp. In conclusion, theoretical physicist Gruner conceded that although "es mag dem Naturforscher schwer fallen," his field cannot satisfy "das Sehnen nach absoluter Wahrheit" which fills contemporary academic youth. "Dem blossen Denken und Beobachten der Naturwissenschaften ist dieses intuitive Schauen versagt." Natural science can tell us nothing of the meaning of the world and our life; the disaster we are experiencing today is due to this intellectual advance without ethical-religious foundation. These are, of course, precisely the charges which theoretical physicist von Laue, taking Rudolf Steiner as his ostensible target, was trying to rebut a few months later. (See note 15 and text thereto.)

for a new spirit, a new life-feeling . . . this epoch, at whose beginning we unquestionably find ourselves today, is fed up with this rationalistic attitude. Whether we direct our attention toward expressionism in art, or to more recent philosophical tendencies, which in many ways have not yet emerged entirely distinctly, or to any other area of life and thought whatsoever, we find everywhere an ever stronger *aversion* for *that* spirit which believed that it had to express, and that it could express, everything whatsoever in dry words, in one formula—an aversion deriving from the unconscious feeling: *this* path has never and will never lead us to the *essence* of things, we must try to get "nearer" to the object, to transfer ourselves inside of it itself. Whether the new path leads to the goal, or whether it can only get us closer, may be left undecided. Here my intent was only to point out in the *domain of natural science itself,* which has served as a model for so many others, that the *mathematical* treatment of the material of experience does not begin to impart information about the essence of the world, that is, to yield true cognition.[126]

And still Doetsch was not quite finished. After this tirade against his discipline he quoted Hegel's dictum that mathematics is "kein Begreifen," and clinched his case by observing that if Hegel "should not be regarded as the proper person to bring applied mathematics to a correct estimation of itself, then I refer to the words of our most brilliant contemporary mathematician, Hermann Weyl, in whose famous work, *Raum-Zeit-Materie.* . . ." It was, of course, rather daring of Doetsch to speak his mind *so* freely—although his general academic audience must have been very happy indeed to hear their views confirmed by a mathematician. It was, however, foolhardy of Privatdozent Doetsch to publish such sentiments in the journal of the German Society of Mathematicians, where it was read by, and had necessarily to offend, senior and influential colleagues. Indeed it may well have cost him a chair.[127]

---

126. G. Doetsch, "Der Sinn der angewandten Mathematik," *Jahresbericht der Deutschen Mathematiker-Vereinigung, 31* (1922), 222-233, on 231-232. "Antrittsvorlesung gelegentlich der Umhabilitierung von der T. H. Hannover an die Universität Halle a/S. am 27. Januar 1922."

127. *Ibid.,* p. 233. Otto Blumenthal to Theodor von Karman, 8 July 1923: "Das Ergebnis Ihrer Anfragen betreffs papabler Mathematiker finde ich im höchsten Masse betrübend. Ich habe von Doetsch nicht die Ansicht, dass er für uns brauchbar ist. C. Müller hat uns seiner Zeit ein sehr vernünftiges und abfälliges Gutachten über ihn gegeben. Auch habe ich von einem Vortrag, den er in Halle über das Wesen der angewandten Mathematik (D.M.V.) gehalten

And finally, to place in evidence an example from the latter part of our period, consider the picture of "The Peculiar Nature of the Mathematician's Mind" which Max Dehn, Professor of Pure and Applied Mathematics, held up before his assembled university in January 1928. Painted in pure Spenglerian style, the characteristic mental tone of the contemporary [German] mathematician is skepsis, mistrust of reason, self-inculpation, pessimism, and resignation:

> This somewhat sceptical attitude of many a contemporary mathematician is reinforced by what is going on in the neighboring field of physics. Here it appears to be the case that the physical phenomena no longer admit of being construed consistently [widerspruchslos] in a mathematical four-dimensional space-time manifold. Up to now we were able to provide physics with sufficiently freely built scaffolding for its ever bolder constructions. Now, however, in certain reflections arising from important investigations of the finest structure of matter, physics is perhaps in the process of cutting itself loose from mathematics. [Dehn is, *inter alia*, two years behind the times—see sections III.4–6.]
>
> All this has impelled many of us to be somewhat sceptical in more general questions as well. The fundamental conviction of every philosopher that the world can be comprehended consistently [widerspruchslos] by the human reason is, for the mathematician, no longer certain. ... This attitude is, to be sure, not entirely original; it is reminiscent of the thought of the later Eleatics at the time of the foundation crisis in ancient Greece.
>
> Out of this skepsis there develops a certain resignation, a kind of mistrust for the power of the human mind in general.
>
> ... because of the boundedness of human intellectual power a limit is set to abstraction, to the departure from the intuition. Beyond this limit no further development is possible. But contemporary mathe-

---

hat, einen ungünstigen Eindruck erhalten, müsste mir allerdings den Artikel noch mal durchsehen" (Karman Papers, California Institute of Technology Archives, Box 13). By 1924 Doetsch had become reconciled to his discipline, and had even persuaded himself that although mathematics and natural science—"immediately casting everything into *rational* schemata"—"can never uncover the real meaning of the world and its interconnections, ... nonetheless the mind [Geist] at least comes ever nearer to that which it would really like to grasp." ("Sinn der reinen Mathematik und ihrer Anwendungen," *Kant-Studien, 29* (1924), 439-459, on 458-459; "Vorträge ... Januar 1924.") And later that same year Doetsch received his first chair.

matics is by no means dead, and naturally, even in topology, for example, a man can and hopefully will come who simplifies the processes so much . . . that a new development sets in. . . . Such achievements will, however, scarcely arise in the course of an organized routine. But if the mathematician is already complaining for this very reason—that in consequence of the modern development finally even the pursuit of his science has become organized—then he must properly say to himself: *mea maxima culpa*. For through mathematics the constructive power of the human being first unfolded, and thus brought forth the age of technology. And if, confronted by this disaster which he has brought about, the mathematician is seized with despair, then, for the third time, resignation saves him.[128]

The foregoing examples—especially the cases of von Mises and Doetsch—demonstrate most clearly that there were mathematical physicists who went so far in assimilating the values and mood of their intellectual milieu as to effectively repudiate their own discipline. They show, moreover, that this process of ideological adaptation to the intellectual environment was, either explicitly or implicitly, in large measure a capitulation to Spenglerism. These cases are extreme, of course, and as such atypical. Yet the stages by which von Mises advanced to this extreme, and the readiness of even a Sommerfeld to flirt with the very antiscientific tendencies he deplored, makes it difficult to avoid the conclusion that most German mathematicians and physicists largely participated in, or accommodated their persona to, a generally Spenglerian point of view.

This conclusion is supported by the combination of ample evidence that Spengler's book was read by many, if not most, German physicists and mathematicians and the remarkable paucity of public criticism by representatives of these disciplines. In reviewing the

---

128. M. Dehn, *Über die geistige Eigenart des Mathematikers. Rede anlässlich der Gründungsfeier des Deutschen Reiches am 18. Januar 1928*, Frankfurter Universitätsreden 28 (Frankfurt, 1928), 20 pp., on 15, 18. This address by one of Hilbert's oldest students (Ph.D. Göttingen, 1899) seems to have drawn considerable attention. Otto Neugebauer quoted it, without citing it, in concluding his exposition of the elaborate installations of the new Göttingen mathematical institute (*op. cit.*, note 8). Although Neugebauer's final lines were in defense of "organisation," he did not fail to concede to Dehn that "Gewiss lässt sich auch für eine solche Auffassung viel ins Feld führen." One may therefore suppose that Hilbert's remarks in September 1930 (*op. cit.*, note 94) were meant as much for his colleagues—and former students—as for his lay audience.

Untergang in 1919 Troeltsch had emphasized the desirability of such criticism, but reported that "to be sure, when I asked one of our most eminent mathematicians and physicists [Planck?] to give his opinion of the book, and briefly described Spengler's principal theses, he refused to read any part of it."[129] But that reaction was either untypical or changed very quickly, for I have seen explicit references to Spengler, either suggesting or demonstrating acquaintance with his book, by Max Born, Albert Einstein, Franz Exner, Philipp Frank, Gerhard Hessenberg, Pascual Jordan, Konrad Knopp, Richard von Mises, Friedrich Poske, Hermann Weyl, and Wilhelm Wien.[130] This list, which I expect could be substantially lengthened, is already of such extraordinary length as to make it virtually certain that Spengler's theses, and not merely the public enthusiasm for them, were generally known to the Weimar physicists and mathematicians. And yet they did or said remarkably little to oppose them. Reviewing the literature of the "controversy over Spengler" in 1922 Manfred Schroeter found that "both the cornerposts of the book, the first and sixth chapters, mathematics and physics, have remained almost unanswered." Indeed, Schroeter was able to find very few criticisms by mathematicians, and only one by a physicist—Wilhelm Wien.[131]

Where and when a physicist or mathematician came forward to attack Spengler it was almost invariably in defense of that most basic tenet of the scientific ideology, the autonomy, objectivity, and

129. E. Troeltsch, *Ges. Schr.* (*op. cit.*, note 32), *4*, 682.
130. Born-Einstein, *Briefwechsel* (*op. cit.*, note 14), p. 44; F. Exner, *op. cit.* (note 93), preface; P. Frank, *op. cit.* (note 43), p. 54; G. Hessenberg, *op. cit.* (note 66); P. Jordan, *Anschauliche Quantentheorie* (Berlin, 1936), p. 279; K. Knopp, *op. cit.* (note 94), p. 208, cf. note 131; R. v. Mises, *op. cit.* (note 117), p. 32; F. Poske, "Anschauliche und abstrakte Begriffsdefinitionen im physikalischen Unterricht. Vortrag, Naturforscherversammlung, Nauheim, September 1920," *Zeitschr. f. den physikal. u. chemisch. Unterricht, 34* (1921), 97-103 ("Den immerhin unvermeidlich intellektualistischen Charakter der theoretischen Physik hat neuerdings Oswald Spengler in seinem Buche, *Der Übergang* [sic] *des Abendlandes*, hervorgehoben," and Poske reluctantly agreed, without an inkling of criticism); H. Weyl, "Das Raumproblem," *Jahresber. d. Dtsch. Mathematiker-Vereinigung, 31* (1922), 205-221, reprinted in Weyl's *Gesammelte Abhandlungen*, ed. K. Chandrasekharan, 4 vols. (Berlin, 1968), *2*, 332; W. Wien, *op. cit.* (note 100), pp. 36-39.
131. M. Schroeter, *Der Streit um Spengler. Kritik seiner Kritiker* (Munich, 1922), pp. 56-57, 70. The only scientist critics he knows are W. Wien, O. Neurath [sociologist], Leonard Nelson [philosopher], G. Hessenberg, and K. Knopp, who published a "sehr absprechend" review.

universality of scientific knowledge.[132] This notion Spengler claimed to have exploded by demonstrating that there are no immanent, invariant criteria of knowledge, that the science of a period is dependent in toto upon its *Lebensgefühl*. Yet for every opponent of Spengler's thesis one can cite another exact scientist who, more or less explicitly and more or less fully, identified himself with this doctrinal touchstone of Spenglerism.[133] And once again von Mises provides evidence

132. Thus the rebuttals cited in note 93.
133. Thus, for example, Gustav Mie, *Das Problem der Materie. Öffentliche Antrittsrede* [as Professor of Physics at the University of Freiburg i. Br.], *gehalten am 26. Januar 1925* (Freiburg in Baden, 1925), pp. 23-24: "Ich bin der Überzeugung, dass in der Geschichte der geistigen Bewegungen überall ein Zusammenhang der verschiedensten Gebiete des Geisteslebens zu beobachten ist. Der Atomismus ist ein Kind des 18. Jahrhunderts und des Rationalismus. . . . Deswegen sind wir im Begriff, uns einem andern Weltbild zuzuwenden. Ich glaube, dass dieses neue Weltbild gewisse charakteristische Züge trägt, die auch sonst im Bilde des modernen Geisteslebens auffallen, ich meine, das Suchen nach einer grossen Einheit und nach einem allgemeinen Zusammenhang im physikalischen Geschehen, der sehr gegen das Auseinanderfallen in die einzelnen Atome kontrastiert. Es ist eine interessante Beobachtung, dass auch die streng an experimentelle Erfahrungen gebundene Physik auf Bahnen geführt wird, die zu den Bahnen der geistigen Bewegungen auf anderen Gebieten durchaus parallel verlaufen."
Again R. Courant, Sept. 1927 (*op. cit.*, note 112), p. 90: "It is surely no accident that the *Umschwung* in the orientation of mathematics from naive productivity to rigorous scientificness is temporally parallel with the great intellectual and social transformations which the European world has undergone since the beginning of the French revolution."
Again E. Madelung, 1931 (*op. cit.*, note 108), pp. 4-6: "I can say without exaggerating that I have not the least interest in the world, but only in the picture [Bild] that I possess of it. Out of this picture I draw my joys and sufferings, my fear and hope, my feeling of comfort and of sorrow. . . . By means of language a communal world-picture is created as a convention. . . . We designate today as '*die Wissenschaft*' our stock of knowledge [Besitz an Wissen] which is codified by the written word and sanctified by convention."
And once again, E. Schrödinger, *Über Indeterminismus in der Physik. Ist die Naturwissenschaft milieubedingt? Zwei Vorträge zur Kritik der naturwissenschaftlichen Erkenntnis* (Leipzig, Barth, 1932), pp. 38-39: "Unsere Kultur bildet ein Ganzes. Auch wer das Glück hatte, die Forschung zu seinem Hauptberuf zu machen—ganz abgesehen davon, dass dies nicht die einzigen sind, die sie fördern,— ist doch nicht nur Botaniker, nur Physiker, nur Chemiker. Vormittags auf dem Katheder spricht er wohl der Hauptsache nach bloss von seinem Fach. An demselben Abend sitzt er in einer politischen Versammlung, hört und spricht ganz andere Dinge,—steht ein andermal im Kreis einer Weltanschauungsgemeinde, wo wieder von anderem die Rede ist. Man liest Romane und Gedichte, geht ins Theater, treibt Musik, macht Reisen, sieht Bilder, Skulpturen, Architektur— und vor allem, man liest und spricht viel über diese und andere Dinge. Kurz wir alle sind Mitglieder unseres Kulturmilieus. Sobald bei einer Sache die Einstellung unseres Interesses überhaupt eine Rolle spielt, muss das Milieu, der Kulturkreis, der Zeitgeist oder wie man es sonst nennen will, seinen Einfluss

that the repudiation of this tenet of the scientific ideology was, in some cases and to some extent, a capitulation to the *Untergang, per se*. Thus in February 1920 von Mises had still been a good enough positivist to deny the influence of political and social conditions or the associated *Lebensgefühl* on the quantity, vitality, direction, or content of the higher intellectual productions. By September 1921, however, he had, as we have seen, gone over to Spengler in this respect as well.[134]

II.4. A Craving for Crises

The exploration of the forms and extent of the ideological adaptation by the physical scientist to his environment must not stop at that unmarked and undefinable frontier where motivation and metaphysics end and the scientific activity itself begins. For to the ideology belongs not merely the general conceptions of the nature and goals of scientific activity, not merely the morale and esprit of the scientist, but also the scientist's perception of the state of his discipline, his hopes, fears, and expectations for its future development. Here, then, we return to the notion and mood of crisis, the conviction of a crisis of culture and of science, which was an essential component of the persona of the Weimar academics.

But before inquiring how far the German mathematical-physical community was likewise infected by this mood, how far a craving for crises affected the exact scientist's perception of the significance and bearing of specific scientific problems, it is worthwhile to emphasize how ready the mathematicians and physicists were to serve themselves with the crisis rhetoric when addressing a general academic audience. For as the notion of crisis became a cliché, it also became an entrée, a ploy to achieve instant "relevance," to establish rapport between the scientist and his auditors. By applying the word "crisis"

üben. Es werden sich auf allen Gebieten einer Kultur gemeinsame weltanschauliche Züge und, noch sehr viel zahlreicher, gemeinsame stilistische Züge vorfinden, in der Politik, in der Kunst, in der Wissenschaft. Wenn es gelingt, sie auch in der exakten Naturwissenschaft aufzuweisen, wird eine Art Indizienbeweis für Subjektivität und Milieubedingtheit erbracht sein." Translated very freely by James Murphy in Schrödinger, *Science, Theory, and Man* (New York, 1957), pp. 98-100.

134. R. v. Mises, *op. cit.* (note 117), pp. 3, 32.

to his own discipline the scientist has not only made contact with his audience, but has *ipso facto* shown that his field—and he himself —is "with it," sharing the spirit of the times. A presumption is thus insinuated, and often explicitly stated, that in the course of this crisis his science will shed all those characteristics which the academic audience finds most objectionable.[135,136]

But now, unless we are willing to charge duplicity and suppose that the physicists and mathematicians were engaged in a cynical manipulation of their image, I think we must allow that their accommodation to the intellectual environment penetrated deeper

---

135. For example: Walter Schottky, "Das Kausalproblem der Quantentheorie als eine Grundfrage der modernen Naturforschung überhaupt. Versuch einer gemeinverständlichen Darstellung," *Naturwiss.*, 9 (1921), 492-496, 506-511, opening paragraph: "Darstellung der Krisis in der sich die heutige Physik befindet"; M. H., "Ein Vortrag Einsteins: 'Neue Ergebnisse über die Eigenschaften des Lichtes'," *Neue Züricher Zeitung* (20 June 1922), nr. 808, p. 2: "The reason for the choice of precisely this theme was offered by the circumstance that in respect to the problem of the nature of light physics finds itself today in a severe crisis [sich heute . . . in einer schweren Krise befindet]"; Leo Graetz, *Alte Vorstellungen und neue Tatsachen der Physk. Drei Vorlesungen* (Leipzig, 1925), p. 1; Wilhelm Wien, "Kausalität und Statistik," *Illustrierte Zeitung* (Leipzig), Nr. 4169 (February 1925), pp. 192-196: "Das Aufwerfen dieser Fragen hat die ganze theoretische Physik in eine Krisis gebracht"; Erhard Schmidt, *Über Gewissheit in der Mathematik. Rede zum Antritt des Rektorats der . . . Universität zu Berlin am 15. Oktober 1929* (Berlin, 1930), p. 12: "so steht in der Tat die mathematische Gewissheit in einer Krise"; Richard Gans, *Die Physik der letzten dreissig Jahre. Rede, gehalten bei der Reichsgründungsfeier am 18. Januar 1930*, Königsberger Universitätsreden 7 (Königsberg, 1930), p. 1: ". . . eine Krisis in der Physik heraufbeschworen hat, wie sie in unserer Wissenschaft vorher nicht bekannt war"; once again, Gans, "Der Zufall in der Physik," *Schriften der Königsberger Gelehrten Gesellschaft, Naturwissenschaftliche Klasse*, 4 (1927), 113-125, opens: "I believe, however, that I may be permitted to claim your interest insofar as I want to try to show you what things excite the physicist today, and how he exerts himself to wind his way out of the most serious crisis in which our science has ever found itself"; Hans Hahn, ed., *Krise und Neuaufbau in den exakten Wissenschaften: Fünf Wiener Vorträge* (Leipzig-Vienna, 1933), Vorwort: "The growing interest of ever wider circles for the exact sciences is surely above all a seeking after one of the regions which are removed from the world of crises. . . . In truth the exact sciences are by no means secure from crises and precisely in recent decades, from theoretical physics on out into logic, they have been shaken by severe crises."

136. One can find prewar precedents for this "crisis talk" in popular essays and academic addresses. So, for example, Paul Ehrenfest's inaugural lecture as Professor of Theoretical Physics in Leiden, *Zur Krise der Lichtaether-Hypothese* (Leiden-Berlin, 1913), reprinted in Ehrenfest's *Collected Scientific Papers*, ed. M. J. Klein (Amsterdam, 1959), pp. 306-327. Just how widespread this was, and just what its significance was, I am not able to say, but it is my impression that it was then more common in France (Poincaré, Abel Rey) than in Germany.

than the rhetoric. Indeed, the rhetoric itself reacts back upon the persona of the scientist, upon his view of the conceptual situation in his science, of the extent and character of the reconstruction necessary or desirable. In fact, in this period, both mathematics and physics—but above all *German* mathematicians and physicists—went through deep and far-reaching crises, whose very definitions showed the most intimate relation with the principal currents of the Weimar intellectual milieu.

"The New Crisis in the Foundations of Mathematics" proclaimed by Hermann Weyl was precipitated virtually out of thin air in the two or three years following Germany's defeat. With extraordinary suddenness the German mathematical community began to feel how insecure were the foundations upon which the entire structure of mathematical analysis rested, how dubious the methods by which that edifice had been erected. Now, with quasi-religious enthusiasm, considerable numbers of German mathematicians rallied to L. E. J. Brouwer's standard calling for a complete reconstruction of mathematics, a redefinition of the enterprise, which, appropriately enough, went under the name "intuitionism."[137,138] The seriousness of this movement and its consequences may be judged by the vehemence of David Hilbert's counterattack in the spring of 1922. "If Weyl notices an 'inner untenability of the foundations upon which the construction of the empire [Reich] rests,' and worries himself over 'the threatening dissolution of the polity [Staatswesen] of analysis,' then he is seeing ghosts." Weyl and Brouwer are trying "to erect a repressive dictatorship [Verbotsdiktatur]"; to follow "such reformers" is to risk losing the most valuable treasures of mathematics. "No, Brouwer is not, as Weyl believes, the revolution, but only the repetition, with old means, of an attempted putsch . . . and now

137. H. Weyl, *Das Kontinuum* (Berlin, 1918; reprinted New York, 1962); "Der circulus vitiosus in der heutigen Begründung der Analysis," *Jahresber. d. Dtsch. Mathematiker-Vereinigung, 28* (1919), 85-92, reprinted in K. Chandrasekharan, ed., *Gesammelte Abhandlungen von Hermann Weyl* (Berlin, 1968), 2, 43-50; "Über die neue Grundlagenkrise der Mathematik," *Math. Zeitschr., 10* (1921), 39-79, reprinted in *Ges. Abhl., 2,* 143-180.
138. For description, bibliography, and documents of intuitionism: Abraham A. Fraenkel and Y. Bar-Hillel, *Foundations of Set Theory* (Amsterdam, 1958), especially pp. 203-204; Jean van Heijenoort, *From Frege to Gödel. A Source Book in Mathematical Logic, 1879-1931* (Cambridge, Mass., 1967); Constance Reid, *Hilbert* (New York, 1969).

especially, with the government [Staatsmacht] so well armed and secured by Frege, Dedekind, and Cantor, condemned to failure from the start."[139]

Can one read this rhetoric and not suppose that both Weyl and Hilbert at the very least saw close parallels between the crisis in mathematics and the political crises then wracking Germany, that their sense of the significance of the mathematical issues was colored by their perceptions of the political issues, that perhaps this crisis in mathematics depended for its very existence upon the social-intellectual atmosphere in the aftermath of Germany's defeat? Looking back thirty years afterward, Weyl almost conceded as much, and in fact the "crisis" itself, never resolved, eventually simply ceased to be felt.[140]

Turning to physics one finds once again a notable internal crisis. This is the "crisis of the old quantum theory" which gripped atomic physicists—first and foremost the Germans—in the years before the introduction of the quantum mechanics in 1925/26.[141] I have myself devoted some effort to the intriguing problem of isolating the particular difficulties and frustrations which led at a particular moment to a conviction that "the whole system of concepts

139. D. Hilbert, "Neubegründung der Mathematik. Erste Mitteilung," *Abhandlungen aus dem Math. Seminar der Hamburgischen Universität, 1* (1922), 157-177, reprinted in Hilbert's *Gesammelte Abhandlungen, 3* (Berlin, 1935; reprinted New York, 1965), 157-177, on 159-160. Hilbert had delivered this tirade against his most brilliant pupil as a lecture at a number of universities before printing it. And it is important to note that the adherents of the intuitionist movement not only admitted its destructive impact, but seem almost to have welcomed that consequence in a spirit of abnegation and resignation: "That proceeding from this standpoint only a part, perhaps only a wretched [kümmerlich] part, of classical mathematics is tenable is a bitter but inevitable fact." (Weyl, "Diskussionsbemerkung zu dem zweiten Hilbertschen Vortrag über die Grundlagen der Mathematik," *ibid., 6* [1928], 86-88, reprinted in Weyl, *Ges. Abhl., 3,* 147-149, and translated in Heijenoort, *Source Book,* pp. 482-484.)

140. H. Weyl, "Nachtrag, Juni 1955" to "Über die neue Grundlagenkrise . . . ," *Ges. Abhl., 2,* 179: "Only with some reluctance do I acknowledge [bekenne ich mich zu] these lectures, whose occasionally quite bombastic style reflects the mood of an agitated period—the period directly after the First World War." One might even hear in Hilbert's rhetoric a quite literal warning to Weyl and his other friends against the political attitudes of the leader whom they had chosen to follow. See Schröder, *op. cit.* (note 52), pp. 219-220; Reid, *Hilbert,* p. 188.

141. Thomas S. Kuhn, "The Crisis of the Old Quantum Theory, 1922-1925," address delivered at the American Philosophical Society, April 1966. Friedrich Hund, *Geschichte der Quantentheorie* (Mannheim, 1967), p. 103.

of physics must be reconstructed from the ground up," as Max Born asserted in the summer of 1923.[142] And while it is undoubtedly true that the internal developments in atomic physics were important in precipitating this widespread sense of crisis among German-speaking Central European physicists, and that these internal developments were necessary to give the crisis a sharp focus, nonetheless it now seems evident to me that these internal developments were not in themselves sufficient conditions. The *possibility* of the crisis of the old quantum theory was, I think, dependent upon the physicists' own craving for crises, arising from participation in, and adaptation to, the Weimar intellectual milieu.

Of this predisposition to perceive the state of physics as critical, we have many examples between the summer of 1921 and the summer of 1922, which is to say in the year immediately preceding that in which the crisis of the old quantum theory was precipitated. Taking only those cases in which the crisis is proclaimed in the title itself, there is Richard von Mises's lecture "On the Present Crisis in Mechanics" of September 1921, Johannes Stark's pamphlet on *The Present Crisis in German Physics* of June 1922, Joseph Petzoldt's remarks "Concerning the Crisis of the Causality Concept" of July 1922, and Albert Einstein's popular article "On the Present Crisis in Theoretical Physics," dated August 1922.[143] Very roughly speaking each of these physicists is pointing in the same direction, viz. toward the quantum theory. There, of course, the agreement ends; each is putting his finger upon a largely, or completely, different "problem." But that very circumstance—the widespread but initially poorly focused application of the word and notion of a crisis —suggests most strongly that the crisis of the old quantum theory,

142. M. Born, "Quantentheorie und Störungsrechnung," *Naturwiss.*, *11* (1923), 537-542, quoted in my essay on "The Doublet Riddle and Atomic Physics *circa* 1924," *Isis*, *59* (1968), 156-174.
143. R. v. Mises, "Über die gegenwärtige Krise der Mechanik," *Zeitschr. f. angewandte Math. u. Mech.*, *1* (1921), 425-431, reprinted in *Naturwiss.*, *10* (1922), 25-29, and reprinted once again in v. Mises' *Selected Papers* (Providence, R.I., 1964), *2*, 478-487, lecture, Math.-Phys. Congress, Jena, September 1921; J. Stark, *Die gegenwärtige Krisis in der Deutschen Physik* (Leipzig, 1922), 32 pp., preface dated "Anfang Juni 1922"; J. Petzoldt, "Zur Krisis des Kausalitätsbegriffs," *Naturwiss.*, *10* (1922), 693-695, dated 2 July 1922, to which Walter Schottky replied 6 October 1922 under the same title, *ibid.*, p. 982; A. Einstein, "Über die gegenwärtige Krise der theoretischen Physik," *Kaizo* (Tokyo), *4* (Dec. 1922), 1-8, dated August 1922.

far from being forced upon the German physicists, was more than welcome to them.

And here again, as with "intuitionism" in mathematics, one cannot help but be struck by the extraordinary convenience of the chief slogan of this crisis: the failure of mechanics. However appropriate this slogan may have been as a diagnosis of the internal difficulties in theoretical atomic physics, it certainly was *most* appropriate as a code word signaling the physicists' intent to rid their discipline of its most obnoxious elements. Conversely, the almost universal conviction among the German atomic physicists that this crisis was going to last a long, long time—although in fact it was "resolved" within two or three years by the discovery of the quantum mechanics—can be understood in part as a reluctance to contemplate giving up their fashionable and praiseworthy plight, but also in part as an expression of a Spenglerian pessimism: "in my heart I am once again convinced that this quantum mechanics"—which I, Werner Heisenberg, have just discovered—"is the answer, for which reason Kramers accuses me of optimism."[144]

III. "DISPENSING WITH CAUSALITY":[145,146] ADAPTATION OF KNOWLEDGE TO THE INTELLECTUAL ENVIRONMENT

III.1. Introduction. The Concept of Causality

Composing his article "On the Present Crisis in Theoretical Physics" for a popular audience in August 1922, Einstein began with a definition of the aim and structure of physical theory. "It is

---

144. ". . . mich des Optimismus anklagt. . . ." W. Heisenberg to Wolfgang Pauli, 29 June 1925, as quoted by B. L. van der Waerden, *Sources of Quantum Mechanics* (Amsterdam, 1967; reprinted New York, 1969), p. 27. Heisenberg's paper propounding this quantum mechanics, which was indeed "schon richtig," appeared as "Über quantentheoretische Umdeutung kinematischer und mechanischer Beziehungen," *Zeitschr. f. Phys.*, *33* (18 Sept. 1925), 879-893, received 29 July 1925; it is translated in v. d. Waerden, *op. cit.*, pp. 261-276, and reprinted in M. Born, W. Heisenberg, P. Jordan, *Zur Begründung der Matrizenmechanik*, Dokumente der Naturwissenschaft, Abteilung Physik, Band 2, ed. Armin Hermann (Stuttgart, 1962), pp. 31-45.

145. This is the title of the opening chapter of Albrecht Mendelssohn-Bartholdy, *The War and German Society* (New Haven, 1937). The eminent emigré legal scholar there asserted that "War canceled causality. It seemed to do so, at least, to the German people . . . the people as a whole, regardless of their interest in politics, their state of education, or their profession and walk in life,

the goal of theoretical physics," Einstein maintained, "to create a logical [!] conceptual system, resting upon the smallest possible number of mutually independent hypotheses, which allows one to comprehend causally [!] the entire complex of physical processes."[147] On the whole the conception of theoretical physics expressed by this definition is neither unfamiliar nor surprising; it was probably shared by most physicists of the period.[148] Yet two of Einstein's restrictions upon a physical theory seem either superfluous (of course a *logical* conceptual system, what else?) or gratuitous (why must one comprehend physical processes *causally*?). It is precisely these seemingly superfluous and gratuitous additions to what is otherwise the common creed of his colleagues which signal issues Einstein evidently regarded as crucial to theoretical physics as an enterprise. The first of these issues—a logical, as opposed to, say, an intuitive structure of physical theory—I will leave largely aside.[149]

---

realize the change quite clearly, long before it could be measured by historians or sociologists" (p. 20).

146. Earlier studies touching upon the pre-1926 movement to dispense with causality: Victor F. Lenzen, "The Philosophy of Nature in the Light of Contemporary Physics," *University of California Publications in Philosophy,* 5 (1924), 27-48; Alois Gatterer, S.J., *Das Problem des statistischen Naturgesetzes,* Philosophie und Grenzwissenschaften, herausgegeben vom Innsbrucker Institut für scholastische Philosophie, 1. Band, 1. Heft (Innsbruck, 1924), 70 pp.; Stefan Kis, *Das Kausalitätsprinzip in der Physik,* doctoral diss., U. Greifswald (Greifswald, 1925), 35 pp.; Hugo Bergmann, *Der Kampf um das Kausalgesetz in der jüngsten Physik,* Sammlung Vieweg, Heft 98 (Braunschweig, 1929), 78 pp.; Philipp Frank, *Das Causalgesetz und seine Grenzen,* Schriften zur wissenschaftlichen Weltauffassung, Band 6 (Vienna, 1932); Ernst Cassirer, *Determinism and Indeterminism in Modern Physics. Historical and Systematic Studies of the Problem of Causality,* trans. O. T. Benfey (New Haven, 1956); Max Jammer, *The Conceptual Development of Quantum Mechanics* (New York, 1966).

147. A. Einstein, *op. cit.* (note 143), p. 1.

148. The conception is essentially that stated most fully and forcefully by Pierre Duhem, *La Théorie physique: son objet, sa structure,* 2nd ed. (Paris, 1914), trans. by P. P. Wiener as *The Aim and Structure of Physical Theory* (Princeton, 1954), pp. 19, 52, 107, *et passim.*

149. There are two aspects to this issue: 1) a rationalist-irrationalist opposition which is reflected in attitudes toward causality, and which I will therefore touch upon in passing; 2) an opposition, cutting across this first alignment, between intuitiveness and abstractness, *Anschaulichkeit* and *Unanschaulichkeit,* as desirable or necessary in a physical theory. The demand for *Anschaulichkeit* was, once again, very closely connected with the predilections and antipathies characteristic of the Weimar intellectual environment. This opposition played an important role in physics and mathematics in Germany in the Weimar period (see notes 235 and 237), and especially in the Nazi period. I do not, however, attempt here to deal with this difficult problem.

It is the second issue, the "violent dispute over the significance of the law of causality"—that is the way Max Planck saw the situation in February 1923[150]—which I will describe and analyze.

But what is meant by "causality," or, rather, what notion did the physicists and the philosophers who stood closest to them associate with this term at this time? That notion was, in a word, lawfulness. "The principle of causality," Moritz Schlick explained in 1920, "is ... the general expression of the fact that everything which happens in nature is subjected to laws which hold without exception."[151] (Cf. Spengler's "Causality is coextensive with the concept of law. There are only causal laws."[152]) And even ten years later Heisenberg, in his *Physical Principles of the Quantum Theory,* could still broach this question by declaring that "the resolution of the paradoxes of atomic physics can be accomplished only by further renunciation of old and cherished ideas. Most important of these is the idea that natural phenomena obey exact laws—the principle of causality."[153] Of course Heisenberg immediately improved upon this formulation by introducing distinctions which were made and employed only after the development of quantum mechanics. In the period 1919–1925, however, both physicists and philosophers held this essentially Kantian notion of causality as conformity to law, so that, as Hans Reichenbach put it in 1920, "if there is cognition of nature, then the principle of causality is valid, for, without this principle, cognition, by its very meaning, is impossible."[154]

150. M. Planck, *op. cit.* (note 18), p. 140: "Seit langem ist über die Bedeutung des Kausalgesetzes in der Natur- und Geisteswelt . . . nicht so heftig gestritten worden wie in unseren Tagen. . . . Fast hat es den Anschein, als ob die denkende Menschheit bezüglich dieser Fragen in zwei getrennte Lager gespalten ist."

151. M. Schlick, "Naturphilosophische Betrachtungen über das Kausalprinzip," *Naturwiss., 8* (11 June 1920), 461-474; opening paragraph. This was still the general view five years later when *Kis* (*op. cit.,* note 146), p. 3, maintained that "Das Causalprinzip ist das Allgemeinste unter den Prinzipien der Naturwissenschaften. Es besagt, dass die Naturvorgänge nach strengen Gesetzen ablaufen, es ist mit den Worten Machs unsere Zuversicht in die Gesetzmässigkeit der Natur."

152. Note 77. Cf. Walter Rauschenberger, reviewing Ernst Berg, *Das Problem der Kausalität* (Berlin, 1920) in *Kant-Studien, 26* (1921), 174: "Verfasser ist also strenger Determinist. Sein Kampf gegen das Kausalitätsgesetz erscheint deshalb nicht recht verständlich. Kausalität und Gesetzmässigkeit wurden bisher stets als gleichbedeutend angesehen." I.e., determinism = causality = lawfulness.

153. W. Heisenberg, *The Physical Principles of the Quantum Theory* (Chicago, 1930), p. 62.

154. H. Reichenbach, "Philosophische Kritik der Wahrscheinlichkeitsrech-

But now, if such was the physicist's notion of causality, how indeed could he even contemplate dispensing with this principle? The point was made in 1924 by Alois Gatterer, philosopher and Jesuit, as "a sort of *argumentum ad hominem* which ought to give pause to especially those physicists who, like [Franz] Exner, take pleasure in trying to degrade all physical and chemical laws to statistical laws, and nonetheless at the same time, full of confidence and pride, pursue magnificent researches on the constitution of the chemical atom. How, I ask, can one approach this research with hope of success, and devote oneself actively to it, if one secretly nurtures the conviction that the elementary processes proceed, at least in part, lawlessly at random . . .?"[155] Gatterer's question is a good one, and his appeal—given the physicists' own notion of causality—ought to have been a strong one; and yet, evidently, it simply was not cogent. Thus if we find physicists repudiating causality—and taking pleasure in doing so—without making any attempt to critically analyze and revise the notion itself, then I think we must construe such repudiations as directed against the sort of cognitive enterprise in which physicists theretofore had understood themselves to be engaged.

Precedents for such a reaction against the cognitive enterprise of physics do exist, and, as Stephen Brush has made us recognize, such sentiments provided much of the steam behind the positivist movement at the turn of the century.[156] Where Mach was content to challenge the universal validity of the laws of mechanics, the radical fringe around the positivist-monist standard advanced to the denial of *any* exact laws for atomic processes with the intent of making room for "an element of indeterminacy, spontaneity, or

---

nung," *Naturwiss, 8* (20 February 1920), 146-153: "Dass eine funktionelle Form existiert, garantiert die Kausalität. . . . Allerdings ist es *denkbar,* dass das Naturgeschehen ohne funktionelle Abhängigkeiten verliefe; aber wenn es eine *Erkenntnis der Natur* gibt, dann gilt das Kausalprinzip, denn ohne dieses ist Erkenntnis ihrem Sinne nach nicht möglich. . . . Wir könnten durchaus zu einer physikalischen Erkenntnis kommen, wenn das Energiegesetz nicht gilt; die Gleichungen würden dann eben anders lauten; aber ohne Geltung des Kausalgesetzes wäre Erkenntnis unmöglich, weil wir überhaupt keine quantitativen Funktionalbeziehungen aufstellen könnten."

155. A. Gatterer, *Das Problem des statistischen Naturgesetzes* (*op. cit.,* note 146), pp. 45-46.

156. S. G. Brush, "Thermodynamics and History," *The Graduate Journal, 7* (1967), 477-565.

absolute chance in nature," as C. S. Peirce urged in *The Monist* in 1892.[157] Taking acausality in Peirce's sense—and that is, fundamentally, the sense in which it was understood in the early 1920's—I do not know of any other notable physicist who publically advocated this doctrine in the following quarter century, i.e., before the end of the First World War.[158]

> 157. Quoted by Brush, *ibid.*, p. 531. In this, Comte's latter day disciples were but accepting with relish what Comte believed but dreaded, namely that "the natural laws . . . could not remain rigorously compatible . . . with a too detailed investigation." Quoted by Emile Meyerson, *Identity and Reality*, trans. Kate Loewenberg (1930; reprinted New York, 1962), p. 20. It should be noted, however, that the editor of *The Monist*, Paul Carus, was scandalized: "Mr. Charles S. Peirce's Onslaught on the Doctrine of Necessity," *Monist*, 2 (1892), 560-582.
>
> 158. The persistence of a subterranean anticausality current is however suggested by some public refutations of the notion in the intervening period. On 3 August 1914, addressing a founder's day convocation at the close of his term as rector of the University of Berlin, Max Planck acknowledged that "this dualism" between causal and statistical laws which has arisen as a result of the introduction of the statistical point of view into physics "is regarded by many as unsatisfying." Consequently attempts have been made to deny that there are any dynamical (= causal) laws whatsoever and to regard all regularity as statistical: "the concept of an absolute necessity would be lifted from physics entirely. Such a view must, however, very quickly show itself to be an error as disastrous as it is shortsighted." (Planck, "Dynamische und statistische Gesetzmässigkeit," *Physikalische Abhandlungen und Vorträge* [Braunschweig, 1958], *3*, 77-90, on 86.) In a posthumously published contribution to the issue of *Die Naturwissenschaften* dedicated to Planck on his sixtieth birthday (23 April 1918), Marian von Smoluchowski pointed to "the tendency which holds sway today to reduce the totality of physical laws . . . to the statistics of hidden elementary events," and he regarded it as "perfectly possible" that in time Lorentz' theory of electrons, relativity, and the law of conservation of energy will also be subjected to this program. Yet Smoluchowski's aim was to show that "chance—in the sense of the word ordinarily used in physics—can perfectly well result from exactly defined lawbound [gesetzmässig] causes," and he emphasized that the calculus of probabilities is thus not to be regarded as a new principle of research but "merely a simplifying statistical schematization of certain functional interconnections which arise very frequently." (Smoluchowski, "Über den Begriff des Zufalls und den Ursprung der Wahrscheinlichkeitsgesetze in der Physik," *Naturwiss.*, *6* [1918], 253-263.)
>
> Although neither Planck nor Smoluchowski identifies the physicists they mean to be refuting, it may well be that both are thinking especially of Franz Exner, whose views Smoluchowski surely knew from his years in Vienna, and Planck may have learned from Exner's *Rektoratsrede* at the University of Vienna: *Über Gesetze in Naturwissenschaft und Humanistik. Inaugurationsrede, gehalten am 15. X. 1908* (Vienna, 1909), 45 pp. This publication, which came to my attention only after the final draft of the present essay was completed, does indeed maintain that "Alles Geschehen in der Natur ist das Resultat zufälliger Ereignisse" (p. 42), and that "if we were capable of slowing down the molecular motion so enormously that we could follow the individual molecular processes, then we would perceive nothing but a chaos of chance events, in which we would seek in vain for any regularity" (p. 13). Exner then went on to sketch a "unified

It must, of course, be acknowledged that in precisely this period Mach himself, the positivist movement in general, and even neo-Kantians like Cassirer were waging a campaign against quite a different concept of causality, so that by 1918 Friedrich Poske could well observe that "it has recently become the fashion among those concerned with the theory of scientific method to throw the causal concept onto the scrap heap."[159] At issue here was not, however, the notion of conformity to law, but rather the "metaphysical," "animistic," "fetishistic" doctrine of cause and effect (*Ursache und Wirkung*) as an ontological assumption, which Mach and his allies wished to replace by the mathematical conception of function.[160] "This putting together of functional interconnections is what theoretical physics is really all about," Wilhelm Wien explained to a lay audience in 1914; "causality—i.e., *der Satz von Ursache und Wirkung*—has nothing to do with the business."[161] And by 1918 this point of view had become almost a matter of course among physicists and the philosophers closely associated with them—

---

and comprehensive world picture" in which all law is but the expression of the law of large numbers, and the want of laws in the *Geisteswissenschaften* is due neither to the peculiar nature of their subject matter, nor to "Das Lebendige," nor to free will, but results simply from the relatively small number of equally chance events underlying the phenomena they study.

It is noteworthy, moreover, that P. Frank, *Kausalgesetz* (*op. cit.*, note 146), pp. 56-58, treated "Die energetische Naturauffassung" as one of the "Kausalitätsfeindliche Strömungen," and asserted that "at the time when the ideas of the Ostwaldian *Naturphilosophie* appeared to be dominant among those active in natural science and even among the majority of the laity interested in natural science, one could regard the notion of mechanical causality in Laplace's sense as having been disposed of, and the introduction of soul-like factors seemed necessary." This, like so much of Frank's book, is largely blather; nonetheless, it is suggestive.

159. F. Poske, *Zeitschr. f. den physikal. u. chem. Unterricht*, 31 (March 1918), 39.

160. E. Mach, *The Analysis of the Sensations*, trans. from the 5th edition, 1906, by S. Waterlow (reprinted New York, 1959), p. 81; Joseph Petzoldt, "Naturwissenschaft," *Handwörterbuch der Naturwissenschaften* (Jena, 1912), 7, 50-94, on 79-80; Paul Volkmann, *Einführung in das Studium der theoretischen Physik . . . mit einer Einleitung in der Theorie der physikalischen Erkenntnis*, 2nd enlarged ed. (Leipzig-Berlin, 1913), pp. 385-398; Ernst Cassirer, *Substance and Function*, trans. from the German edition, 1910, by W. C. and M. C. Swabey (Chicago, 1923; reprinted New York, 1953), Ch. 5; Hans Kelsen, *Society and Nature* [German title: *Kausalität und Vergeltung*]: *A Sociological Inquiry* (Chicago, 1943), p. 381.

161. W. Wien, "Ziele und Methoden der theoretischen Physik [1914]" (*op. cit.*, note 90), p. 156.

resisted only by a few arch-reactionaries like Ernst Gehrcke[162]—so that "causality," stripped of all ontologic overtones, was taken as equivalent to functional determination. Often, but not always, causality was further specified as the "Laplacian" conception of the necessary and sufficient conditions for such complete determination, viz. a cross section of the "world" at a given moment in time;[163] such a conception followed not merely from classical dynamics but equally from the very notion of a field theory.

Beginning with causality as the postulate of the lawfulness of natural processes, we have ended with causality as rigorous determinism. One might object that there is room for several distinct positions between these two conceptions. The possibility of satisfying a (weaker) postulate of lawfulness without demanding that every detail of every natural process be unambiguously determined did not entirely escape physicists in the years before the discovery of a quantum mechanics having this general character. Nonetheless, the essential point is that in the period treated in this paper every such suggestion of a relaxation of complete determinism was advanced as, and regarded as, a failure or abandonment of *causality*. In fact, as we proceed we will occasionally find the word "causality" being used in several senses narrower than, not wider than, "determinism"—as equivalent to the laws of classical mechanics, to the conservation of energy and momentum, to visualization in space and time, to the absence of action at a distance, to action by contact, or to description by differential equations. And again, in many instances these special definitions of causality were advanced in conjunction with, and as the justification for, an assertion of the invalidity of the law of causality. In every instance, however, such special definitions of causality, and a fortiori the general requirement of unambiguous

162. E. Gehrcke, *Physik und Erkenntnistheorie,* Wissenschaft und Hypothese 12 (Leipzig-Berlin, 1921), pp. 43-51.
163. E.g., Rudolf Carnap, "Dreidimensionalität des Raumes und Kausalität," *Annalen der Philosophie,* 4 (1924), 105-130, who, laying the groundwork preliminary to demonstrating his principal thesis that "die Dreidimensionalität des Raumes (gleichbedeutend mit der Vierdimensionalität des Weltgeschehens) ist die logische Folge der Gesetzmässigkeit des Geschehens," notes: "Die Geltung der Kausalität im Sinne der Physik besagt: in der physikalischen Welt herrschen determinierende Gesetze, und zwar sind alle Vorgänge eindeutig bestimmt, wenn die Gesamtheit der Vorgänge eines beliebig kleinen Zeitabschnittes bestimmt ist. Die Begriffe 'bewirken,' 'Ursache' u. dgl. haben also mit dem physikalischen Begriff der Kausalität nichts zu tun."

determination, were held to be equivalent to the assumption of the comprehensibility of nature, and repudiated or defended as such.

III.2. The First Intimations of an Issue, 1919-1920

If one examines the annual indices to German books and periodicals in the first decades of the century, one finds a remarkable number of articles and tracts with the word "causality" in their title. Most striking, however, is the spate of such tracts in the five years 1918-1922.[164] Typically, these are short answers to the riddle of the universe, the revelations of enthusiasts rather than the ruminations of academics. (They show, *inter alia*, that Spengler was not alone in seeing causality as the key to that riddle.) Yet the German academics too were anxious not to be left out of this discussion; in 1915 the Prussian Academy of Sciences had offered a prize for the best history of the causal problem since Descartes, awarding it in 1919 to a devoted student of the noted determinist Benno Erdmann.[165]

It is also at just this time—I know of no example earlier than 1919—that intimations of this issue appear in the private correspondence and public addresses of German physicists. In June 1919, replying to a lost letter from Max Born, Einstein asked ironically, "Is a hardboiled x-brother and determinist allowed to say with tears in his eyes that he has lost faith in humanity. Precisely the instinctive behavior of our contemporaries in political matters is suited to maintain a vivid belief in determinism."[166] Here Einstein is, on the one hand, gently ridiculing Born for feeling sorry for himself and his country by reminding Born of the public image of the theoretical physicist—hardboiled determinist—and, on the other hand, Einstein is making a small joke which can only be to some point if it were a recognized fact that the law of causality was under attack in the social sphere and under discussion among physicists.

Einstein could still joke about the matter with Born in early

---

164. *Deutsches Bücherverzeichnis*, 6 (1915-1920), 770; *10* (1921-1925), 1298, lists ten such.
165. Else Wentscher, *Geschichte des Kausalproblems in der neueren Philosophie* (Leipzig, 1921), preface.
166. Einstein-Born, *Briefwechsel* (*op. cit.*, note 14), pp. 29-30.

December,[167] but at the end of January 1920 his tone had become most serious, for in the meantime Born, in a long letter also lost, had evidently confessed to Einstein that he was willing to entertain the idea of acausality, supporting himself upon arguments of his subordinate, Einstein's former student, Otto Stern.[168] "That business of causality plagues me a great deal too," Einstein conceded, shaken by Born's defection but also anxious not to give offense by too categorical an assertion of his own "very very great reluctance to forgo *complete* causality." Yet even more interesting than these remarks themselves is the association of ideas which their precise location in Einstein's long letter—clearly a point by point reply to Born's— reveals. They occur toward the end, immediately following not unsympathetic remarks on Oswald Spengler, whose *Decline of the West,* published the year before, included barbs directed at both Einstein and Born. "Sometimes in the evening," Einstein allowed, "one likes to entertain one of his propositions, and in the morning smiles about it."[169]

167. *Ibid.,* p. 38. Born had published a short popular exposition of general relativity, "Raum, Zeit und Schwerkraft," *Frankfurter Zeitung,* 23 Nov. 1919, Nr. 876, pp. 1-3, in which, among other jabs at Kant, he remarked that "Wer diese Entwicklung [of relativity theory] miterlebt hat, der wird sich des Zweifels am a priorischen Character auch anderer Kategorien des Denkens nicht erwehren können." Born was then singled out for attack by Robert Drill, "Ordnung und Chaos. Beiträge zur Geschichte von der Erhaltung der Kraft," *Frankfurter Zeitung,* 30 Nov. 1919, Nr. 895, pp. 1-2; 1 Dec. 1919, Nr. 899, p. 1, a wild, fanatical Kantian who sought to prove the a priori character of the [metaphysical, ontological] concept of causality using the very concrete example of our anticipation of the taste of a piece of *Wurst*. This tickled Einstein: "Sein Nachweis der Causalität a priori ist wahrhaft erhebend."

168. Einstein to Born, 27 Jan. 1920, *Briefwechsel,* pp. 42-45. It would be most interesting to know just what Born's and Stern's views were. Some indication of them was probably contained in a talk on "Wahrscheinlichkeit und Kausalität in der Physik" which Born delivered to the Physikalischer Verein in Frankfurt on 27 July 1920 (*Jahresbericht,* 1919-1925, p. 107), but which appears to have remained unprinted.

169. Einstein, 27 Jan. 1920, *loc. cit.*: "Der Spengler hat auch mich nicht verschont. Man lässt sich gern manchmal am Abend von ihm etwas suggerieren und lächelt am Morgen darüber. Man sieht, dass die ganze Monomanie aus der Schullehrer-Mathematik kommt. Euklid-Cartesius ist sein Gegensatz, den er nun in alles hinarbeitet, aber—wie man gern zugibt—mit Geist. Solche Dinge sind amüsant, und, wenn morgen einer mit dem nötigen Geist das Gegenteil sagt, so ist es wieder amüsant, und was *wahr* ist, weiss der Teufel!"

"Spengler didn't spare me either" is a puzzle, for, so far as I can see, in the original edition (cf. note 65) relativity is handled rather unpejoratively as "die letzte Form der faustichen Natur," pp. 599-601, while Born is not mentioned

What proposition might Einstein have had in mind? Might he not have been thinking of Spengler's most fundamental proposition, the axis of the system, "the opposition of the destiny-idea and the causality-principle"? Is the juxtaposition in Einstein's letter of Spengler's *Untergang* and the issue of causality in atomic physics pure chance? Is it not more likely that in Born's mind and/or in Einstein's mind there is an intimate, although perhaps not fully conscious, association between the physicists' new and sudden inclination to forgo causality and Spengler's enormously popular culture-criticism in which the physicist "whose entire mental existence is founded upon the principle of causality" symbolizes the late and decadent fear of the irrational, the incomprehensible?[170]

Such an association between Spengler and the issue of acausality is certainly explicit in Wilhelm Wien's public lecture on "The Connections of Physics with Other Disciplines," which he delivered just one month later, at the end of February 1920, in the Prussian Academy of Sciences. Previously I used this lecture to illustrate a chameleonlike adaptation of the physicists' ideology to changes in the intellectual environment. Here, however, it stands as the first of a series of attempts to draw a clear line between that environment and physics as a cognitive enterprise. The apparent contradiction reflects rather the distinction I made in Part II between the peripheral and the central features of scientific ideology. Although Wien was ready to advance a new conception of the wellsprings and social-intellectual function of scientific activity designed to make the enterprise seem worthwhile in the public's eye, he was unwilling to compromise those ideological tenets which he regarded as essential to the scientific method and its cognitive goals. The true source and value of natural science lies, to be sure, in "an inner need of the human mind," but that need is for a particular kind of knowledge;

---

at all. It is then only in the revised edition (1923) that relativity is described as "a ruthlessly cynical working hypothesis," and the space given it much reduced, pp. 544-545 (Eng. ed., p. 419), while Born, mistaken for a chemist ignorant of mathematics, receives his just deserts in a footnote on pp. 205-206 (Eng. ed., p. 156).

170. See notes 77 and 78 and texts thereto. P. Frank, *op. cit.* (note 146), devoted a chapter to "Kausalitätsfeindliche Strömungen." He suggested that such currents were widespread in the general intellectual milieu of the Weimar period, but gave only one example: Spengler.

it is a "longing to comprehend the causality of the course of phenomena [die Kausalität des Geschehens]."

Wien's motivation for incorporating causality in the very definition of natural science becomes quite evident when, at the end of his lecture, he comes to the *Decline of the West*. While conceding that there is ample evidence of the accuracy of Spengler's characterization of our present cultural situation, Wien rejects in principle the notion of historical laws, of any necessary course of history. All such laws can be and are repeatedly violated "by irrational expressions of the human spirit." Turning Spengler against Spengler, Wien emphasizes that if we suppose a generally valid law of aging of cultures and use it to predict the future of our own, "then we reintroduce a covert causality into history." But to adapt Kant's well-known epigram, "I would like to assert that there is so much the more true historical science in history the less it contains of physics. . . . Causality is the foundation of the physical world picture, but it is a category [Denkform] of our mind [Geist] and cannot be employed again for the analysis of the same spirit [Geist], whose effects it is the task of history to portray."[171]

Embracing causality in order to effect a separation between his discipline and his milieu, Wien has then the task of repulsing Spengler's attempt to make physics culture-bound, and especially Spengler's contention that contemporary physics, as its nerve failed, was renouncing causality. It is nature which has compelled the physicists to resort increasingly to the use of statistics; it is neither a sign of "decadence" nor of any renunciation of causality. On the contrary, every utilization of statistics "postulates causality," but because of the great complexity the causal interconnections cannot be traced in detail. Where in 1914 Wien stressed that of all the natural sciences it is theoretical physics in which the personality has the greatest scope and importance, both in constructing theories and influencing the course of scientific development, now, contra Spengler, he is at pains to emphasize that "however strongly the shaping of physical modes of thought depends upon the constitution of the physicist, it is nonetheless decisively determined by the nature of the things themselves." Archimedes' results accord entirely with

---

171. W. Wien (*op. cit.*, note 100); the quotations are from pp. 20, 35, and 38-39, respectively.

our own, and our results will in all probability be utilizable by physicists of a later culture.[172]

The attention which Wien gives to Spengler, his focus upon the issue of causality, and his consistent effort to isolate physics—as a cognitive enterprise for which causality is the defining characteristic—from its acausal, irrational historical milieu, all suggest that he sensed an intimate connection between the treasonable murmurings against causality among his colleagues and Spengler's brilliant expression of certain powerful currents in the contemporary milieu.

III.3. Conversions to Acausality, 1919–1925

a. The Earliest Converts: Exner and Weyl

It was not until 1919, when Franz Exner was a full seventy years old, that his *Lectures on the Physical Foundations of the Natural Sciences* were printed.[173] And although the crucial concluding lectures on laws of nature may have been worked out long before in the mind and the conversation of the distinguished Viennese spectroscopist, they were probably not even included in the course as delivered to the public before the war.[174] The argument begins with the assertion that *none* of our laws of nature is exact. From this postulate—and here perhaps is the link with the late nineteenth-century positivist-monist repudiations of causality—Exner jumped to the conclusion that "causality" does not obtain, that if examined sufficiently closely during sufficiently short time intervals the motion of a falling body would be found to be perfectly random, directed up as often as down.[175] The apparent lawfulness which we discover at

---

172. *Ibid.*, p. 37; W. Wien, "Ziele und Methoden der theoretischen Physik. Festrede . . . 1914" (*op. cit.*, note 90), pp. 152-154.

173. F. Exner, *Vorlesungen über die physikalischen Grundlagen der Naturwissenschaften* (1st ed., Vienna, 1919; 2nd enlarged ed., Leipzig and Vienna, 1922). The crucial 86th-94th *Vorlesungen* are in all essential respects identical in both editions. The following quotations are from the preface to the first edition and from the 93rd and 94th *Vorlesungen*.

174. Internal evidence in these concluding lectures on laws of nature and Exner's reference to them in the preface as an "Anhang" argue that they were never delivered as such, but synthesized during the war when the book was written. See, however, note 158.

175. Exner claimed (86th *Vorlesung*, p. 658 in 2nd ed.) to have obtained Ludwig Boltzmann's assent to this proposition.

the macroscopic level is then "explained" by Exner's second thesis that all macroscopic natural laws are essentially statistical in character, the regularity arising in some unspecified way out of the collaboration of the random motions. The speculation that *all* macroscopic laws are essentially statistical, that none is exact, was by no means unprecedented. What is novel is the leap from that supposition to the conclusion that causality fails. For this leap no justification is offered, and the problem of how perfectly acausal microscopic motions result in statistical regularities is not even raised by Exner.

Exner the experimentalist takes a radical nominalist-empiricist stance: the absolutely rigorous laws "are a creation of man and not a piece of nature." Nor have we the right to postulate even "the existence of an absolute causality," least of all on the grounds that it is necessary in order for us to understand nature. "Nature does not inquire at all whether men understand her or not, and we are not to construct a nature adequate to our understanding, but our task is simply to come to terms with that which is given as best we can." Although Exner cannot consistently maintain his empiricist posture and also categorically deny the existence of causality at the microscopic level, he wants very much to do so in order "to arrive at a unified world picture" in which *all* law is purely statistical, a world of pure chance. He therefore does his best to convince his (lay) readers of the implausibility of the existence of such a causal substratum, switching back and forth between, and largely confounding, the question of the validity of the laws of classical mechanics in the atomic domain and the validity of the principle of causality in the same domain.

Influential as Exner's lectures indeed were, they have in many respects an archaic air. Exner is a curious mixture of the philosophical currents of the two preceding generations, a self-confessed mechanist-materialist yet clearly also a positivist in his view of scientific constructs. Of the *Lebensphilosophie* and existentialism which will figure so prominently in most of the following conversions to acausality there is scarcely a hint. Radioactivity and Brownian motion are the most recent developments in physics to which he refers; in his efforts to cast doubt upon the causal character of atomic processes he omits to serve himself with a "quantum" of any sort. Thus the first of the calls for a renunciation of causality was clearly independent of the problems raised by the quantum theory of the atom or radiation.

Apart from Exner, the earliest to speak out against causality was Hermann Weyl. Weyl was a phenomenologist of quite a different sort from his Machian ex-brothers. As Privatdozent at Göttingen shortly before the war Weyl had fallen under the influence of Edmund Husserl's program of "pure phenomenology." This Platonizing phenomenology of the mind, based upon intense introspection, had originated in epistemological concerns but in this period was degenerating into existentialism. Dating from 1917 is the first avowed intrusion of Weyl's philosophical outlook into his scientific work—his own attempt to place the continuum on an intuitionist foundation.[176,177] But as I indicated in Section II.4, Weyl soon

---

176. H. Weyl, "Erkenntnis und Besinnung (Ein Lebensrückblick)," *Studia Philosophia* (1954), as reprinted in Weyl's *Ges. Abhl., 4,* 631-649, recalled that in his student years at Göttingen, 1906-1910, his adolescent Kantianism was converted to positivism—he read Mach, Poincaré, and F. A. Lange—and only shortly before his departure for Zurich in 1913 "was it Husserl, then, who led me out of positivism to a freer view of the world once again." The contact with his colleague was mediated by one of Husserl's numerous enthusiastic students, Helene Joseph, whom Weyl married in 1913. Explicit citations of Husserl first appeared in *Das Kontinuum* (Berlin, 1918), written in 1917, and in the introduction to *Raum-Zeit-Materie,* 1st ed. (Berlin, 1918), preface dated Easter 1918. Weyl's extraordinary deference to Husserl is evident in the repeated quotations, always with full approbation, in his *Philosophy of Mathematics and Natural Science* (Princeton: Princeton University Press, 1949), translated from the German edition of 1927. In return the Husserl school was happy to lean upon Weyl and claim him for one of their own: Oskar Becker, "Beiträge zur phänomenologischen Begründung der Geometrie und ihrer physikalischen Anwendungen," *Jahrbuch für Philosophie und phänomenologische Forschung, 6* (1923), 385-560, on 387-388. By 1928, however, the success of formalist metamathematics had shaken Weyl's allegiance: "If Hilbert's view prevails over intuitionism, as appears to be the case, *then I see in this a decisive defeat of the philosophical attitude of pure phenomenology,* which thus proves to be insufficient for the understanding of creative science even in the area of cognition that is most primal and most readily open to evidence—mathematics." (Weyl, *op. cit.,* note 139). See also Peter Beisswanger, "Hermann Weyl and Mathematical Texts," *Ratio, 8* (1966), 25-45.

177. An admirable and uniquely intelligible account of this remarkable intellectual phenomenon is given by Herbert Spiegelberg, *The Phenomenological Movement: A Historical Introduction,* 2nd ed., 2 vols. (The Hague, 1965), which may be supplemented by Joseph J. Kockelmans and Theodore J. Kisiel, eds., *Phenomenology and the Natural Sciences: Essays and Translations* (Evanston, Ill., 1970).

became the principal champion of Brouwerian intuitionism in Germany. That Weyl saw an intimate connection between intuitionism in mathematics and acausality in physics emerges quite clearly from his initial manifesto against causality, "The Relation of the Causal to the Statistical Approach in Physics," printed in August 1920.[178,179] "Are statistics merely a shortcut to certain consequences of causal laws," Weyl asks, "or do they imply that no rigorous causal interconnection governs the world and that, instead, 'chance' is to be recognized alongside law as an independent power restricting the validity of the law? The physicists are today entirely of the first opinion." And yesterday, in the spring of 1918, Weyl had been too, having in his proposed extension of general relativity made an "attempt," as he admits, "at carrying through the idea of a pure

178. H. Weyl, "Das Verhältnis der kausalen zur statistischen Betrachtungsweise in der Physik," *Schweizerische Medizinische Wochenschrift, 50* (19 August 1920), 737-741, reprinted in Weyl's *Ges. Abhl., 2,* 113-122. Weyl had prepared an address with this title for a symposium on "the Significance of Probability for Natural Science and Medicine" which had been organized by Heinrich Zangger, professor of forensic medicine in Zurich, for the annual congress of the Schweizerische Naturforschende Gesellschaft in Lugano in September 1918. The congress was canceled, however, due to the grippe epidemic, so that the earliest statement we have of Weyl's position is the 500-word abstract of the address Weyl delivered the following year, 8 September 1919; *Schweizerische Naturforschende Gesellschaft, Verhandlungen* (1919), 2. Teil, pp. 152-153. This abstract has the same general structure as the printed paper, includes Husserlian phenomenological-existentialist jargon ("das nur im Willen erlebte 'Grund-sein,' " etc.), and concludes with an affirmation of Weyl's belief "that at the basis of statistics there lies an independent principle which is not to be reduced to causality." Nonetheless, it suggests very strongly that in the fall of 1919 Weyl had not yet advanced as far as his position of August 1920, and that, in particular, he had not yet made the connection with intuitionism in mathematics, the repudiation of causality and "der reinen Gesetzesphysik" being based solely upon their incompatibility with the "für unser ganzes Erleben fundamentale Einsinnigkeit der Zeit."

179. Although he failed to mention Hermann Weyl in this connection, A. d'Abro, *Decline of Mechanism* (New York, 1939), reprinted as *The Rise of the New Physics,* 1 vol. in 2 (New York, 1952), pp. vii, 212, justified inclusion of a chapter treating "The Controversies on the Nature of Mathematics" in his historical exposition of the quantum theory on the grounds that "In our opinion these controversies originate from the same psychological differences which appear to be responsible for the current controversy concerning the principle of causality in physics." In concluding that chapter d'Abro suggested "It might even be said that modern physics is witnessing the same crisis that we have been discussing in mathematics: the quantum theorists occupy the position of the intuitionists while Einstein and Planck occupy that of the formalists." I think d'Abro's conjecture is essentially correct.

physics of law for the entirety of the world."[180] But now Weyl has changed his mind and is placing himself in opposition to the prevailing opinion. Why? He has certain dissatisfactions with classical statistical mechanics and the treatment of fluctuation phenomena, but the real issue, he admits is that

> finally and above all, it is the essence of the continuum that it cannot be grasped as a rigid [starr] existing thing, but only as *something which is in the act of an inwardly directed unending process of becoming.* . . . In a *given* continuum, of course, this process of becoming can have reached only a certain point, i.e. the quantitative relations in an intuitively given piece S of the world [regarded as a four-dimensional continuum of events] are merely approximate, determinable only with a certain latitude, not merely in consequence of the limited precision of my sense organs and measuring instruments, but because *they are in themselves afflicted with a sort of vagueness.* . . . And only "at the end of all time," so to speak, . . . would the unending process of becoming S be completed, and S sustain in itself that degree of definiteness which mathematical physics postulates as its ideal. . . . Thus the rigid [starr] pressure of natural causality relaxes, and there remains, without prejudice to the validity of natural laws, *room for autonomous decisions* [*Entscheidungen*], *causally absolutely independent of one another,* whose locus I consider to be the elementary quanta of matter. These "decisions" are what is *actually real* in the world.[181]

I have quoted Weyl at some length, both because he goes on at some length and because a mere ascription of such radically existentialist views and motives would very likely be dismissed as incredible. Yet, clearly, these motives are primary. Weyl has resolved to abandon the ideal of a pure field physics—for which he had labored so hard and achieved such striking success—and adopted matter, or rather its free will, as the ultimate reality. The field and its laws, like geometry before Einstein, were now a mere backdrop. Why? Because it seemed necessary in order to escape the determinism which the field conception involved. Here, in the fall of 1919 and the summer of 1920 Weyl says not a word about Planck's quantum

---

180. H. Weyl, *Ges. Abhl.*, 2, 116-117; "Gravitation und Elektrizität, *Preuss. Akad. der Wiss., Berlin, Sitzungsber.* (30 May 1918), 465-480, reprinted *Ges. Abhl.*, 2, 29-42, and trans., with additional notes, in H. A. Lorentz, *et al., The Principle of Relativity* (London, 1923; reprinted New York, 1952), pp. 201-216.
181. H. Weyl, *Ges. Abhl.*, 2, 121-122.

of action. It has evidently not yet occurred to him that the quantum theory could be dragged in to provide an ostensible physical basis for his existentialist repudiation of causality. It was only in the fall of 1920, when preparing the fourth edition of *Space-Time-Matter,* that Weyl seized upon the quantum theory as compelling him to say "clearly and distinctly that physics in its present state is simply no longer capable of supporting the belief in a closed causality of material nature resting upon rigorously exact laws." There Weyl also added that crucial existentialist consideration which had been with him for some time—the repudiation of determinism restores the unidirectionality of time, "the most fundamental fact of our experience of time," which field physics denied us a priori.[182] Thus, "not only is matter restored to its old claim to reality, but also the genuine idea of causality, of *Verursachung,* as we experience it most immediately in our will, awakes to new life. Branded as fetishism by Mach . . ." etc., etc.[183]

It seems pretty clear—and indeed it is characteristic of the acausalists—that the sort of primary reality which Weyl would have matter enjoy is simply not a sort of reality which is accessible to physical cognition. Thus by the summer of 1924, in carrying his "Leibnizian agent-theory of matter" to its logical conclusion, Weyl was led back to the field as the primary *physical* reality:

> the material particle itself is not even a point in space, but is something entirely outside the category of extension. . . . It is analogous to the Ego, whose actions, despite the fact that it is itself nonextensional, always have their origin, through its body, at a definite place in the world continuum. Yet whatever this field exciting agent may be in its inner essence—perhaps life and will—in physics we consider it only in terms of the field actions which are excited by it and we are able to characterize it numerically (charge, mass) only by virtue of these field actions.[184]

182. H. Weyl, *Raum-Zeit-Materie,* 4th ed. (Berlin, 1921), pp. 283-284; *Space-Time-Matter,* trans. from 4th ed. by H. L. Brose (London, 1922; reprinted New York, 1952), pp. 310-312. The preface to this edition is dated November 1920. Again, with a more precise statement of "causality," in the 5th ed. (Berlin, 1923), pp. 286-287.
183. H. Weyl, "Feld and Materie," *Annalen der Physik,* 65 (1921), 541-563, received 28 May 1921; reprinted in Weyl's *Ges. Abhl.,* 2, 237-259, on 255.
184. H. Weyl, "Was ist Materie?" *Naturwiss.,* 12 (11, 18, 25 July 1924), 561-569, 585-593, 604-611; *Ges. Abhl.,* 2, 486-510, on 510.

Weyl was now able to reconcile himself to this resurrection of the field because he thought he had finally found an escape from the proposition that the classical field theories embody and impose the Laplacian conception of causality. In a semipopular article in the form of a dialogue, Weyl argued that "according to the general theory of relativity the concept of the relative motion of several bodies with respect to one another is just as little tenable as that of the absolute motion of a single body." Consequently, the principle of causality cannot involve these untenable states of motion, and so reduces to the assertion that "the world of events only depends upon, and must be unambiguously determined by, the charge and mass of all material particles. Since this is obviously absurd ... that principle of causality must be abandoned."[185]

b. 1921, Summer and Fall: von Mises, Schottky, Nernst, *et al.*

The quasi-religious conversions to acausality, of which Weyl's is the earliest example, became a common phenomenon in the German physical community during the summer and fall of 1921. As if swept up in a great awakening, one physicist after the other strode before a general academic audience to renounce the satanic doctrine of causality and to proclaim the glad tidings that the physicists are about to release the world from bondage to it. The cases known to me are: Walter Schottky in June, Richard von Mises in September, Walther Nernst in October.[186]

The conversion of von Mises to acausality is particularly interesting not only because it shows the suddenness with which this regeneration could take place and its essential independence of the difficulties encountered in atomic physics, but also because it provides *prima facie* evidence of a direct connection between the repudiation of causality by a loyal scion of Austrian positivism and

---

185. H. Weyl, "Massenträgheit und Kosmos. Ein Dialog," *Naturwiss.*, 12 (14 March 1924), 197-204; *Ges. Abhl.*, 2, 478-485.
186. W. Schottky, "Das Kausalproblem der Quantentheorie als eine Grundfrage der modernen Naturforschung überhaupt. Versuch einer gemeinverständlichen Darstellung," *Naturwiss.*, 9 (24 and 30 June 1921), 492-496, 506-511; R. von Mises, "Über die gegenwärtige Krise der Mechanik" (*op. cit.*, note 143); W. Nernst, *Zum Gültigkeitsbereich der Naturgesetze* (Berlin, 1921), 26 pp., reprinted in *Naturwiss.*, 10 (26 May 1922), 489-495. This is Nernst's inaugural lecture as rector of the University of Berlin, 15 October 1921.

his capitulation to the *Weltschmerz* of Spengler's *Decline of the West*. In von Mises' inaugural (and farewell) address as delivered in February 1920 at the Technische Hochschule Dresden, and as printed in August 1920, causality was still handled unself-consciously and unpejoratively as equivalent to physical explanation. "We see now in our time, how a new and simply enormous field of phenomena, the multiplicity of the chemical elements, is drawn into the realm of causal explanation." And von Mises takes it for granted that the goal of atomic physics, as of all natural science, is and must be "to explain all these phenomena on the basis of a very few principles, to reveal their causality."[187] But when one turns to the thoroughly Spenglerian appendix which von Mises added in September 1921 to the republication of this lecture, one finds his attitude toward causality—as toward so much else—entirely transformed. Every electrical, every thermal, every optical process is a statistical phenomenon and as such fundamentally incompatible with the concept of causality. So long as we base ourselves upon that concept "the quantum theory and everything connected therewith must appear as an insoluble riddle. Whoever traces back the history of physical cognition cannot help but recognize that here an essential *alteration of our mode of thinking,* of the entire scheme of 'physical explanation,' is inexorably demanded and is gradually being prepared."[188]

Admittedly, von Mises has invoked the quantum theory as the occasion for the repudiation of causality. But he was not willing that it be *more* than the occasion, that, in particular, his own discipline of applied classical mechanics remain saddled with the stigma of causality. In this same month, September 1921, at the first of the annual German physics-mathematics congresses, von Mises read his colleagues a lecture—or better, made a public confession before an assembly of his peers—regarding "The Present Crisis in Mechanics."

> Stated in the briefest form, this question—in whose negative answer I discern the crisis in the present state of mechanics—runs thus: can we still assume that all phenomena of motion and equilibrium which we observe in visible bodies are explicable within the framework of the

187. R. von Mises, *Naturwissenschaft und Technik der Gegenwart* (*op. cit.,* note 118), p. 19.
188. *Ibid.,* p. 30.

Newtonian axioms and their extensions. In other words, can the temporal course of every motion of an arbitrarily delimited portion of mass be unambiguously determined by specifying the initial state and assuming some appropriate force law to be acting? . . . All that I want to try to show here is that the accumulated facts which we possess today make it evident that it is highly improbable that this goal of classical mechanics could ever be attained, and that other, perfectly definite and no longer unfamiliar, considerations are destined to relieve or to supplement the rigid causal structure [den starren Kausalaufbau] of the classical theory . . . whether the sacrifice be great or small, whether we find it difficult or easy, it seemed to me unavoidable for once clearly and frankly to state that within the purely empirical mechanics there are phenomena of motion and equilibrium which will forever escape an explanation on the basis of the differential equations of mechanics. . . .[189]

One cannot help but be struck by the "me too" tone of von Mises' repudiation of "the stiff causal structure" of classical mechanics and his representation of that renunciation as an act of moral virtue. Yet it is also precisely this tone which suggests that a conversion to acausality carried with it significant social approbation, social rewards so substantial that von Mises could not bear to let the atomic physicists monopolize them.

Although Weyl had already turned to the quantum theory in seeking support and ammunition for his attack on causality, Walter Schottky seems to have been the first atomic physicist to publish an acausal manifesto treating "The Problem of Causality in the Quantum Theory as a Basic Question for Modern Natural Science as a Whole."[190] Schottky's article of June 1921, subtitled "Attempt at a Popular Exposition," is clearly an expanded version of a lecture—very likely an inaugural lecture as Privatdozent for theoretical physics at the University of Würzburg, where he had recently habilitated after several years at the research laboratories of Siemens and Halske in Berlin. Schottky feels sure that inasmuch as one is accustomed to regard the rigorous laws of physics as a model and ideal for "all analytical contemplation of nature," a general and historical

189. R. v. Mises, *op. cit.* (note 143), *Selected Papers, 2,* 482, 487.
190. W. Schottky, *op. cit.* (note 186).

exposition of the "crisis," the "revolution in the basic conception of the form and range of physical laws" which is in preparation, will be welcome to his audience.[191]

In the first installment of the article Schottky builds up to the proposition that the electromagnetic field and its variables are finished, done. For, he argues with impeccable logic, if we don't know the laws of the interaction of atoms with radiation, but yet can only observe the electromagnetic field quantities through their interactions with matter, then these "state variables of the field theory . . . no longer possess any significance whatsoever for scientific research." Allowing that "that is a consequence which to be sure thus far only very few physicists have accepted," Schottky proceeds immediately to ask what sort of observable quantities, and what sort of connections between them, are to be put in place of the electromagnetic field. And the answer: "The law of causality itself, with its complete conditioning of the coming phenomena by the present and past phenomena, appears . . . to be placed in doubt."[192]

So much in the first installment. In the second installment we discover the conventional electromagnetic field variables and equations back at work, and all that remains of the earlier "analysis" is the insistence that any solution to the problem of the interaction of atoms and radiation must cancel causality. Schottky's first proposal is the oft-recurring conjecture that the field equations determine merely the *rate* at which the quantal elementary processes take place. But this "at first sight very attractive way is impassable," for Einstein has told him that because of the inexact fulfillment of the conservation laws, in the course of sufficiently long times a motion with arbitrarily large velocity could arise out of nothing—a point which escaped Bohr, Kramers, and Slater three years later.

Schottky now turns to his own pet idea that there is a direct connection of the emitting with the absorbing atom by retarded action at a distance, so that at the moment when a quantum is emitted it is already predetermined where, when, and by what atom it will be absorbed. But is this not causality with a vengence, a physics à la Calvin? That is certainly how Tetrode, an under-appreciated Dutch theoretical physicist, represented the case when, exactly one year

191. *Ibid.*, p. 492.
192. *Ibid.*, pp. 495-496.

later, he published the outlines of a theory based upon this same conception.[193] Yet such a thought never enters Schottky's mind; all that he sees is a failure of causality arising from the fact that it is no longer possible "to conceive the course of events like a continually and uniformly flowing stream," that because the "unbreakable threads" connecting emission and absorption extend infinitely far towards past and future, it is no longer possible in principle to predict the future from a cross section of the world at a given moment in time. And finally, to make the acausality doubly sure, Schottky asserts categorically—but inexplicably—that these elementary acts of emission and absorption, the precise positions of the beginnings and ends of these unbreakable threads, are indeterminate, "without direct cause and without direct effect," "outside the relation of cause and effect."[194]

Thus Privatdozent Schottky. Is perhaps the demonstration of the failure of causality which Geheimrat Professor Walther Nernst offered four months later in his inaugural lecture as rector of the University of Berlin—and which produced a correspondingly greater stir—less tendentious, less shallow and fallacious? Scarcely. Here again, what is most striking is the author's resolve to sink the law of causality by hook or by crook. And his motive for doing so is clear enough: "But, now, can philosophy and natural science really assert with certainty that, for example, every human action is the unambiguous result of the circumstances prevailing at the moment? If absolutely rigorous laws of nature controlled the course of all events, one would in fact scarcely be able to escape from this conclusion." But philosophy has adopted this position only because it has been tyrannized by the exact natural sciences, whose "conception of the principle of causality as an absolutely rigorous law of nature laced

---

193. H. Tetrode, "Über den Wirkungszusammenhang der Welt. Eine Erweiterung der klassischen Dynamik," *Zeitschr. f. Phys.*, *10* (1922), 317-328, received 14 June 1922. Tetrode is generally critical of "der einseitig gerichteten [!], zum Teil zufallsmässig bedingten Kausalität" to which the modern development of physics, above all the theory of field action, has led. But this conception of causality is not aboriginal in the human mind; therefore why not consider another conception. The result is remarkably like the Bohr-Kramers-Slater theory —the electro-magnetic field becomes unreal, the conservation of energy and momentum is statistical—but with exactly the opposite intent, namely to strengthen rather than relax determinism.

194. W. Schottky, *op. cit.* (note 186), pp. 509-511.

the mind [Geist] in Spanish boots, and it is therefore at present the obligation of research in natural science to loosen these fetters sufficiently so that the free stride of philosophical thought is no longer hindered."[195]

In outline Nernst's argument is that, first, the principle of causality implies the existence of exact natural laws, but none of the natural laws with which we are acquainted is exact, ergo it is possible, even likely, that causality does not obtain. (A debt to Exner is *not* acknowledged.) Second, even if it should be the case that the motions of individual molecules follow exact laws, we may postulate that the fluctuations in the zero-point energy of the æther disrupt these motions. As there are no experimental means for isolating a portion of the æther, the ideal of identically prepared, isolated systems is in principle unrealizable. "The law of causality demands that in the case of identical initial conditions, two different systems will follow identical courses in their changes; now, however, we conclude that two systems of this sort do not admit of being realized at all."[196] Nernst is not, of course, prefiguring a quantum field theory in which the fluctuations of the æther are themselves in principle indeterminate, but rather he implicitly assumes that, as with any classical field, the time, place, and manner of such fluctuations would be completely determined if the state of the *entire* æther could be specified. This possibility Nernst can exclude only on the grounds that "then we come to an infinitely extended system, in the face of which our laws of thought fail."[197]

Thus it is clear that although Nernst wishes with all his heart and soul to renounce causality, he is simply unable to free himself from the implicit assumption that the world *really is* causal. Nernst himself had begun to perceive this by the spring of 1922 when his lecture was republished in *Die Naturwissenschaften*. He then added a postscript pointing out that "most religions maintain that all events

---

195. W. Nernst, *op. cit.* (note 186), pp. 492, 495. The following quotations are from pp. 494-495.
196. Cf. Werner Heisenberg, "Über den anschaulichen Inhalt der quantentheoretischen Kinematik und Mechanik," *Zeitschr. f. Phys., 43* (1927), 172-198, received 23 March 1927: "But in the sharp formulation of the law of causality: 'If we know the present exactly, we can calculate the future' it is not the conclusion but the presupposition which is false. We are unable in principle to get to know the present in all of its determinative elements" (p. 197).
197. In the original: "unsere Denkgesetze versagen."

occur according to the will of a most high intelligence, and thus with complete logic, which is identical to the requirement of the principle of causality." Therefore "it is obviously less a question of whether or not one regards the principle of causality as rigorously valid, but much more a question of whether one conceives the natural processes to be comprehensible or, on the contrary, holds that the human mind is incapable of following these processes down to their last details." This latter is, now, Nernst's position—"only statistical mean values of the course of events are accessible to our natural-scientific cogniton"—and so we see once again that the repudiation of causality is in fact a repudiation of both reason itself and the cognitive enterprise in which physicists had theretofore been engaged.[198]

Apart from their common theme of *ignorabimus,* the three cases just examined—von Mises, Schottky, Nernst—show a remarkable temporal coincidence, suggesting a wave of conversions to acausality. And if one recalls that there were at just this moment no specific developments in physics which could plausibly be regarded as the source of such acausal convictions, then one can scarcely escape the conclusion that what we are dealing with is, essentially, a capitulation to those intellectual currents in the German academic world which we charted in Part I. Moreover, I am inclined to regard this capitulation as a very widespread phenomenon precisely because of the lack of negative evidence. The only other general academic lecture by a theoretical physicist at this moment with which I am acquainted contains, to be sure, no explicit renunciation of the principle of causality, but the clearest indications that it is a controversial issue: upon assuming the rectorate of the University of Berne, Paul Gruner made the most strenuous efforts to hang the opprobri-

---

198. Nernst is really quite old fashioned in his physical conceptions, and tries only to draw modish conclusions from them. His postulate that the motions and interactions of a *sub*-atomic mechanical system (the æther) perturb those of atomic-molecular mechanical systems, so that the laws of motion of a single gas molecule would express only mean values, had been entertained by Ludwig Boltzmann a quarter century earlier—without, of course, any failure of causality having been seen therein. (Boltzmann, *Vorlesungen über Gastheorie* [Leipzig, 1896-1898], trans. by Stephen G. Brush as *Lectures on Gas Theory* [Berkeley, 1964], p. 449.) In proposing that the fluctuations of the zero-point energy of the æther are responsible for triggering the decay of radioactive atoms Nernst is, in truth, adopting a causal explanation and mechanism for this prime example of an apparently acausal natural process.

ous epithet "causal" upon the mechanistic-materialistic world view and to sink the two together.[199]

c. Later Notable Conversions: Schrödinger and Reichenbach

In the fall of 1921, Erwin Schrödinger came to Zurich as professor of theoretical physics at the University, and so also came into contact with Hermann Weyl. Schrödinger had earlier been in close personal contact with Franz Exner as student, assistant, and Privatdozent in Vienna before the war. And when, one year later, he delivered his public inaugural lecture, he too delivered himself of a manifesto against causality bearing much resemblance to those issued on like occasions a year earlier. Schrödinger's manifesto, however, is distinguished not merely by its tight exposition and fine literary form, but also by its stress upon Exner's priority and importance.[200]

The principle of causality is the postulate "that every natural process or event is absolutely and quantitatively determined at least through the totality of circumstances or physical conditions that accompany its appearance." But "in the past four or five decades physical research has demonstrated perfectly clearly that for at least the overwhelming majority of phenomena, the regularity and invariability of whose courses has led to the postulation of general causality, the common root of the observed rigorous lawfulness is—*chance*." Now, insofar as the physical laws are statistical, they do not *require* that the individual molecular events be rigorously causally determined. (It was "Exner who in 1919, for the first time, with complete philosophical clarity" pointed out the groundlessness of the common assumption that molecular processes are in fact causal.) Moreover, Schrödinger finds most unsatisfying the duality in the laws of nature implied by the assumption of rigorous causality in the microcosm. "In the world of visible phenomena"—governed as it is by statistics, and thus by the concept of pure number—"we have

---

199. P. Gruner, *Die Neuorientierung der Physik. Rektoratsrede* (*op. cit.*, note 125), pp. 5, 11.
200. E. Schrödinger, "Was ist ein Naturgesetz?" *Naturwiss.*, 17 (4 Jan. 1929), 9-11; trans. as "What is a Law of Nature?" in Schrödinger, *Science, Theory, and Man* (New York, 1957), pp. 133-147. This was Schrödinger's inaugural lecture as professor of theoretical physics at the University of Zurich, 9 December 1922, which remained unprinted at the time.

clear intelligibility, but behind this a dark, eternally unintelligible imperative, an enigmatic 'must.'" (Compare Spengler: "Out of the principle of causality speaks fear of the world. Into it the intellect banishes the demonic in the form of a continually valid necessity, which rigid and soul-destroying is spread over the physical world picture."[201]) "This duplication of the laws of nature," Schrödinger continues, "reminds one too much of the animistic duplication of natural *objects* for me to believe in its tenability." And he concludes his lecture by asserting that the solution to our difficulties in atomic physics will depend upon "liberation from the rooted prejudice of absolute causality."

But here again the most striking features of the manifesto are, on the one hand, the quasi-moral terms in which causality—"ein dunkles, ewig unverstandenes Machtgebot"—is repudiated and, on the other hand, the frivolousness with which the objections to dispensing with causality are dismissed. And so once again there seems good reason to regard the conversion as a form of accommodation to the intellectual environment—especially good reason inasmuch as Schrödinger himself was prepared to admit the Spenglerian thesis that physical theory is an expression of, and thus conforms itself to, the *Zeitgeist*.[202]

I am acquainted with one further clear and dramatic example of a quasi-religious repudiation of causality in the years before quantum mechanics: Hans Reichenbach's conversion in the fall of 1925. In 1924, when he wrote his *Axiomatization of the Relativistic Theory of Space and Time,* Reichenbach still adhered firmly to the ideal of causality.[203] And even as late as August, or possibly September, 1925, Reichenbach could open a popular article on "Probability Laws and Causal Laws" by asserting that the law of causality, "this supreme law," is the precondition for the application of mathematics to physics and thus for physics to be an exact science.[204] But the

201. See note 77 and note 158.
202. Schrödinger, *op. cit.* (notes 133, 228, and 235).
203. H. Reichenbach, *Axiomatik der relativistischen Raum-Zeit-Lehre* (Braunschweig, 1924), trans. by Maria Reichenbach as *Axiomatization of the Theory of Relativity* (Berkeley-Los Angeles, 1969), p. 15.
204. H. Reichenbach, "Wahrscheinlichkeitsgesetze und Kausalgesetze," *Die Umschau, 29* (3 October 1925), 789-792.

further one reads in this article the clearer it becomes that Reichenbach's allegiance to causality is beginning to waver. "Will we one day see the old ideal of physics realized, and comprehend the atomic world perfectly rigorously? Many researchers, including the most significant, believe this. . . . Others, on the contrary, and also among them significant researchers, are of the opinion that here perhaps there is an intrinsic limit to all explanation whatsoever." And Reichenbach concludes: "One is not permitted to say that under any circumstances it must be possible to find a causal explanation at the atomic level. Rather, the decision on this question must be reserved to physics itself, and cannot be made by philosophy."

Thus far our logical empiricist Reichenbach, in August or September 1925. Consider now the paper on "The Causal Structure of the World" which the notorious existentialist Reichenbach wrote in the following month or two.[205] The opening section—which carries a most curious subtitle: "Determinism and the Problem of the 'Now'"—begins: "It has become usual to regard the hypothesis of causality in physics as so self-evidently necessary that one no longer even thinks of subjecting it to criticism. And for the most part one does not notice at all to what a high degree this hypothesis is an extrapolation above and beyond the facts of experience. The assertion that without the hypothesis of causality no exact knowledge of nature is possible exhausts the customary defense of this standpoint." Here one searches in vain for the anticipated citation of Reichenbach's own earlier publications. "In what follows it will be shown that even without the hypothesis of rigorous causality it is possible to give a quantitative description of the course of nature which does everything that physics can possibly do. . . ." From a brief analysis of the concept of causality there then emerges very quickly, and essentially without argument, the "conclusion" that causality in the sense of determinism is an unjustified and needless extrapolation: "For physics the hypothesis of determinism is completely empty." It is therefore to be discarded, and in its place is set the concept of probability, taken as fundamental and irreducible.[206]

205. H. Reichenbach, "Die Kausalstruktur der Welt und der Unterschied von Vergangenheit und Zukunft," *Bayerische Akad. d. Wiss., München, math.-naturwiss. Abteilung, Sitzungsber.* (1925), pp. 133-175. Presented by C. Carathéodory in the session of 7 November 1925.
206. *Ibid.*, pp. 133, 136.

What is the occasion, the motive, the driving force behind this revolution? Is it perhaps that the decision which Reichenbach the logical empiricist had reserved for physics has suddenly fallen? Of any such developments we hear not a word. Rather we are assured that "It is the demand for a minimum of assumptions which compels us to renounce rigorous causality." Which is to say that existentialist philosophy, disguised as logical empiricism, has preempted the decision. But at this point Reichenbach the existentialist strips off his disguise: various investigations, notably those of Reichenbach the logical empiricist, have shown that the idea of a causal chain is closely connected with the topology of time, that is, with the fundamental concepts "earlier," "later," and "simultaneously."

> But [our existentialist Reichenbach stresses] what these investigations could not resolve is the problem of the "now" ... the "now-point" as experience [Erlebnis] of the boundary between the past and the future. ... An "earlier" and "later" exist also for determinism, but there is no "now"; there is no distinguishable point in time. And the feeling that my own existence is a reality, whereas Plato's life only throws its shadow into reality, must be an error. That, however, contradicts the entire orientation of our existence; we have a completely different attitude towards the future than towards the past. And unless one wants to regard every single one of our actions, every thought which accompanies us in the ordering of our daily life, as a single huge error, then determinism must be false. ... If one renounces it, the contradiction with our elemental life-feeling can be avoided. Of course such a feeling must not be decisive if reason speaks cogently against it—let one therefore first analyze reason to see if the maintenance of determinism is necessary. And that it is not.[207]

In the suddenness of the conversion to acausality, in its explicit independence of recent developments in atomic physics, and in its perfectly manifest connection with a capitulation to existentialist *Lebensphilosophie,* Reichenbach's case is certainly extreme. Yet every one of the cases I have examined—and most especially those of Weyl, von Mises, and Schrödinger—share these characteristics to some extent. Excepting Exner, all of them have the qualities of a quasi-religious experience, of a rebirth, of contrition for past sins—in a word, of a conversion. When our converts attempted to demon-

---

207. *Ibid.,* pp. 138-141.

strate the necessity for this renunciation of causality, their arguments, as often as not, ought logically to have led to the opposite conclusion. From this I think one must infer that they fully anticipated that *any* argument advanced by a physicist as a demonstration of the failure of causality would be received by their audience with uncritical applause. And when one recalls that the audiences for most of these renunciations of causality were, in the first instance, the whole body of a university assembled on a ceremonial occasion, then I think it reasonable to construe such renunciations as attempts to alter, or at least receive a special dispensation from, an unbearably opprobrious public image of the theoretical physicist as a "hard-boiled determinist."

III.4. Unregenerates against the Tide, 1922–1923

The wave of conversions to acausality in the latter part of 1921 prompted a series of public demonstrations in support of causality by "the most significant" theoretical physicists. Planck and Einstein—Nernst's and von Mises' colleagues at the University of Berlin—were quite disturbed; they felt that their colleagues were (unwittingly) betraying their calling, and carrying fuel to the antiscientific fires then raging in Germany. In 1922 and 1923 they both came forward to rebuke such rashness and to defend the principle of causality in physics and beyond.

Among the first, however, to raise his voice was Mach's old bulldog, Joseph Petzoldt, who in a long letter to the editor of *Die Naturwissenschaften* "Concerning the Crisis of the Concept of Causality" lectured Schottkey and Nernst like schoolboys.[208] The questions which they have dragged up were thoroughly considered and disposed of more than two decades ago. To Schottky he pointed out that temporal action at a distance is quite as compatible with the Machian concept of causality as is spacial action at a distance. To Nernst he declared firmly that while it is *conceivable* the regularity of nature could fail, "there is no limit to the 'understanding' [des 'Begreifens']." Only Schottky replied to Petzoldt, and his rebuttal was weak, vague, and disingenuous—"it goes without saying that the

208. J. Petzoldt, "Zur Krisis des Kausalitätsbegriffs," *Naturwiss., 10* (11 August 1922), 693-695, dated 2 July 1922.

physicists too are not glad . . . to renounce the assumption that *all* events are tied together by laws."[209] One thus sees how little prepared the converts were to meet criticism, how disconcerting they found it, and how readily they could be silenced by it.

On 29 June 1922, some weeks before the publication of Petzoldt's letter, Max Planck, as secretary of the Prussian Academy, took the occasion of the annual public session in honor of their spiritual founder Leibniz to affirm the transcendental character of the law of causality and to reprimand academician Nernst—naturally, without naming him—for his irresponsible talk.[210] When the quantum hypothesis shall have been developed sufficiently so that one can properly speak of a quantum theory, that will be the proper moment to consider its consequences for our scientific-causal thought. "Meanwhile groping speculation offers itself the most various possibilities, whose rich profusion admonishes critical caution all the more as precisely at the present time not inconsiderable dangers to the sure advance of scientific work have arisen from various sides." Chief among these dangers is penetration by a "lively, but basically unfruitful dilettantism," confusing and fusing science and religion, seeking "directly and relatively effortlessly to pluck the golden fruits of knowledge and bliss from the rich tree of life, in contrast to the so-called school or guild science, which only in hard, protracted, specialized studies is able to gather one tiny little grain after another into its barn. Today it cannot yet be foreseen when and where these colorfully iridescing foam bubbles will finally burst. . . . Vis-à-vis such intellectual currents the academies find themselves in a substantially better protected situation than their sister institutions the universities, which have to stand far more directly against the shifting surge of the waves of public life."[211] Evidently then, Planck, too, saw, or at least sensed, an intimate connection between an anticausal manifesto by a rector of the University of Berlin and that constellation of attitudes which made the Weimar intellectual milieu seem to the theoretical physicist so hostile to his enterprise.

Early in the following year on 17 February 1923, Planck devoted

---

209. W. Schottky, "Zur Krisis des Kausalitätsbegriffes," *Naturwiss.*, *10* (1922), 982, dated 6 October 1922.
210. M. Planck, "Ansprache des vorsitzenden Sekretärs" (*op. cit.*, note 17).
211. *Ibid.*, pp. 46-48.

an entire public lecture, again in the Prussian Academy, to a most uncompromising and courageous reaffirmation of allegiance to the principle of causality—not merely in the natural sciences, but in the *Geisteswissenschaften* too.[212] Planck knew full well that in this "violent dispute" over causality, "splitting the intellectuals into two camps," one of reason and one of feeling, the bulk of his audience lay within the latter camp, that much of what he said would "provoke" them and might even appear "a blasphemy, as cheap as it is intolerable."[213] Nonetheless he proceeded to tell his audience that "the assumption of a causality without exception, of a complete determinism, forms the presupposition and the precondition for scientific [*wissenschaftlich*] cognition." And anticipating precisely the issues which the uncertainty principle and complementarity were to raise, Planck knew well in advance what position he would adopt: "But has it then—one could now certainly ask—any sense whatsoever to continue speaking of a definite causal interconnection when no one in the world is capable of actually comprehending that causal interconnection as such? . . . Absolutely. . . . For causality is . . . transcendental, it is entirely independent of the constitution of the inquiring intellect, indeed it would retain its significance even in the complete absence of a knowing subject."[214]

Again in the summer of 1923 Planck took the opportunity offered by his contribution to the issue of *Die Naturwissenschaften* commemorating the tenth anniversary of the Bohr atom in order to warn his colleagues against those "eminent physicists"—unnamed, of course, but evidently Exner, Nernst, Schrödinger, and, yes, Bohr himself—"who want to allow the principles of the classical theory basically only a statistical significance. . . . Such a conception seems to me, however, to shoot far over and beyond the target, if only because with the abandonment of classical dynamics they simultaneously pull out the foundations of every rational statistics."[215]

212. M. Planck, *Kausalgesetz und Willensfreiheit* (*op. cit.*, note 18).
213. *Ibid.*, pp. 140, 160. See note 150.
214. *Ibid.*, p. 161. In his address of 3 August 1914 (*op. cit.*, note 158), pp. 78, 88-89, Planck had asserted these same propositions equally categorically but without any suggestion that his views were unwelcome to his audience.
215. M. Planck, "Die Bohrsche Atomtheorie," *Naturwiss.*, *11* (6 July 1923, Bohr Heft), 535-537; reprinted in Planck's *Physikalische Abhandlungen und Vorträge* (Braunschweig, 1958), *2*, 543-545. Note that, like Exner, Planck, too, confuses the validity of classical dynamics in the atomic domain with the validity

On this issue of causality Planck and Einstein were in complete agreement, and their stand together against the rising tide of acausal sentiment contributed to the preservation of a close personal bond between these two men despite the wide divergence in their political and social views. Writing to Einstein on 22 October 1921, a week after Nernst's *Rektoratsrede*, Planck, as president of the Gesellschaft deutscher Naturforscher und Ärzte for 1922, appealed to Einstein's "fine feeling for causal interconnections," and so succeeded in overcoming Einstein's resolve to boycott an organization which, he felt, had treated him meanly the previous year.[216]

Apart from insisting that the adjective "causal" occupy a conspicuous place in his definitions of the goal and function of scientific activity, Einstein was not given to dogmatizing publicly and popularly on this issue.[217] His own efforts were devoted to searching in the

---

of causality. Bohr had recently associated himself with "the view, which has been advocated from various sides, that, in contrast to the description of natural phenomena in classical physics in which it is always a question only of statistical results of a great number of individual processes, a description of atomic processes in terms of space and time cannot be carried through in a manner free from contradiction by the use of conceptions borrowed from classical electrodynamics...." ("Über die Anwendung der Quantentheorie auf den Atombau. I. Die Grundpostulate der Quantentheorie," *Zeitschr. f. Phys., 13* [ca. 1 Feb. 1923], 117-165, on 157; English translation in *Cambridge Philosophical Society, Proceedings* [1924], supplement, 42 pp., on 35.) It was, however, only in 1924 that Bohr spoke of "a causal description in space and time."(*Op. cit.*, note 226, p. 790.) By 1927 Bohr had ceased to regard "causal" descriptions and "space-time" descriptions as equivalent, and saw them, rather, as "complementary." (*Op. cit.*, note 241.)

216. "Es ist doch sonst bei Ihrem feinen Gefühl für Kausalzusammenhänge nicht Ihre Art, bei sachlichen Überlegungen allgemeinen Gefühlsstimmungen den entscheidenden Einfluss zu gewähren." Planck to Einstein, 22 October 1921, Einstein Collection, Institute for Advanced Study, Princeton. The experience at Nauheim, September 1920, had left a very bad taste in Einstein's mouth; nonetheless he agreed to deliver a major lecture at the following, hundredth anniversary congress in Leipzig. That summer, however, in the aftermath of Rathenau's assassination, Einstein felt compelled to withdraw from public life, and from Germany, for a time.

217. See note 147 and text thereto; cf., Einstein, "Das Kompton'sche Experiment. Ist die Wissenschaft um ihrer selbst willen da?" *Berliner Tageblatt*, 20 April 1924, Nr. 189, I. Beiblatt (Readex Microprint edition of the publications of A. Einstein, Nr. 147), where Einstein maintains that the great educational task of science "darin besteht, das Streben nach kausalem Erkennen in der Gesamtheit zu wecken und wach zu erhalten." Addressing a popular audience in June 1922 on "New Results Regarding the Nature of Light," *op. cit.* (note 135), "Einstein in conclusion gave expression to his opinion that considering the great advances in our knowledge of nature one can count upon a future solution of this problem also, and that the human consciousness possesses the necessary capabilities [Voraussetzungen] for the comprehension of the natural processes."

field-theoretic apparatus of general relativity for a super-causal solution to the quantum problem by means of over-determined systems of differential equations.[218] Any program—e.g., Tetrode's—to solve the problem by tightening rather than loosening the causal interconnections he greeted most enthusiastically, while efforts in the opposite direction—e.g., the Bohr-Kramers-Slater theory—he received most cooly and critically.[219] Einstein was convinced, and rightly so, that his fellow physicists were rushing to embrace a failure of causality without having made any serious attempt to explore the possibilities for a causal solution. In order to advertise this point,

218. How long had Einstein been pursuing this program? Russell McCormmach, "Einstein, Lorentz, and the Electron Theory," *Historical Studies in the Physical Sciences*, 2 (1970), 41-88, especially 83-84, raising the general problem of Einstein's conversion to a field approach, locates that reorientation in the years 1907-1909, and sees Einstein as aiming thenceforth at "a field theory with quantum solutions, not a quantum 'mechanics.'" Einstein's own statements in the 1920's of his quantum-theoretical program are quite consistent with this early date. Thus in January 1920 he wrote Born: "I believe now as before [nach wie vor] that one must seek an overdetermination by differential equations in such a way that the *solutions* no longer have the character of a continuum. But how??" (*Briefwechsel* [*op. cit.*, note 14], p. 43.) And again on 28 June 1929, receiving from Planck's hands the second Planck Medal of the German Physical Society—the first had gone to Planck himself—Einstein implied that it had *always* been and always would be his program: "There were two ideas, especially, around which my ardent exertions grouped themselves. The evolution of the world [das Naturgeschehen] seems to be so largely determined that not only the temporal course, but also even the initial state is largely bound by law. To this idea I believed I had to give expression by finding overdetermined systems of differential equations. The postulate of general relativity as well as the hypothesis of the unified structure of physical space, or the field, were supposed to serve as guideposts in this search. There the goal stands, unattained. And there was scarcely a fellow physicist to be found who shared my hope of arriving by this route at a deeper understanding of reality. What I found on the subject of quanta are only chance insights [Gelegenheitseinsichten] or, to a certain extent, fragments, which broke off in the course of my fruitless exertions upon the great problem. I am ashamed now to receive for this so high an honor.

"Despite the fact that I believe strongly that we will not remain stuck at a subcausality, but rather, ultimately, we will even arrive at a super-causality in the sense indicated, nonetheless I most highly admire the contributions of the younger generation of physicists which are comprised under the name "quantum mechanics," and I believe in the deep truth-content of this theory; only I believe that the restriction to statistical laws will be only temporary." (Einstein, "Ansprache . . . an Prof. Planck," *Forschungen und Fortschritte*, 5 [1929], 248.)

219. Writing to Paul Ehrenfest late in August 1922 (SHQP Microfilm Nr. 1; the letter is undated), Einstein recommended "eine sehr geistvolle Arbeit von Tetrode über das Quantenproblem. Vielleicht hat er Recht; jedenfalls zeigt er sich durch diese Arbeit als Kopf ersten Ranges. Schon lange hat mich nichts mehr so elementar gepackt." For his reactions to the Bohr-Kramers-Slater theory: Martin J. Klein, "The First Phase of the Bohr-Einstein Dialogue," *Historical Studies in the Physical Sciences*, 2 (1971), 1-39, on 32-33.

Einstein published in December 1923 a sketch of his own program, despite the fact that he had made essentially no progress with it.[220]

Altogether, were one unacquainted with the overwhelming anticausal sentiment in the Weimar intellectual environment and the social pressures to which a physicist stepping before a general academic audience was exposed, one would have to be surprised at just how few physicists came forward to defend causality, and take issue with their colleagues who were, in fact, repudiating physics as a cognitive enterprise. It seems reasonable to suppose, however, that although few had the courage to brand themselves publicly as determinists, many a senior and influential colleague let it be known with what displeasure he viewed these capitulations to antiscientific currents.[221] And such intimidation may well have been responsible for the decline in the number of full-scale manifestoes against causality by physicists after the end of 1921.

III.5. The Situation circa 1924

Although public silence seems to have been imposed fairly effectively—only in 1929 did Schrödinger allow his manifesto to be printed, while Reichenbach's was interred in the proceedings of the Munich academy—the tide against causality was not stemmed. There are numerous indications that privately the question continued to be "much discussed,"[222] and it was the impression of a contemporary

---

220. A. Einstein, "Bietet die Feldtheorie Möglichkeiten für die Lösung des Quantenproblems?" *Preuss. Akad. d. Wiss., phys.-math. Kl., Sitzungsber.* (13 Dec. 1923), pp. 359-364, published 15 January 1924. It must be said, however, that Einstein gave a rather different impression of the origins of his paper and his intentions in publishing it when mentioning it to H. A. Lorentz, 25 December 1923 (for source see note 114): "Ich sehe eine Möglichkeit den Quantentatsachen von der Feldtheorie aus beizukommen unter Preisgabe der mechanischen Gleichungen. Das mechanische Verhalten der Elektronen (Singularitäten) soll durch überbestimmte Feldgleichungen mitbestimmt werden. Leider sind die mathematischen Schwierigkeiten für meine Kräfte zu gross. Ich habe deshalb durch eine kurze Abhandlung das Interesse der Fachgenossen auf die Methode zu lenken versucht."

221. For intimations of such a distaste for "polemics" see my "Doublet Riddle" (*op. cit.*, note 142), p. 171.

222. Wolfgang Pauli, "Quantentheorie," *Handbuch der Physik*, Band 23: *Quanten* (Berlin, 1926), p. 11: the moment of transition of a single excited atom "appears, according to the present state of our knowledge, to be determined solely by chance. It is a much discussed but still undecided question whether we have to regard this as a fundamental failure of the causal description

observer that considerable sympathy for, and more or less explicit avowals of belief in, acausality were to be met with "ever more frequently."[223] Where in 1922 Friedrich Poske was simply shocked by Nernst's renunciation of causality, a year later he "warmly recommended" the second edition of Exner's lectures.[224]

And in seeking the grounds for this relatively undramatic but quite definite drift away from causality circa 1924 one can finally point to specific recent developments in atomic physics. For, as I discussed in Section II.4, in 1923 and 1924 the atomic physicists were becoming convinced of the fundamental inadequacy of the extant quantum theory of the atom—which supposed classical mechanics to be valid for motions within the stationary states—and were beginning to doubt the reality of the visualizable atomic models to which that theory had been applied. I argued there that the Weimar intellectual milieu at the very least facilitated the precipitation of a generalized conviction of a crisis of the old quantum theory, and

---

of nature, or only as a temporary incompleteness of the theoretical formulation." (The article was written in 1924-1925.) Compare the remarks which H. A. Kramers added in this same connection to the German translation of his popular account of the Bohr theory, originally written in Danish jointly with Helge Holst: Kramers posed for the first time the question whether the probability laws have an underlying causal mechanism or "das physikalische Kausalitätsgesetz in Wirklichkeit nicht gilt." He then warned against stamping this latter conception as an epistemological impossibility, and added that "for the moment it is certainly rather a matter of taste which alternative one prefers, and perhaps will remain so forever. The actual choice affects the methods of physical research far less than one would at first perhaps like to believe." (*Das Atom und die Bohr'sche Theorie seines Baues* [Berlin, 1925], p. 139. The preface is dated March 1925.)

223. A Gatterer, *op. cit.* (note 146), p. 47; also p. 36. Although Nernst seems to have refrained from printing anything further on the question, he was not entirely silent. On 11 February 1925 he delivered one of those popular lectures to the lay members of the Kaiser Wilhelm Gesellschaft which were "supposed to serve to give them an insight into the scientific work of the institutes" under the title "Causalgesetz und neuere Naturforschung." It was noticed in the *Mitteilungen der Gesellschaft Deutscher Naturforscher und Ärzte, 2* (April 1925), 10.

224. F. Poske, *Zeitschr. f. den physikal. u. chem. Unterricht, 35* (July 1922), 188-189, emphasized that Nernst's parallel between loosening of the causal principle and "certain theological doctrines . . . makes clear how earth-shattering his conception, if it were accepted, would have to be for the entire *Weltanschauung*." In March of 1923 Poske, *ibid., 36,* 133-134, merely described the position taken in Exner's "especially noteworthy" final chapter, observing that "this conception is closely related to other recently expressed views according to which the role of the law of causality has been played out, and causeless chance governs."

I emphasized how very apt, from the point of view of an adaptation to the intellectual environment, the principal diagnosis—"the failure of mechanics"—indeed was. Yet however much this crisis and rallying cry themselves owed to precisely the same intellectual currents which were driving the reaction against causality, in the present connection the important fact is that at this moment the antimechanical and anticausal movements coalesced, reinforcing one another. The confluence and synergy of these movements appears all the more intelligible if one recalls the persistent tendency, evidenced by such diverse figures as Exner and Planck, to confuse and confound the validity of the laws of classical mechanics and the validity of the law of causality.

Now, finally, after all the posturing before popular audiences, we find the first attempts to *do* a little acausal physics. The earliest of these, appropriately enough, we owe to a quasi-crank, Hans Albrecht Senftleben. The program advanced in his paper "On the Foundation of the 'Quantum Theory' " of November 1923 included such prescient postulates as that "natural phenomena generally are to be regarded as statistical effects of totalities of elementary molecular processes which are themselves not subject to the requirement of causality," and that "Planck's constant h limits in principle the possibility of describing a process in space and time with arbitrary accuracy." Moreover, Senftleben was not entirely ignored.[225]

But when it comes to attention, few papers could compare with that published by Bohr, Kramers, and Slater in the spring of 1924. In January John Clarke Slater, fresh from the two Cambridges, had carried to Copenhagen a semi-deterministic space-time picture of light quanta traveling along the Poynting vector of a virtual radiation field which—and this was novel—Slater assumed to be continually emitted by atoms throughout their existence in stationary states. "When this view was presented to Professor Bohr and Dr. Kramers," Slater recalled not long afterward, "they pointed out that

---

225. H. A. Senftleben, "Zur Grundlegung der 'Quantentheorie,' " *Zeitschr. f. Phys.*, 22 (March 1924), 127-156, received 13 November 1923. Quotations from pp. 129-131.

Kis, *op. cit.* (note 146), discussed Senftleben quite seriously. In the summer of 1924 Kramers visited him in a sanatorium in Denmark. (Letters to Bohr and Kramers of 23 August and 8 October 1924 in the Archive for History of Quantum Physics.)

the advantages of this essential feature would be kept, although rejecting the corpuscular theory, by using the field to induce a probability of transition rather than by guiding corpuscular quanta. . . . Under their suggestion, I became persuaded that the simplicity of mechanism obtained by rejecting a corpuscular theory more than made up for the loss involved in discarding conservation of energy and rational causation [n.b.], and the paper . . . was written."[226] And it is, I think, only by reference to the widespread acausal sentiment that one can understand the immediate and widespread assent which the theory received in Germany, even though it was in fact hardly a theory at all but rather a vague suggestion of how, renouncing causality, one might try to give a "formal" account of the interaction between atoms and radiation.[227]

---

226. N. Bohr, H. A. Kramers, and J. C. Slater, "Über die Quantentheorie der Strahlung," *Zeitschr. f. Phys., 24* (ca. 20 May 1924), 69-87, received 22 February 1924, dated January 1924. The publication of the paper was probably delayed in order that it not appear earlier than the English version in the May issue of the *Philosophical Magazine, 47* (1924), 785-802. J. C. Slater, "The Nature of Radiation," *Nature, 116* (1925), 278, dated 25 July 1925; quoted by van der Waerden *op. cit.* (note 144), pp. 13-14, who also reprints "On the Quantum Theory of Radiation," pp. 159-176. The very real difference between Slater's original notion and the view to which Bohr and Kramers persuaded him suggests a distinction between *probabilistic* and *acausal* approaches. Thus the guiding field approaches to the problem of light quanta, which had long been commonplace, and de Broglie's suggestion of a wave as a guiding field for material particles, were probabilistic, but only by anachronistically imposing Heisenberg's uncertainty principle can one say that they abandoned causality. Their proponents did *not* suppose it impossible to get behind these probabilities to the determinants of the individual events. An *acausal* theory, on the contrary, is one which excludes this possibility in advance. Thus the Bohr-Kramers-Slater interpretation was formed from Slater's original proposal by precluding in principle "rational causation" in the interaction of atoms and radiation. This feature was then made more palatable by stressing the "formal character" of their description of the interaction, in contrast, one might add, with Slater's "physical" picture.

227. Interesting for its testimony to the extent and strength of belief which the "theory" received in Germany—as for much else—is W. Pauli to H. A. Kramers, 27 July 1925 (Archive for History of Quantum Physics, SHQP Microfilm Nr. 8, Section 9): emphasizing that he does not want to be mistaken for one of the "true believers," "Ich halte es überhaupt für ein ungeheures Glück, dass die Auffassung von Bohr, Kramers und Slater durch die schönen Experimente von Geiger u. Bothe sowie durch die kürzlich erschienenen von Compton so schnell widerlegt worden ist. Es ist zwar natürlich richtig, dass Bohr selbst, auch wenn diese Experimente nicht gemacht worden wären, nicht mehr an dieser Auffassung festgehalten hätte. Aber viele ausgezeichnete Physiker (wie z.B. Ladenburg, Mie, Born) hätten daran festgehalten und diese unglückselige Abhandlung von Bohr, Kramers und Slater wäre vielleicht für lange ein Hemmnis des Fortschrittes der theoretischen Physik geworden!"

Certainly that same essentially moral feeling underlying Schrödinger's repudiation of causality predominated in his response to Bohr, Kramers, Slater. Having demonstrated that the "Exner-Bohr" conception of statistical conservation of energy involves an unbounded random walk of the energy content of a closed system, Schrödinger did *not* conclude that the theory is impossible, but rather, "clutching at it with both hands," he saw in it a demonstration that "a certain stability in the course of the world sub specie aeternitatis can only subsist through an *interconnection* of every individual system with the entire rest of the world. . . . Is it idle play with ideas," Schrödinger asked rhetorically, "if one is, in this connection, struck by the similarity with social, ethical, cultural phenomena?" Clearly Schrödinger thought one ought to be, and that that recognition should be decisive.[228]

III.6. Causality's Last Stand, 1925–1926

We are now approaching the end of the development which I have been trying to trace, that is of the rise of a will to believe that causality does not obtain at the atomic level *before* the invention of an acausal quantum mechanics. With the introduction of Heisenberg's matrix mechanics in the fall of 1925 and of Schrödinger's wave mechanics in the spring of 1926, physicists realized relatively quickly that that belief no longer had to rest primarily upon ethical considerations or to involve a purely gratuitous renunciation of the possibility of exact knowledge of atomic processes. The grounds of argument and belief were thereby substantially altered. I will not attempt here to treat the growing realization of this new situation

---

228. E. Schrödinger, "Bohrs neue Strahlungshypothese und der Energiesatz," *Naturwiss.*, 12 (5 September 1924), 720-724. Schrödinger, peculiarly, seems to have seen in the Bohr-Kramers-Slater proposal an attempt to rid the quantum theory of discontinuities, a goal he then pursued in and through the wave mechanics which he began to develop late in 1925 on the basis of de Broglie's ideas. Writing to Wilhelm Wien on 18 June 1926, Schrödinger observed: "Es scheint ja, dass zur Zeit nicht auf allen Seiten die Ueberzeugung besteht, dass eine Abkehr von den grundsätzlichen Diskontinuitäten unbedingt zu begrüssen ist, *wenn* es damit geht. Ich aber habe immer mit Inbrunst gehofft, dass das möglich sein wird und würde mit beiden Händen zugegriffen haben—wie ich bei Bohr-Kramers-Slater mit beiden Händen zugriff—auch wenn der Zufall nicht gerade mir selbst den ersten (mit Rücksicht auf de Broglie muss ich richtiger sagen: den zweiten) Zipfel in die Hände gespielt hätte." (Archive for History of Quantum Physics.)

in any detail, but only emphasize once again how conscious the physicists were of the fact that they were playing before an audience hostile to causality=mechanism=rationalism, and how anxious many were to play up to that audience.

Not all, however, did so. During this period it was Wilhelm Wien who assumed again the role of champion of causality. In January 1925 he had taken his case to the general public through the pages of the Leipzig *Illustrierte Zeitung* where his denial that the quantum theory has, will, or could lead to an abandonment of the law of causality threaded its way among pictures of cabinet meetings and catastrophes, opera balls and carnival costumes. "The notion that nature is comprehensible . . . is identical with the conviction that all natural processes can be reduced to causality, to invariably valid natural laws." Of all purely philosophical notions the concept of causality has had the greatest impact on the development of humanity. It is responsible for the suppression of superstition, for modern natural science, and for the revolutions in technology and industry (n.b., the audience was nonacademic). Although the problem of the interaction of atoms and radiation "has brought all of theoretical physics into a crisis which will occupy it for a long time," the present form of the quantum theory can only be transitional, for "a statistics without a causal foundation will never be recognized by physics as something final."[229]

During the academic year 1925–1926 Wien fully exploited the platform available to him as rector of the University of Munich, speaking out in defense of causality in both his official addresses.[230]

229. W. Wien, "Kausalität und Statistik," *Illustrierte Zeitung* (Leipzig), Nr. 4169 (Feb. 1925), pp. 192, 194, 196. Max Planck was by no means silent: "Physikalische Gesetzlichkeit im Lichte neuerer Forschung," *Vorträge und Erinnerungen* (Stuttgart, 1949), pp. 183-205, especially pp. 184, 194-196; also reprinted in Planck, *Physikalische Abhandlungen* (Braunschweig, 1958), *3*, 159-171. This lecture was delivered on 14 February 1926 in Düsseldorf, and again on 17 February in the Auditorium Maximum of the University of Berlin. (*Forschungen und Fortschritte*, 2 [15 March 1926], 50.)

230. W. Wien, *Universalität und Einzelforschung. Rektorats-Antrittsrede gehalten am 28. November 1925*, Münchener Universitätsreden, Heft 5 (Munich, 1926), 19 pp.; *Vergangenheit, Gegenwart und Zukunft der Physik. Rede gehalten beim Stiftungsfest der Universität München am 19. Juni 1926*, Münchener Universitätsreden, Heft 7 (Munich, 1926), 18 pp. In his one other published academic address, *Goethe und die Physik. Vortrag gehalten in der Münchener Universität am 9. Mai 1923* (Leipzig, 1923), 39 pp., on p. 5, Wien had made a point of owning his allegiance to causality: "Accustomed to seek the law of causality everywhere, the physicists ever and again give themselves great pains to uncover the reasons which led Goethe to his unfavorable attitude toward physics."

Although his inaugural lecture of November 1925 contained no reference to the current situation in physics, Wien nonetheless took the opportunity, as we saw in Section I.1, to stress the historical importance of causality, equating it once again with the conviction that nature can be comprehended by the logical force of the human intellect, and then went on to criticize Langbehn, Chamberlain, and Spengler for their antirationalism and pessimism. The slightly equivocal tone of this lecture had, however, disappeared entirely in June 1926 when, towards the end of his term as rector, Wien spoke at the annual founder's day ceremonies on "The Past, Present, and Future of Physics," or, more accurately, on causality in the past, present, and future of physics. The theme first appears on page 4 of the printed text as the capacity of the human intellect to grasp the causality of natural processes, continues on pages 6–8 where it is emphasized that, even when the laws are statistical, causality must reign at the level of the elementary processes, and reaches a climax on pages 10 and 11 where Bohr is attacked directly and by name.

Here one must recall that, supporting himself in part upon Heisenberg's discovery of a way to do atomic physics while renouncing the goal of a detailed picture of intra-atomic motions and mechanisms, Bohr had recently been expressing far more openly and categorically his hope and belief that such pictures were impossible in principle, that physics was faced "with an essential failure of the pictures in space and time on which the description of natural phenomena has hitherto been based."[231] Quoting these words, Wien then sought to reprimand and silence Bohr and all others of like convictions with that same demand for self-censorship which Planck had advanced so successfully in 1922: "The physicists have always openly displayed before all the world the difficulties with which they have to contend. . . . But we must be very careful with pronouncements whose significance extends far beyond the limits of the field of physics." And Wien then went on to assert in the strongest terms that there is no physical field which is closed

---

231. W. Heisenberg, *op. cit.* (note 144). N. Bohr, "Atomic Theory and Mechanics," lecture at the Sixth Congress of Scandinavian Mathematicians, 31 August 1925, and revised before publication in *Nature, 116* (5 December 1925), 845-852; reprinted in Bohr, *Atomic Theory and the Description of Nature* (Cambridge, 1934), pp. 25-51; quotation from 34-35. The German text, "Atomtheorie und Mechanik," appeared in *Naturwiss., 14* (January 1926), 1-10.

to our understanding, and that physicists will not rest until they have subjected atomic processes to the law of causality.[232]

At this point, having dealt with Bohr and causality, Wien turned upon his colleague, the Professor of Theoretical Physics, Arnold Sommerfeld—without, of course, naming *him*. Although Wien had readily adapted his justifications for doing physics to the changing public values, he had nonetheless been concerned to shield the enterprise itself from the influence of the Wiemar cultural milieu. Sommerfeld's "Atomystik," on the contrary, dressed up for the public with pythagorean numerical harmonies and number mysteries, was not merely an attempt to use the quantum theory to play up to the ambient antirationalism, but represented an actual research program. "The number mysticism," Wien hoped and expected, "would be supplanted by the cool logic of physical thought; not perhaps to everyone's joy. For mysticism often exerts upon many minds a greater force of attraction than the cold and sober physical mode of thought. It is far from my intent to attack mysticism as such. There are many areas of the life of the soul from which mysticism cannot be excluded; but in physics it does not belong. A physics in which mysticism governs, or even collaborates, relinquishes the ground from which it draws its strength, and ceases to deserve its name." Wien then concluded his lecture by reaffirming once again his confidence that "insight into the causal interconnections of natural processes will continue to be possible," suggesting that those who express doubts on this score are just suffering from mental exhaustion, and perhaps also on that account are inclined to harken to pessimistic words about the *Untergang des Abendlandes* or the *Zusammenbruch der Naturwissenschaft*.[233]

The confidence and corresponding aggressiveness which Wien manifested on the issue of causality in the spring of 1926 derived

---

232. W. Wien, *Vergangenheit, Gegenwart und Zukunft der Physik*, (*op. cit.*, note 230), p. 10. Cf. note 221.

233. *Ibid.*, pp. 15, 18. We may perhaps read this as a veiled allusion to the breakdown which Bohr suffered in 1921 and which often threatened to recur. Wien's hostility toward Sommerfeld's "Atomystik" and his agressiveness due to confidence in Schrödinger's wave mechanics is corroborated by Werner Heisenberg's recent memoirs, *Der Teil und das Ganze: Gespräche im Umkreis der Atomphysik* (München, 1969), trans. [often quite inaccurately] as *Physics and Beyond: Encounters and Conversations* (New York, 1971), pp. 104-105, and 72-73, respectively.

chiefly from Erwin Schrödinger's papers on wave mechanics which Wien was then publishing in his journal, the *Annalen der Physik*. Having repudiated causality for social-ethical reasons in 1922-1924, by the fall of 1925 Schrödinger had converted back to causality for what were most probably personal-political reasons.[234] He now conceived and developed the wave mechanics as a causal space-time description of atomic processes in opposition to the Copenhagen-Göttingen matrix mechanics. To accept their contention that such a description is not possible "would be equivalent to a complete surrender." For, Schrödinger argued in February 1926 in his second paper, "we really cannot change the forms of thought, and what cannot be understood within them cannot be understood at all. There are such things—but I do not believe that the structure of the atom is one of them."[235]

Yet at just that moment in June 1926 when Wien, armed with Schrödinger's theory, was striking out so vigorously, the anticipated victory was being transformed into defeat by Max Born's statistical interpretation of the wave function, building an abandonment of causality right into the foundations of the wave mechanics.[236] "The

234. V. V. Raman and Paul Forman, "Why Was It Schrödinger Who Developed de Broglie's Ideas?" *Historical Studies in the Physical Sciences, 1* (1969), 291-314. A collection of seventeen letters from Schrödinger to W. Wien, December 1925-November 1927, which has recently come to light, strengthens the case advanced in that publication. Xerox copies of these letters have been deposited in the Archive for History of Quantum Physics.

235. E. Schrödinger, "Quantisierung als Eigenwertproblem (Zweite Mitteilung)," *Ann. d. Phys., 79* (April 1926), 489-527, reprinted in Schrödinger, *Die Wellenmechanik,* Dokumente der Naturwissenschaft, Abteilung Physik, Band 3, ed. Armin Hermann (Stuttgart, 1963), pp. 25-63; on 509 and 45, respectively. A partial, and occasionally quite erroneous, translation is included in Gunther Ludwig, *Wave Mechanics, Selected Readings in Physics,* ed. D. ter Haar (Oxford, 1968), pp. 106-126, on 120-121.

On 25 August 1926, Schrödinger wrote W. Wien that: "Ich möchte aber heute nicht mehr gern mit Born annehmen, dass solch ein einzelnes Ereignis [e.g., the interaction of an electron with an atom] 'absolut zufällig' d.h. vollkommen undeterminiert ist. Ich glaube heute nicht mehr, dass man dieser Auffassung (für die ich vor vier Jahren sehr lebhaft eingetreten bin) viel gewinnt. . . . Bohrs Standpunkt, eine räumlich-zeitliche Beschreibung sei unmöglich, lehne ich a limine ab. Die Physik besteht nicht nur aus Atomforschung, die Wissenschaft nicht nur aus Physik und das Leben nicht nur aus Wissenschaft. Der Zweck der Atomforschung ist, unsere diesbezüglichen *Erfahrungen* unserem übrigen Denken einzufügen. Dieses ganze übrige Denken bewegt sich, soweit es die Aussenwelt betrifft, in Raum und Zeit." (Archive for History of Quantum Physics.) One thus meets once again (cf. note 228) Schrödinger's insistent demand that scientific views conform with world views.

236. Max Born, *Zur statistischen Deutung der Quantentheorie,* Dok. der Naturwiss., Abt. Physik, Bd. 1, ed. Armin Hermann (Stuttgart, 1962).

true state of affairs," Heisenberg declared in the spring of 1927, "can be characterized thus: Because all experiments are subject to the laws of quantum mechanics, . . . quantum mechanics establishes definitively the fact that the law of causality is not valid."[237] And once again, when one sees how rapidly this failure of causality was accepted by physicists not merely as a definitive feature of the theory, but equally of reality, one can scarcely escape the conclusion that such a result, far from being regretted, was greeted with relief and satisfaction. The atomic physicists had fulfilled the obligation which Nernst—and their social-intellectual milieu—had laid upon them.

That conclusion is surely also suggested by the physicists' general anxiousness to carry the good news to the educated public—Heisenberg published a popular article retailing his conclusions even before his "technical" paper was printed[238]—but also from the terms

---

237. W. Heisenberg, "Über den anschaulichen Inhalt . . .," (*op. cit.*, note 196), p. 197, received 23 March 1927. A full year earlier Senftleben, *Physikal. Berichte*, 7 (April 1926), 520, had pointed to Heisenberg's paper initiating the matrix mechanics (*op. cit.*, note 144) as an example of the recent tendency to accept "to a certain degree" the view he had advanced in 1923 (*op. cit.*, note 225). Presumably Senftleben would have regarded the principle of indeterminacy which Heisenberg now propounded as merely the consummation of that process of acceptance of his own views.

238. W. Heisenberg, "Über die Grundprinzipien der 'Quantenmechanik'," *Forschungen und Fortschritte, 3* (10 April 1927), 83: "so scheint durch die neuere Entwicklung der Atomphysik die Ungültigkeit oder jedenfalls die Gegenstandslosigkeit des Kausalgesetzes definitiv festgestellt." Considered biographically, this enthusiasm is not unexpected. Heisenberg stresses repeatedly in his memoirs (*op. cit.*, note 233) that when he entered upon the study of theoretical physics at the University of Munich in the Fall of 1920 he had been active in the German *Jugendbewegung* for some years, and he continued so for some years afterward. Although Heisenberg is studiously vague about the particular organization in the politically variegated youth movement to which he belonged—W. Z. Laquer asserts that Heisenberg was a *Weisser Ritter*, and the following observations are especially applicable to this generally rightist group—the intellectual orientation of the movement as a whole has been well characterized by Theodor Wilhelm: "the *Jugendbewegung* is firmly and deeply embedded in that glorification of undivided life, which was intoned by Nietzsche, systematized in the *Lebensphilosophie* of the beginning of the century, paraphrased by the movements for reform in art and pedagogy, and from which the Hitler movement, too, profited in its own way." In fact it was the most radical antagonists of the exact sciences among the vulgar *Lebensphilosophen*—Ludwig Klages, Hermann Keyserling, Rudolf Steiner—who had the greatest following and exerted the strongest influence within the *Jugendbewegung*. Laquer quotes the leader in charge of the youth movement's career counseling office—never mind that he was a communist—contending in November 1918 that some professions were "without value for our future community and its plans to conquer the world"; heading that list was, naturally, physics, followed by chemistry, medicine, and engineering.

in which they presented these glad tidings. In a public lecture at the University of Hamburg early in 1927 Arnold Sommerfeld raised "the question which is discussed so much these days, whether the rigid pattern [starre Form] of causality which we have inherited from the 18th century"—read enlightenment, utilitarianism, materialism, etc.—"and from the rationalistic science of mechanics, is appropriate to our contemporary body of experience."[239] And when the question is posed in this form there is no doubt either about the answer which his audience wished to hear. Or again, consider

---

Although this orientation is never permitted to appear explicitly in Heisenberg's memoirs, it may be read between the lines. Thus Heisenberg represents himself (pp. 19, 27) as being forced to defend his decision to make a career of theoretical physics—which, interestingly, he claims to have done on the grounds that theoretical physics has "thrown up problems that challenge the whole philosophical basis of science, the structure of space and time, and even the validity of causal laws." Refusing to choose between theoretical physics and the *Jugendbewegung*, during his first two years at the University Heisenberg divided himself between "two quite different worlds. . . . Both worlds were so filled with intense activity that I was often in a state of great agitation, the more so as I found it difficult to shuttle between the two." The nature and intensity of that agitation becomes clearer if one recalls, on the one hand, that the youth movement organizations of the Weimar period, the *Bunde*, unlike present or Anglo-Saxon organizations, demanded a total commitment—as Theodor Wilhelm says, "Man verschrieb sich seinem Bund ganz"—and also notes, on the other hand, that Heisenberg's monitor in his second world, Wolfgang Pauli, was the very epitome of all that the youth movement detested: unathletic, hedonistic, indifferent to nature, addicted to urban night life, sarcastic, cynical, incisively critical, and Jewish to boot. (Walter Z. Laquer, *Young Germany: A History of the German Youth Movement* [New York, 1962], pp. 34, 102, 116, 141; Theodor Wilhelm and Wilhelm Ehmer in Werner Kindt, ed., *Grundschriften der deutschen Jugendbewegung* [Düsseldorf-Köln, 1963], pp. 12, 232.)

239. A. Sommerfeld, "Zum gegenwärtigen Stande der Atomphysik. Vortrag, gehalten auf Einladung der naturwissenschaftlichen Fakultät zu Hamburg," *Physikalische Zeitschr., 28* (1927), 231-239, received 18 February 1927, reprinted in Sommerfeld's *Gesammelte Schriften* (Braunschweig, 1968), *4*, 584-592, on 588. In the pre-quantum mechanical period, too, although never willing to renounce the full and unique determination of physical processes, neither could Sommerfeld resist the temptation to play up to the anticausal sentiments of a popular audience. Thus in addressing a general session at the Innsbruck Naturforscherversammlung, September 1924, he passed over the use of transition probabilities, the Bohr-Kramers-Slater paper, etc., without comment, but took the structure of the semi-empirical formulas for the relative intensities of spectral lines as the occasion for opening the prospect of a "teleologische Umbildung der Kausalität." Sommerfeld, "Grundlagen der Quantentheorie und des Bohrschen Atommodelles," *Naturwiss., 12* (21 November 1924), 1047-1049; *Ges. Schr., 4*, 535-543. Cf. note 31 and text thereto.

the terms in which Max Born discussed the same question in the *Vossische Zeitung,* Berlin's highbrow liberal newspaper, in the spring of 1928. After defining causality as determinism, and adding that all previous laws of physics had that characteristic, Born observed that "such a conception of nature is deterministic and mechanistic. There is no place in it for freedom of any sort, whether of the will or of a higher power. And it is that which makes this view so highly valued by all 'good rationalists.' " But happily physics has now discovered new laws which give it an entirely different character.[240] That character, Bohr had stressed repeatedly in his lectures at Como and at the Solvay Congress the previous fall, is an "inherent 'irrationality' "; indeed "the inevitability of the feature of irrationality characterizing the quantum postulate" was accepted most willingly by Bohr, who showed no sympathy for

240. M. Born, *Vossische Zeitung,* 12 April 1928, as quoted at length by H. Bergmann, *Der Kampf um das Kausalgesetz (op. cit.,* note 146), pp. 34-37. Born is reviewing Emanuel Lasker's *Die Kultur in Gefahr* (Berlin, 1928), 64 pp. Lasker, himself much provoked by the professional physicists over the theory of relativity, adopted a very provocative tone: "The old axiom 'from nothing comes nothing' is refuted by the new discovery that the principle of causality is not valid. It's hard to say from whom the genial notion came. Inspired by the spirit of the age of c [the velocity of light] the prophets of the new doctrine had this bright idea which is destined to make world history. Long it grew in secret, carefully weighed and considered, until it has now celebrated in the *Handbuch der Philosophie* [i.e., Weyl, *op. cit.,* note 177] its entrance into the realm of science. . . . The new result runs: in physics and chemistry the principle of causality holds only probably. The old idea of the necessity, unambiguity, and regularity of the laws of nature is ridiculous. The pattern for a law of nature is the lottery. Until further notice. It depends upon what we decide. We believe in principle in the power of experiment. Our council decides the meaning of the experiment—by majority decision. . . . Unfortunately there are a few experimentalists who don't understand the meaning of their own experiments. They still struggle for the old, outmoded view. Rigid habits of thought! The interpretation of an experiment is reserved solely to those who understand experiments and at the same time have a high flying, world embracing imagination. The opinion of those who do not satisfy both these conditions doesn't count. The physicist who is content to measure remains an artisan. He becomes an artist only when he is also a philosopher. The philosopher in turn is negligible if he isn't stamped as an experimental physicist. The physicist-philosopher alone is permitted to interpret and evaluate experiments. . . . The true instrument of the physicist-philosopher is illumination. . . . We are prepared to debate with anyone who is both physicist and philosopher and accepts our methods. To debate with other people would be a waste of time, and we have quite enough work to do turning science into new pathways. Just at this moment we have our hands full replacing the principle of causality by another which we will postulate, and which we will then impose upon the philosophers" (pp. 20-22).

Schrödinger's attempt "to remove the irrational element expressed in the quantum postulate."[241]

It is true that Sommerfeld himself, even as he raised the question of "the rigid pattern of causality," stressed that it was not his intent to call into question "the lawlike definiteness of the physical processes," and elsewhere, as we saw in Section I.1, was at this time actually writing against the less academic forms of the contemporary romantic reaction. But it seems to me that this circumstance only strengthens the inference that an acausal quantum mechanics was particularly welcome to the German physicists because of the irresistible opportunity it offered of improving their public image. Now they too could polemicize against the rigid, rationalistic concept of causality and hope to recover lost prestige thereby.

III.7. Conclusion

In an interview with Einstein in 1932, James Gardner Murphy, an Irish literary man with wide acquaintance among the German theoretical physicists, remarked that "it is now the fashion in physical science to attribute something like free will even to the routine processes of inorganic nature." "That nonsense," Einstein replied, "is not merely nonsense. It is objectionable nonsense. . . . Quantum physics has presented us with very complex processes and to meet them we must further enlarge and refine our concept of causality." Murphy: "You'll have a hard job of it, because you'll be going out of fashion . . . scientists live in the world just like other people. Some of them go to political meetings and the theater and mostly all that I know, at least here in Germany, are readers of current literature. They cannot escape the influence of the *milieu* in which they live. And that *milieu* at the present time is charac-

---

241. N. Bohr, "The Quantum Postulate and the Recent Development of Atomic Theory," *Nature, 121* (14 April 1928), 580-590, reprinted in Bohr, *Atomic Theory and the Description of Nature* (Cambridge, 1934), 52-91, on 580, 586, 590, and 54, 75, 91, respectively; German translation in *Naturwiss., 16* (1928), 245-257. Cf. Philipp Frank, "Gibt es ein irrationales Moment in den Theorien der modernen Physik?" *Neue Züricher Zeitung* (17 December 1928), Nr. 2355, who is at pains to combat this notion which had already been seized upon gleefully by Adolf Koelsch, "Die Verpersönlichung des Elektrons," *ibid.* (20 October 1928), Nr. 1910.

terized largely by a struggle to get rid of the causal chain in which the world has entangled itself."[242]

Murphy's assertion of the inescapability of the influence of the milieu is the more worthy of our attention as it is but a paraphrase of a passage from a lecture by Schrödinger, "Is Natural Science Conditioned by the Milieu?," published earlier that year.[243] Murphy's own contribution is the specific identification of hostility toward causality as the dominant characteristic of the contemporary milieu, and the implication that the scientist's attitude toward this particular concept had virtually been determined thereby.

Schrödinger's and Murphy's analysis is, as the foregoing investigation has shown, remarkably accurate, at least for the German-speaking Central European physicists. Their craving for crises, their readiness to adapt their ideology to the values of their social-intellectual environment argue a substantial and largely indiscriminate participation in the attitudes of their academic milieu, a readiness to swim along in the intellectual currents of the day. This circumstance is the more surprising if one bears in mind that the values characteristic of these intellectual currents which set in so strongly after Germany's defeat were fundamentally antithetical to the scientific enterprise. Indeed the mathematical physicist, the personification of analytical rationality, was often singled out as the prime exemplar of a despicable way of grasping the world. Above all, with astonishing unanimity, it was the physicist's attempt to subject the world to the rigid, dead hand of the law of causality— to use the rhetoric Spengler made so popular—which was taken to epitomize all that was most detestable in the scientific enterprise. These two circumstances—hostile environment and accommodation to its values—were then found to be linked by much direct and indirect evidence suggesting that the accommodation was in response to the hostility. Stated in terms of Karl Hufbauer's distinctions: suddenly deprived by a change in public values of the approbation and prestige which they had enjoyed before and during World War I, the German physicists were impelled to alter their ideology

---

242. "Epilogue: A Socratic Dialogue. Planck-Einstein-Murphy," in Max Planck, *Where Is Science Going?* trans. James Murphy (New York, Norton, 1932), pp. 201-221, on 201-205.
243. Schrödinger, *op. cit.* (note 143).

and even the content of their science in order to recover a favorable public image. In particular, many resolved that one way or another, they must rid themselves of the albatross of causality.

In support of this general interpretation I illustrated and emphasized the fact that the program of dispensing with causality in physics was, on the one hand, advanced quite suddenly *after* 1918 and, on the other hand, that it achieved a very substantial following among German physicists *before* it was "justified" by the advent of a fundamentally acausal quantum mechanics. I contended, moreover, that the scientific context and content, the form and level of exposition, the social occasions and the chosen vehicles for publication of manifestoes against causality, all point inescapably to the conclusion that substantive problems in atomic physics played only a secondary role in the genesis of this acausal persuasion, that the most important factor was the social-intellectual pressure exerted upon the physicists as members of the German academic community.

And here, saving perhaps the case of Hermann Weyl, it was not a question of "philosophical" influences in any serious intellectual sense. By far the single most influential "thinker" was Spengler, and that only because the *Untergang des Abendlandes,* the concentrated expression of the existentialist *Lebensphilosophie* that was diffused through the intellectual atmosphere, was read with attention by most German mathematicians and physicists on account of the prominent role Spengler had given their sciences. Thus, excepting Franz Exner, the philosophical theses of the latter nineteenth century to which Jammer has drawn attention, while they may perfectly well have some ultimate responsibility for the ideational content of the *Lebensphilosophie* of the Weimar period, played, *per se, an sich,* a negligible role in the sudden rise of anticausal sentiment among German physicists after the First World War. Rather, it was only as and when this romantic reaction against exact science had achieved sufficient popularity inside and outside the university to seriously undermine the social standing of the physicists and mathematicians that they were impelled to come to terms with it.

There are, moreover, many indications that this accommodationist strategy met with considerable success. The "objectionable nonsense" about the free will of electrons which philosophers, aided and abetted by physicists, were talking in the late 1920's, constituted in fact a very favorable press. Although distasteful to Einstein, this image of modern physics was exactly suited to the taste of the educated public of the Weimar period. And I would emphasize that much of the nonsense announced with great fanfare by philosophers in the late 1920's owed nothing whatsoever to the quantum mechanics discovered in 1925–1926, but was based wholly and solely upon the manifestoes against causality issued by physicists before that date. Such, for example, were the articles which Ludwig von Bertalanffy published in 1927 gloating over the fact that "in physics itself views are coming to be accepted which in biology would be designated as vitalistic. . . . The causal world picture of the physicist is dissolving—into its place steps one which recognizes individuality, even for the molecular process. . . . Indeed that allusion of Nernst's to the freedom of will of the theologians can even be employed to support one of Spengler's most controversial ideas: that modern physics, renouncing rigorous causality and exact laws of nature, will give way to a new mysticism."[244]

244. L. v. Bertalanffy, "Über die Bedeutung der Umwälzung in der Physik für die Biologie," *Biologisches Zentralblatt, 47* (Nov. 1927), 653-662; on 653-656. Likewise, Bertalanffy, "Über die neue Lebensauffassung," *Annalen der Philosophie, 6* (Sept. 1927), 250-264. A somewhat more delicate picture, again based solely upon pre-1925 sources, was painted by Karl Joël, "Überwindung des 19. Jahrhunderts im Denken der Gegenwart," *Kant-Studien, 32* (1927), 475-518, especially 482-487.

Not every *Lebenspilosoph* reacted in this way; in fact it appears that many a transcendental idealist, *lebensphilosophisch*-existentialist academic philosopher resented the attempts of the physicists to escape from the stocks of causality and usurp their role of national *Seelsorger*. A very early example is Kurt Riezler who in 1923 noted that recent developments in physics "have induced a number [*Reihe*] of natural scientists to express the hope, or at least to hint, that utilizing these and perhaps other discoveries still to be made the concept and the range of validity of natural laws will be transformed in such a way as would allow the bridge from natural processes to historical processes to be espied, and the gulf which appears to separate necessity and freedom to be closed." Riezler aims, therefore, "to probe the question whether, how far, and by what route this bridge to the world of the spirit and of freedom, which some natural scientists believe they espy, can and may be sought." His conclusion is that the cobbler should stick to his last, that "the second presupposition of natural science, determination, is likewise invariable. The natural scientist—[Nernst cited here]—who wants to see 'the bands of the law of causality' loosened, saws off the branch upon which he sits," while the philosopher on the contrary is not restricted to so narrow a conception of the world. ("Über das Wunder gültiger Naturgesetze. Eine naturphilosophische Studie," *Dioskuren: Jahrbuch für Geisteswissenschaften, 2* [1923], 238-274, on 238, 257.) By 1925, however, Riezler was no longer quite so categorical: "Die Hypothese der Kausalität," *Die Akademie: eine Sammlung von Aufsätzen aus dem Arbeitskreis Erlangen 4* (1925), 116-146, esp. 143.

One must admit that Bertalanffy's equation of the renunciation of causality with mysticism is not wholly unjustified. For as we saw, the manifestoes by physicists against causality before 1925 were issued not in spite of, but much rather because of, the general belief that "an abandonment of determinism would signify a renunciation of the comprehensibility of nature." Far from engaging in any critical analysis of the concept of causality, directed toward the relaxation of determinism without renouncing *a priori* the comprehensibility of nature, these physicists actually reveled in that consequence, stressed the failure of analytical rationality, implicitly repudiated the cognitive enterprise in which physics had theretofore been engaged.[245]

For this reason the acausality movement could not but arouse opposition within the German physics community. Indeed one has here the most characteristic difference between those physicists who hastened to renounce causality and those who clung to it even after the discovery of quantum mechanics. For Exner, Schottky, Nernst, and Bohr the failure of causality was essentially a failure of the human intellect; Weyl, von Mises, and Reichenbach went even further, expressing an existentialist revulsion against intellectuality. On the other hand those few physicists—strikingly and significantly few—who came forward to publicly oppose dispensing with causality all based their cases upon the value of rationality and their faith in the capacity of the human intellect to comprehend the natural world: so Einstein, Petzoldt, Planck, Schrödinger (after his reconversion), and W. Wien (vis-à-vis inorganic nature). And for this reason also I have not been able to, nor indeed wished to, maintain a perfectly neutral stance in my exposé. Although a readiness to view atomic processes as involving a "failure of causality" proved to be, and remains, a most fruitful approach, before the introduction of a rational acausal quantum mechanics the movement to dispense with

245. The quotation is from S. Kis, *op. cit.* (note 146), p. 33. The first serious attempt to reanalyze the concept of causality in order to determine just how much of it is required for the comprehensibility of nature appears to be Eino Kaila, *Der Satz vom Ausgleich des Zufalls und das Kausalprinzip. Erkenntnislogische Studien* (Turku [Åbo], 1924 = *Annales universitatis fennicae Aboensis*, Series B, Tom. 2, No. 2), to which we may perhaps add Reichenbach's, *op. cit.* (note 205).

causality expressed less a research program than a proposal to sacrifice physics, indeed the scientific enterprise, to the *Zeitgeist*. My sympathies have consequently been with the conservatives in their defense of reason, rather than with the "progressives" in their denigration of it.

But if this social-intellectual phenomenon is to be comprehended, in part at least, by means of a dichotomy between progressives and conservatives, then correlations might be anticipated between a physicist's position on the causality issue and his general intellectual-political orientation. And in fact, paralleling Ringer's observation that early in the Weimar period the "modernist" academics tended to be "methodologically adventurous," one finds that, by and large, those physicists who were readiest and earliest to repudiate causality had either distinctly "progressive" political views by the standards of their social class and the German academic world, and/or had an unusually close interest in, or contact with, contemporary literature. Nernst, who in his youth had wished to become a poet and who retained his interest in literature throughout his life, was also one of the few German physicists who publicly associated themselves with the cause of parliamentary democracy. Von Mises, although politically conservative and nationalistic, was on his way to becoming the foremost authority on the young Rilke. Born and Weyl were both well disposed toward the German republic, at least at its birth —in itself a sufficiently unusual sentiment in the German academic world—and both had literary wives. On the other hand, with the notable exception of Einstein, those who defended causality tended to be highly principled political conservatives and/or interested in classical literature. Such were Planck, Schrödinger, and Max von Laue—who kept their knowledge of Greek well polished. Standing to their right was W. Wien. And finally to the causalist camp one may add the outright reactionaries: Ernst Gehrcke, Erwin Lohr, Philipp Lenard, and Johannes Stark.[246]

246. F. A. Lindemann and F. Simon, "Walther Nernst, 1864-1941," *Obituary Notices of Fellows of the Royal Society of London,* 4 (1942), 101-112. Nernst had subscribed to a manifesto in the summer of 1917 calling for democratic political reforms in Prussia, and he had participated in the 1926 meeting of republican-parliamentary university professors: Klaus Schwabe, *Wissenschaft und Kriegsmoral: Die deutschen Hochschullehrer und die politischen Grundfragen des*

This very circumstance—that the alignment within the German physics community over the issue of causality correlates closely with the intellectual and political temper of the individual physicist— reminds us, however, that the "sociological" model employed in this paper cannot be the whole truth. It provides a general framework, and seems to work especially well in certain extreme cases. But in order to account for its special applicability to some physicists and its special inapplicability to others one must invoke precisely those factors which are excluded from the model—individual personality and intellectual biography. The mechanism advanced for the entrainment of the German physicists and mathematicians by the *Zeitgeist* is thus clearly not sufficient. And it may be that

---

Ersten Weltkrieges (Göttingen, 1969), p. 264, note 229; Wilhelm Kahl, et al., *Die deutschen Universitäten und der heutige Staat: Referate erstattet auf der Weimarer Tagung deutscher Hochschullehrer am 23. und 24. April 1926* (Tübingen, 1926), pp. 38-39. Von Mises' nationalistic attitudes are reflected in his repeated promotion of political boycotts by German scientists of international congresses. For example: Th. von Kármán to von Mises, 11 December 1923, and P. Debye to von Kármán, 1 May 1926, in the Kármán Papers, California Institute of Technology Archives, but especially the correspondence between von Mises and L. E. J. Brouwer in 1928 in the von Mises Papers, Niels Bohr Library, American Institute of Physics, New York. There are many indications of Max Born's political attitudes in his correspondence with Einstein (*op. cit.*, note 14), where specimens of Hedwig Born's poetry are also published. Hermann Weyl's initial attitude toward the German republic may be glimpsed in a letter to Einstein of 16 November 1918 in the Einstein Collection, Institute for Advanced Study, Princeton, and in the Weyl Nachlass, Bibliothek der Eidgenössischen Technischen Hochschule, Zurich; his continuing attachment to democracy is suggested in the talk by which he introduced himself to the Göttingen mathematics students in 1930, quoted at length in his "Rückblick auf Zürich aus dem Jahre 1930," *Schweizerische Hochschulzeitung, 28* (1955), 180-189, reprinted in Weyl's *Ges. Abhl., 4,* 650-654. There he also described his decision to accept the call to Göttingen as resulting from discussions, carried on in his imagination, with Jacob Burkhardt and Hermann Hesse. Weyl's wife Helene, the former disciple of Husserl, translated Ortega y Gasset into German. Planck's political conservatism is well known, but not quite so deep as it is often represented: he and von Laue were both members of the Deutsche Volkspartei in the Weimar period. (Wilhelm Westphal, "Der Mensch Max von Laue," *Physikalische Blätter, 16* [1960], 549-551; Friedrich Herneck, *Bahnbrecher des Atomzeitalters* [Berlin, 1969], pp. 303-304). Schrödinger's conservative nationalism is implicit in his correspondence with Wilhelm Wien, cited in note 234, most strongly in a letter of 26 April 1927 describing his emotions at the sight of the German countryside upon returning from the United States. Schrödinger frequently used Greek and Latin titles for his research notebooks (Archive for History of Quantum Physics); von Laue frequently used quotations from these languages in his publications.

examination of other episodes of entrainment in the late nineteenth and early twentieth centuries will prove that it is also not necessary. But be that as it may, it seems difficult to deny that the shifts in scientific ideology and the anticipated shifts in scientific doctrine exposed in this paper were *in effect* adaptations to the Weimar intellectual environment. Moreover, whatever similarities one may find in the mental posture of non-German exact scientists in this same period, there is one feature which cannot, I think, be found outside the German cultural sphere: a repudiation of "causality."

Quantum-Relativistic Retrospection and the History of Classical Physics: Classical Rationalism and Nonclassical Science[1]

BY BORIS KUZNETSOV*

In the history of philosophy and science, periods of upheaval sometimes coincide with the appearance of new retrospective evaluations connected with radical changes in the content and style of science. When new and fundamental principles appear in science, science cannot evaluate or even name them without associating them with past principles, without contrasting them with the past. These associations, rapprochements, and oppositions take the form of historical excursions in which new concepts become the measure of the past and, from time to time, the old principles play the role of criteria for the new ones. What Niels Bohr said of our time, "in order to be justified scientific theory has to be sufficiently foolish," is, to some extent, true for each era; scientific progress removes the aura of paradox and foolishness from each concept, deriving it in a natural and logically irreproachable way from new and more general assumptions. In this respect Einstein considered scientific progress as a *running away from miracles*. Here the word "miracles" means more or less the same thing as what Bohr means by "foolishness." In this respect scientific progress is impossible

---

* Institute for the History of Science, Academy of Sciences, U.S.S.R., Staropanski, p. 1/5, Moscow.

1. This article was translated from the Russian by Dr. Petros Odabashian with the assistance of Dr. R. McCormmach.

without reevaluation, revision, and reinterpretation of not only the arsenal of science but its pantheon as well.

Nonclassical science made detailed revision an integral and definite part of its endeavors. Modern science is nonclassical, but not because it relativized and generalized the axioms of Newton, axioms that claimed to play the role of invariable canons of truth analogous to classical canons of beauty in the art of antiquity; rather, modern science is nonclassical because of its style. Its canons are dynamic; the ideas of scientific explanation change during the life of one generation. The dynamics of science and the changes in its fundamental principles have become obvious.

This dynamics extends into the past as well. History claims the right which people refused even to their gods: history can change the past. In light of modern science, we discover that classical science was also dynamic. This is the first result of nonclassical retrospection. It allows us to see that the fundamental principles of classical science were dynamic and that the Victorian era in physics was not too Victorian. But the mobility of the fundamental physical principles of the nineteenth century suggests the movement of the minute and even the hour hand of a watch. Nonclassical physics, by contrast, suggests a certain analogy with the second hand of a watch. It made *obvious* the dynamism of the fundamental principles. The person who has seen for the first time the second hand of the watch has guessed, even before seeing it, that the watch was working. But now the idea of a moving mechanism seems to be a direct impression, the result of unique and ungeneralized observations.

Such an impression is now retrospectively brought about by each major discovery and each major generalization of classical physics. We shall not in any systematic way consider them in this article. It would mean digressing from certain remarks *on* the history of physics, i.e., remarks of an epistemological genre, to the history of physics in its entirety, i.e., to the full historical genre. Let us limit ourselves to two notions of classical physics, notions that will bring us to the characteristics of its general rationalistic spirit and rationalistic sources.

One such notion is that of a field. It is strikingly different from the concepts that classical physics received from mechanics, which, by the way, were not created ready-made and complete as Minerva from the head of Jupiter. Its striking difference lies in the absence or, more precisely, in the meagerness of antecedents. The notion of a field is connected with the idea of contiguous action, with the rejection of action at a distance, with the idea of continuity—with a whole series of quite traditional ideas. But the principal thing that Faraday does, his transformation of the interaction of elements of substance into substance itself, is very nonclassical, nontraditional, and paradoxical in spirit.

The same holds for the notion of entropy. For classical physics, the traditional idea was that of homogeneity. This idea underwent evolution, and the word "traditional" should not be understood in this article in a direct and elementary sense. From homogeneous space (conservation of momentum), we move to homogeneous time (conservation of energy)—this movement shows the emancipation of physics from mechanics and is sufficiently radical. Later this evolution brought about the understanding of the homogeneity of space and time—the pseudo-Euclidean understanding (special theory of relativity), and the understanding of distorted space and time (general relativity). But the notions of entropy and the second law of thermodynamics implied the anisotropy of time, which lay outside the limits of this evolution.

When classical physics is discussed with such a perspective, it is no longer a *system* of answers, but the object of *history,* where the questions cannot be divided from the answers, where every answer means at the same time a question addressed to the future, where we consider not only the positive results of science, but also the meaning of each new discovery and generalization for the tempo and direction of future scientific progress. The criterion of truth and the criteria of velocity and acceleration of our understanding of truth are parts of science. In the final analysis, the history of physics is different from physics in that it has as its object not statements about the structure of the world, but changes of the statements and their derivatives in time, their gradient, their dynamism, and the differential criteria of their dynamism.

The "historicization" of classical science is connected with nonclassical retrospection. In modern, nonclassical science, the epistemological value of scientific discoveries and generalizations is measured not only by new and experimentally proved knowledge, but by new methods, new questions, new forecasts, by all the factors defining the velocity and acceleration of science. At the same time, the utilitarian value of modern science cannot be separated from its epistemological value. The more general and fundamental the new insights, methods, and ideals, the more radical is their effect on economic production—an effect measured by differential indicators and by the velocity and acceleration of technical progress.

By generalizing the movement of science, philosophy long ago accepted the idea of the mobility of its principal categories. This idea grew from the generalization of the *development* of science. On the contrary, natural philosophy in its common meaning—i.e., a generalization of positive and accepted *results* of science—constructed immobile matrices that claimed to possess an *a priori* nature, within which natural philosophy tried to constrict the developing content of science. In this respect, natural philosophy was a continuation of the *identifying* tendency of rationalism, which was to find constant, invariant, and equal interrelations in a continuous stream of existence. This tendency was sometimes separated from an opposing and less official tendency, the search for nonidentity, incompatibility, and paradox. Such a separation was to some extent legitimate as long as the capability of change in the fundamental principles of science remained unclear and analogous to the movement of the hour hand of the watch. But the generalization and the securement of the separation stood in the way of the generalization of the development of science and the idea of the mobility of the fundamental principles of science.

Nonclassical physics opened the era of a nearly continuous, conspicuous transformation of fundamental principles. Consequently, in the eyes of contemporaries, the past of science was also changed. Science saw in its past the less official tendency which we mentioned—i.e., the rejection of identity, the transformation of the very matrices into which new facts are fitted. This was an *inquisitive* tendency. Scientific thought was concentrated on what was not identical, recurrent, invariant; it was looking for generalized schemes, but it could not find answers to the questions that were created during this process; and it addressed these questions to the future, trying to penetrate the future, and, in this respect, it was very prognostic in style.

QUANTUM-RELATIVISTIC RETROSPECTION

The development of science could not take place within the framework of rationalistic thought without the synthesis of the principal "official" characteristic melody and the sensual "inquisitive" accompaniment that violated the established identity of that accompaniment. The understanding of existence—we shall talk about it further on—is made up of the synthesis of universals and of localized, individualized, direct impressions. When rationalistic thought sought to explain not only the arrangement and behavior of bodies, but also their existence, not only the structure of the world, but also its substance, it would move into the world where it would not always find identical definitions and invariant correlations. Here rationalistic thought would invariably separate itself from the positive knowledge of its time, and this is where, in the highest possible degree, it would approximate the future. When the future became the present, when the problem of existence became the central problem of positive science and was given physical contours, having become the problem of experiment, the history of science saw in the evolution of rationalism this constant digression beyond "official" limits, this counteridentifying accompaniment.

The philosophy of Spinoza was the highest reach of rationalistic thought beyond the traditional framework. Descartes approached, but did not overstep, the limits of the framework. His rationalistic understanding of the world only embraced the *behavior* of substance, not its existence. Voltaire put into the mouth of Descartes poems devoted to God: the great rationalist undertakes to create everything that was created by God, once he is given matter and the laws of its motion.

Both the existence of matter and the fundamental laws of its motion are beyond the limits of the physics of Descartes. Within these limits, existence is dissolved into behavior, while the principal laws of existence of matter are considered *a priori* for physics. Spinoza overstepped these limits; he translated the reason for the existence of physical nature from the spiritual world into nature itself. He considered nature as the very reason for its existence (*causa sui*) and thus created the rationalistic concept of *existence*. Not only in the seventeenth century, but during the entire period of classical science, this concept could not find identical physical equivalents. But such equivalents appeared in nonclassical science.

The comparison of nonclassical science with classical rationalism leads not only to new retrospection, to a new understanding of the past, but it also encourages a new understanding of modern ideas, of modern tendencies in science, and of its moving forces and perspectives. It is hard to say whether we submit the seventeenth century to trial by modern times or whether we submit modern times to trial by classical rationalism. Probably both phenomena take place. In any case, in trial procedures both the court and the crime are changed. Classical rationalism turns to us with its "inquisitive side," with its desire to understand existence rationally. In modern science, the first priority is given to its unsolved problems. The historical resonance does not combine the apotheosis of classical ideas with the apotheosis of modern science, but here the opposing, sometimes dissonant, sometimes tragic, chords prevail.

Without the "inquisitive" side, without the feeling of the incompletion of existing theories, without forecast, without a differential approach to the values of modern science, the latter could not arrive at a new retrospection and evaluation of the past. This becomes obvious as soon as we try to examine somewhat more closely the mechanism of this retrospection. Let us start from what could be called "relativistic retrospection," from Einstein's reevaluation of classical physics in his approach to the theory of relativity.

Both historical retrospection and forecast were requisite components of the intellectual drive that led Einstein to relativity theory. He spoke about two criteria for the selection of a physical theory: "external confirmation" and "inner perfection." The first criterion is the agreement of conclusions of the theory with empirical observations. The second is the naturalness of the theory, its logical foundation in more general principles, its strong rejection of *ad hoc* postulates contrived for the explanation of unique observations or series of observations.[2] "External confirmation" required that the theory of relativity, under definite conditions (small energies) when the speed of bodies is not comparable with the speed of light, reduce to classical mechanics, which under these conditions had important empirical confirmations. With this reduction, classical mechanics

2. A. Einstein, "Autobiographical Notes," *Albert Einstein: Philosopher Scientist*, ed. A. Schilpp (New York, 1951), p. 25.

acquires more "inner perfection," becoming a natural conclusion from a more general and precise theory. Along with this, it acquires an historical existence; it loses its absolute and *a priori* characteristics, and it is seen to be connected with a certain historical period in the development of experiment, of production, and of culture. This viewpoint allows us to see in classical theory some inherent contradictions, and it allows us to discuss classical theory not only as a totality of answers to questions put by nature, but also as a summary of questions which did not receive answers and which were readdressed to the future. The history of science, in contrast to the closed "apotheosis" system of knowledge, while analyzing a given period in the development of science, regards and analyzes both questions from the past and questions addressed to the future, and it also analyzes retrospection and forecast. This is why the history of science becomes the basis for the dialectic concept of understanding.

Going back to relativistic retrospection, one should underline that Einstein's theory allowed one to look at classical science in a different way and to recognize this "inquisitive" line; for it included the "inquisitive" component, the forecast, the search, the feeling of its inherent incompletion. Einstein's theory of relativity discovered the incompletion of classical mechanics because it saw in classical mechanics certain limited, approximate modifications of more general principles. These are the principles of the relativity of motion and the homogeneity of space. Einstein spoke of "modifications of Galileo" in referring to the classical correlations of the old and the new meanings of the dynamic variables in the transformation from one inertial system of calculation to another. It is not simply a modernizing phrase that brings together the ideas of Galileo with the ideas of the most modern science. Einstein's view of the ideas of Galileo as an earlier and insufficient form addressed to the future refers to a very general principle. In the *Dialogue on the Two Chief Systems of the World,* Galileo proved that the motion of a system does not produce changes in the internal processes in that system. Consequently, space is homogeneous; in space, there are no privileged places. In Aristotelean physics, these places existed. The center of the earth was considered as the immobile center of universal space. The system of calculation with this center was privileged; for instance, approaching the immobile earth or going away from it was considered "true" and absolute motion.

The classical idea of the relativity of motion, if it is considered in the light of Einstein's theory, seems to be insufficiently general. Bodies moving uniformly submit to this principle; but along with bodies in classical science, there also appears a continuous medium that transmits their interaction as well as transmitting light. It is here that we actually discover the "inquisitive," rebellious, explosive character of the Maxwell-Faraday concept of the field in classical science. The field as a continuous medium enables us to talk of an absolute motion of a body in relation not to other bodies, but to that medium. Such motion, according to the classical rule of the addition of velocities, should reveal itself in different velocities of the propagation of light when the light travels against and when it follows a moving body. But such a conception does not have enough "external confirmation." Light moves with the same velocity in any medium whether this medium is approaching the source of light or is moving away from it. Einstein's theory came out of this discrepancy, having rejected the classical concept of a continuous medium which bodies are immersed in. But from the point of view of this theory, the classical principle of relativity appears to be insufficiently general and to be limited only to mechanical phenomena, for in bodies moving with different velocities not only mechanical processes, but the propagation of light also takes place uniformly. Einstein's theory generalized the principle of relativity, and the classical principle seemed now to be a question to which science in the seventeenth through the nineteenth century did not give a general answer; the question was readdressed to the twentieth century, when a detailed answer was given to it.

But such a retrospective reestablishment of classical relativism required that Einstein's theory itself possess an "inquisitive" and forecasting component addressed to the future. The theory does, in fact, possess such a component. Einstein divides physical objects into two classes, one of which is moving bodies, the other the system of calculation, i.e., the rules and watches that establish the place and time of each event. According to the theory of relativity, the dimensions in the direction of motion of the moving body are reduced in proportion to the velocity; i.e., the length of a rule in a system in which it is immobile is different than the length of a rule in a system in which it is moving. This does not mean that during the motion the length of the rule is reduced relatively to some kind of "normal" or "true" length of the immobile rule. Absolute immobility does not exist, and during the comparison of two rules we can consider each as being at rest (its length not reduced) or as moving (its length shortened). In an analogous way, a clock slows down when it is moving, not with reference to a "correct," immobile clock, but, on the one hand, to a given system of coordinates in which the clock is immobile and, on the other hand, to a given system of coordinates in which the clock moves with some velocity, the systems of coordinates being identical.

This concept explains many observations; it possesses an important "external confirmation." The explanation is based not on postulates of an *ad hoc* character, but on more general postulates, so that the "inner perfection" of the theory is high. But from the point of view of Einstein, it is not sufficiently high, and we can discover in the theory of relativity a certain factor not logically derived from more general principles. This factor is the following: the behavior of bodies appears to be different depending on the behavior of the bodies of calculation—depending on the reduction of the length of rules and the slowing of clocks. But then how is the behavior of the bodies of calculation explained? These bodies—rules and clocks—are composed of moving parts. Do the laws of motion of these parts exist, laws on the basis of which the behavior of macroscopic bodies of calculation can be derived? No, the theory of relativity postulates the behavior of the bodies of calculation as something independent of their motion and structure. "This, in a certain sense, is inconsistent," Einstein wrote. "Speaking strictly, measuring rods and clocks would have to be represented as solutions of the basic equations (objects consisting of moving atomic configurations), not, as it were, as theoretically self-sufficient entities. However, the procedure justifies itself because it was clear from the very beginning that the postulates of the theory are not strong enough to deduce from them sufficiently complete equations for physical events sufficiently free from arbitrariness."[3]

3. A. Einstein, "Autobiographical Notes," *ibid.*, p. 59.

This statement of its own insufficiency is characteristic of the theory of relativity and of nonclassical science as a whole. The latter cannot become classical in the sense of possessing an ideal conclusiveness; but the inconclusiveness itself and the nonclassicism acquire higher rank in the course of time. The evolution of nonclassical science includes increasingly fundamental problems in its "inquisitive" character and in the number of unsolved problems (in contrast to the *obviously* unsolved ones of the past). Consequently, historical reminiscences become more and more profound and distant; modern science revises increasingly fundamental premises, transferring them from the rank of logical or even *a priori* premises to the rank of historical and approximating ones. Such an evaluation goes back further and further, all the way to classical antiquity, when, in first approximation, the most general premises of philosophy were found (rather guessed). Let us note at the same time that modern science, being armed with new, powerful experimental and theoretical resources and methods, in some ways comes closer and closer to the spirit of classical antiquity, which was very nonclassical in its style and which faced radical problems of existence in an *a priori* form and in the form of searches and approaches. Lenin saw this in the *Metaphysics* of Aristotle, the philosopher who later became the most dogmatized thinker of antiquity.[4]

Nonclassical science is characterized by either obvious or nonobvious historical reminiscences. Forecasts are also one of its characteristics, both obvious or nonobvious. The inconclusiveness of the theory of relativity that Einstein noted shows the principal potentialities of a more general concept. It is a more general concept not only in relation to the theory of relativity, but also in relation to quantum mechanics. The theory of relativity presupposes the world as a space and time continuum. In the peripatetic picture of the world, space and time were considered separately. In Aristotelian cosmology and physics, a scheme of existence which is purely spatial, invariable in time, and taken at one moment plays a part; it is the scheme of a world with a center, boundaries, and natural places for bodies. On the other hand, Aristotle considered the nonspatial, purely temporal processes in the qualitative changes of the appearance and disappearance of bodies. For atomists

4. V. I. Lenin, *Complete Collected Works*, 29, 325-332.

(especially Lucretius), and in the mechanical picture of the world, time lost its independent reality; processes taking place in time take place also in space; qualitative changes are restricted to spatial changes. But Newton ascribed to space an independent existence: the propagation of interactions, the propagation of forces, and, in principle, any motion could be instantaneous; consequently, processes taking place in zero time could be real, and simultaneity could be given a physical meaning.

Einstein rejected this fiction. His world was made not from purely spatial, "instantaneous" trajectories (they are preserved as approximations, justified for small velocities of bodies, when the velocity of light and the velocity of the propagation of interactions could be considered infinite), but from four-dimensional world lines, each point of which (world point) has four coordinates, three spatial and the fourth temporal. But does this structure of four-dimensional lines possess a physical existence? What is the difference between the presence of a real particle at a given moment and place, i.e., a real physical event, and its spatial-temporal localization? What is the difference between real motion and its geometrical representation in world lines?

These questions constitute a new modification of the principal problems of Cartesian physics. Moreover, it is a modification of the main problem of classical rationalism, when the latter is considered in its modern retrospection. We will limit ourselves to some preliminary, sketchy explanations.

Descartes identifies space with matter, and he deprives bodies of all qualities other than geometrical ones. Space is given certain physical qualities; its parts are impenetrable, and they move in relation to other parts and constitute material bodies. However, since space and matter actually are one and the same, it is impossible to remove a body from the medium surrounding it, and, consequently, the motion of bodies and their interaction and impenetrability represent premises that lack observable equivalents. The search for such equivalents, the solution of the problem of the existence of matter other than in terms of geometrical premises, is the main "inquisitive" tendency of classical rationalism.

How is this problem solved by modern science?

Here we must interpose a short digression. Quantum mechanics refuses to draw the world line of a particle, that is, to ascribe to it a definite space and time localization independently of an experiment in which the particle interacts with this or that material body. Such a body could be a thin screen with a small opening. When the particle goes through the opening, its spatial coordinates coincide with the already known coordinates of the opening, and the time is defined with the help of a clock. Another body allows us to register the speed of the particle, its momentum, and energy. But interaction has a physical meaning; if it changes the behavior of the body, it changes the space and time localization, the momentum and energy, and the world lines of the interacting bodies. Such an effect which changes the world lines of the interaction is important and noticeable when we deal not with macroscopic bodies, but with elementary particles like electrons. For this reason, the effect was discovered in the study of the microcosm within the limits of atomic physics, a study which explains the structure and qualities of atoms through the behavior of subatomic particles. In this instance, how could the world line of the particle be defined? How could the classical premises of space and time localization of momentum and energy be used in the study of the microcosm? Quantum mechanics solves this problem in the following way. It introduces the notion of a macroscopic body of interaction. This body is considered to be classical, and in relation to it a certain classical unrestrictiveness is shown; it is freed from quantum details; its atomic structure is neglected, as is the change of the world lines of the particles that make up the body during the interaction. When, for instance, a screen with an opening is considered, the absolute immobility of the screen during its interaction with particles is postulated, as is the possibility of a precise definition of the coordinates of the opening. Such a principally macroscopic body of interaction ("classical device") allows us to determine the coordinates of the particle and the moment it passes through the opening, but, at the same time, the momentum of the particle is changed. In its turn, the device which allows us to define the momentum changes the coordinates of the particle. As a result, there appears the possibility of defining one of the variables of the particle's total uncertainty by means of another variable.

Quantum mechanics is based on the fact that the particle by itself, without a macroscopic body of interaction, possesses neither a definite space and time localization nor a definite momentum and energy. The particle possesses only the probability of various meanings of these variables. The probability is measured by *waves of probability*, which can be defined by the solution of a wave equation. The experiment—the interaction of a particle with a macroscopic body, a classical device—allows us to move from probability waves to concrete meanings. But the waves of probability change during the interaction with the device in such a way that the precise definition of the situation of the particle changes its momentum in an uncontrollable way and vice versa. The same holds for the correlation of the uncertainties between the time of an experiment and the energy of the particle: the more precisely we know the moment when the particle interacts with the device, the less precisely we know the energy of the particle. And vice versa, the more precisely we know the energy, the less precisely we know the time.

The correlations of imprecisions or uncertainties connecting the conjugate variables—position and momentum, time and energy—are derived from a more general complementarity principle. The particle possesses the corpuscular qualities of position, momentum, and energy at each moment. But it also possesses wave qualities—frequency and amplitude of oscillation of some variable quantity that could be found at each moment and at each point of the space in which the particle moves. The wave qualities of the particle and its corpuscular qualities are complementary. They lose physical meaning without a complementary factor.

Niels Bohr wanted to extend the complementarity principle into other areas besides atomic physics. Probably the principle could also be extended into the past, becoming the basis for a reevaluation of classical ideas, and into the future, modifying the study of the subnuclear world of elementary particles. The extensions into the past and the future are interconnected in ways we shall soon explain. At the same time, we shall convince ourselves that such extensions transform the very notion of complementarity and the notions connected with it, and that they interact when they approach one another. When we talk about the notions of "contiguity" and "interaction," they seem to be simply metaphorical. But they appear metaphorical only from the viewpoint of a formal, logical tradition, in the limits of which they are grouped and regrouped; without changing their meaning, without being modified, they merge into new combinations, into new and more general notions. One of the results of nonclassical science was the clear demonstration of the flexibility of concepts, a result which became the basis for Hegelian logic. Logical operations are sometimes accompanied by a "strong interaction" of concepts analogous to the strong interaction of particles; the latter interaction not only changes the motion of particles, but also transmutes them into particles of another type.

How does the notion of complementarity change in the transition to the subnuclear world?

It seems that in very small regions of space and time (maybe of the order of 10^{-13} centimeters and 10^{-24} seconds, or even smaller regions), the picture of the constantly moving, self-identical particle ceases to be elementary, stable, and unequivocal. At the same time, the notion of a world line—a continuous sequence of world points, of space and time localizations of a particle—becomes slightly superfluous in very small regions and is introduced from considerably vaster regions. In quantum mechanics, which was created in the 1920's, the world lines already appear to be somewhat vague, because the space and time localization of the particle, its spatial position and the time it occupies this position, cannot be precisely defined simultaneously with the momentum and energy of the particle. Now not only the precision of the world line, but its continuity is threatened. In ultramicroscopic regions, the very existence of a self-identical particle becomes problematic. Here we still do not have a unique picture of events taking place in space-time cells of the order of 10^{-13} centimeters and 10^{-24} seconds, or even smaller ones. But the modification of complementarity is actually nothing else than a generalization of the tendency of science, of its dynamics and its "inquisitive" component. This modification is shown only in prediction, which, with all its nonuniqueness, illustrates the modern dynamic situation in science. We will present one of these illustrations here.

Let us assume that time and space are discrete, that the four dimensional world line of a particle is interrupted or, in other words, that the particle, in a minimal space interval, (for instance, 10^{-13}

centimeters), does not make a transition between two points that are infinitely close, but that the particle is annihilated, becoming a particle of a different type; this latter particle then again becomes a particle of the original type. Such a re-creation takes place within a distance of 10^{-13} centimeters and within an interval of 10^{-24} seconds. Thus, in space-time, there are intervals in which space and time do not appear to be dynamic variables of the moving particles; they lose their physical existence. We focus on these elementary re-creations or motions, in macroscopic approximation; we identify the particle appearing as a result of the re-creation with the original particle; we ascribe self-identity and uninterrupted existence to it, and we draw its continuous trajectory. In order that this trajectory does not vanish during a certain macroscopic time interval, i.e., in order that the particle moves, we need asymmetric probabilities of the elementary motions. If, during re-creation, the particle has an equal probability of moving in any direction, then after a large number of motions—random wanderings—it could find itself close to its original position. If, however, there is an asymmetry according to which a motion to one side has more probability than a motion to the opposite side, then, after a large number of elementary motions, the particle will move forward a certain microscopic distance. The macroscopic speed of the particle is proportional to the degree of asymmetry, and we could identify this with the momentum. The greater the asymmetry, the closer the macroscopic trajectory is to the ultramicroscopic trajectory, which is made up of elementary motions. The speed of the ultramicroscopic trajectory is always the same and equals the elementary translation divided by the elementary interval of time (10^{-13} centimeters: 10^{-24} seconds), i.e., the speed of light. The speed of the microscopic trajectory obviously cannot exceed this constant. The mass of the particle is actually the intensity of the symmetry of space, the symmetry of the probability of elementary motions. This could be explained by the effect on the particle of the homogeneous outer galaxies.[5]

The forecast relative to the transition from atomic physics and quantum theory, created in 1925–1927, to subnuclear physics and a more general theory could also be derived from the above sketch, or

5. B. Kuznetsov, "Complementarity and Relativity," *Philosophy of Science*, 33 (1966), 199-209.

from some other. Such a sketch illustrates two interconnected tendencies of nonclassical physics. One is the unavoidable generalization and modification of the principle of complementarity in forecasting the further evolution of science. The other is the unavoidable generalization of the principle of complementarity in retrospection during the reevaluation of the fundamental, penetrating ideas, which are modified, but not eliminated, in the spiritual history of humanity starting from classical antiquity.

Let us take as an illustration the sketch of the discreteness of time and space discussed above. It may appear that we are talking about the substitution of the mechanical world picture (mechanical in a very general sense: the bricks of the structure of the world are the motions of self-identical bodies) with another, *transmutational* world picture (the bricks of the structure of the world are elementary transmutational particles). In reality, the sketch of the discreteness of time and space illustrates a more radical transformation of the world picture. The very notion of the "bricks of the world-structure" —elementary bodies and processes from which, in the final analysis, the world and its evolution are created—becomes very problematic. What is transmutation? It is the transformation of a particle of one type into a particle of another type bearing a different mass, charge, and life time. But these variables do not have a meaning if there are no world lines, if there is no macroscopic world of constantly moving, self-identical bodies. The mass of the particle and its charge point to its eventual world line, to its behavior in one or another field. We cannot ascribe to the elementary transmutations any physical meaning other than the transformation of one eventual macroscopic world line into another. Without world lines, the elementary motions and the concepts of minimal distance, minimal time, symmetry, and asymmetry lose their meaning.

In their turn, world lines are physically meaningless concepts in the absence of ultramicroscopic processes that abrogate, limit, or reject the constant macroscopic motion. Without this "non-Cartesian" content, world lines become simply a geometrical notion and lose physical meaning. Probably complementarity in a very general sense—the complementarity of concepts that mutually exclude one another and at the same time lose physical meaning in independence of one another—now becomes the original definition of existence, without which it is difficult and probably even impossible to say where science and its fundamental concepts are going.

But such a *prognostic* characteristic of the direction and state of science demands a *retrospective* reevaluation of all of the past evolution of the concept of existence, in particular of the evolution of this concept within the framework of classical rationalism. In light of modern science, the principal road of rationalism becomes the genesis of the understanding of existence as harboring an inconsistency. Consequently, rationalism is no longer regarded as an endeavor to define the world within identifying logical thought. A logical chain of intellectual conclusions of the type, "the notion A is actually the notion B, and the notion B is the notion C, and so on," is inseparable from a sensual accompaniment. Empiricism constantly introduces into this logical chain dissonances, inconsistencies, and nonidentity. But empiricism itself takes part in the evolution of our understanding only after being woven into the logical chain.

This link of rationalism in its traditional meaning with its empirical accompaniment existed during classical times too. But it was hidden. In nonclassical physics, it became obvious. In this respect, nonclassical physics exposed the secret of classical rationalism. But the modern Oedipus, discovering the secrets of the classical sphinx, also discovers his own secrets and, in addition, new lines in his itinerary of further movement. That is why the modern style and the modern problems of science do not correspond at all to a "clean slate" on which one can write new lines only after the old ones have been completely erased. Such a concept, which is close to our understanding of the "history of vacillations," which finally concluded in a newly possessed truth, sometimes takes the character of "nonclassical" haughtiness; i.e., each youth, having read a modern textbook, is immediately attributed with a more profound understanding of the world than that which satisfied the poor fellows from Aristotle all the way to Newton. The whole problem is that they were not entirely satisfied with the thoughts of their own century. "Nonclassical" haughtiness takes its source from a fiction, which was dogmatized and which rested on the laurels of classical science. But classical science was not thus; it did not include *a priori* notions and contradictions, and science, in all probability, could never surpass in courage and paradoxicalness the thought that appeared in antiquity concerning the isotropy of space, concerning antipodes that do not fall "down." The meaning of nonclassical retrospection actually lies in the analysis of the historical tradition of antitraditionalism, a tradition responsible for the transition from classical to nonclassical science, where it finally found an incomparably more obvious form.

We will conclude with a few words about forecasts, which include those of nonclassical retrospection—those of relativistic, quantum, and quantum-relativistic character. The latter allows us to see in modern physics the profound features of classical science and classical rationalism. Such an historic and scientific concept opposes the pessimistic ideas of the failure of the intellect and of science and opposes modern anti-intellectualism. It brings us to gnostic optimism. The change of more and more profound, general, and fundamental principles, in terms of a well-based forecast, follows from the tendencies of nonclassical physics. Science, as far as one can foresee its future, will not move into an asymptotic zone; its turns and changes of path and its digressions from the traditional way will not fade out. On the contrary, the amplitude of the oscillations of "the world line" of science will be ever-increasing. But this does not mean that each radical (constantly more radical) turn of science will erase all of the preceding ones. It will give to the already discovered truths more profound and more precise meaning; it will make more concrete the conditions of its adaptability, and it will derive them from more general principles.

The growth of the amplitude of the oscillations of science, the transformation of its more fundamental principles will become (and is already becoming) a directly acting force of the transformation of production and of civilization as a whole. The transition from gnostic to social optimism was the abnegation of absolute space and absolute time, the abnegation of the medium of space, of the constancy of mass; i.e., the whole sum of positive assumptions of the theory of relativity led to our understanding of the energy produced during the division of atomic nuclei, to the use of this energy in the atomic era. The abnegation of the classical continuity of the electromagnetic field, the abnegation of the classical definition of contiguous variables, i.e., the whole understanding of atomic physics that grew out of the principles of quantum mechanics, led to quantum electronics, to the radical transformation of technology, of communications, of control, and of computing operations, and, finally, to the possibility of the most radical transformation of the nature of labor. Further, post-atomic civilization will realize the potential of subnuclear physics, of the physics of elementary particles.

Our time is the dawn of nonclassical civilization. Maybe a different image is appropriate here, the one used by Alexei Tolstoy in the title of his historical novel about the start of the revolution. He called it *Gloomy Morning*. Nonclassical science and nonclassical civilization are seen through a cloud screen. The clouds do not allow us to see the contours of the unique theory of elementary particles—the ideal of modern science that is escaping from the physicists and is luring them after it. But there is a different, far more threatening cloud, the prospect of a destructive application of science threatening the very existence of civilization. Among the forces which have been called upon to disperse these clouds is the rationalistic pathos of modern science. It is fortified by historical and scientific researches which discover and make obvious in the present, past, and future the irrevocable movement of the intellect toward the truth.

The Growth of Professorial Research in Prussia, 1818 to 1848— Causes and Context[1]

BY R. STEVEN TURNER*

During the nineteenth century the fame of German science rested not only upon its fundamental contributions to natural knowledge, but also upon the unique organizational system which Germany had created for scientific research. Unlike its major European neighbors, Germany pursued science almost entirely within the unlikely framework of its old university system. Within that framework it had constructed by 1875 a costly and highly influential series of university laboratories, seminars, and institutes.[2] The laboratory system gave institutional expression to the ethos of learning which had come to dominate the universities—a professional commitment to research and to the elaboration of research methods and a fundamental concern with training students in the techniques of scientific investigation.[3] By 1830 the universities had already succeeded the various academies as centers of scientific research. By midcentury the career scientist was almost invariably a professor, while the laboratory in which he worked had been only recently annexed to a university structure reaching back to the middle ages.

* Department of History, University of New Brunswick, Fredericton, N.B., Canada.
 1. This paper is drawn from a dissertation in progress on the Prussian universities from 1818 to 1848, with special reference to the growth of scientific research. I am indebted to Professor Thomas S. Kuhn for invaluable criticisms and suggestions during the preparation of the paper, and to Professor Michael S. Mahoney for assistance with various translations from the German.
 2. On the German institutes see W. Lexis, *Die deutschen Universitäten* . . . (Berlin, 1893), vols. 1 and 2.
 3. See D. S. L. Cardwell, "The Development of Scientific Research in Modern Universities . . . ," *Scientific Change,* ed. A. C. Crombie (New York, 1961), pp. 661-663.

That the unexpected, precipitous growth of German science after 1820 occurred within the universities rather than within academies, private laboratories, or newly created institutions for science resulted from an historical development almost unique to the German states. During the early nineteenth century the traditional German professorate was transformed from a predominantly teaching post to a position entailing responsibility for scientific research and publication as well. The professor of the eighteenth century had considered his main duty the transmission of established learning to certain professional groups; in addition to maintaining that goal, his nineteenth-century counterparts tried actively to expand learning in many esoteric fields. Although the roots of that change lay deep in the eighteenth century, the actual redefinition of the professor's duties occurred during the German *Vormärz* era, which stretched from the close of the Napoleonic wars to the March disorders signaling the onset of the Revolution of 1848.[4] During this period, research emerged within university ideology as a fundamental duty of the scholar, and a reputation within one's specialist community beyond the university became more and more a *sine qua non* for even minor university appointments.

The transformation of the professorship occurred not only in the sciences, but also in every discipline of the fourth or philosophical faculty of the universities. The philological and historical disciplines first displayed the intense concern with research and research training. Only later—during the 1830's—were these commitments widely adopted by science professors, often in direct imitation of learned values and institutional models of the humanistic disciplines. In a similar manner these commitments spread even to the professional faculties of law, theology, and medicine, where they brought about a shift in emphasis from the training of practitioners to the education of scholars and research workers in the respective disciplines.

Transcending as it did the internal development of any individual field or faculty, the growth of research as a duty of the professor constitutes a major problem in university historiography. Only

---

4. See Friedrich Paulsen, *Die deutschen Universitäten und das Universitätsstudium,* 2nd ed. (Hildesheim, 1966), pp. 40-75.

rarely, however, have historians of the German universities disentangled this problem from the larger context of university history as a topic for special study. When they have done so, historians have still shown little agreement concerning the causes of this development. Historians writing before 1920 stressed the influence of new ideologies which glorified *Wissenschaft* and creativity. More recently competition and decentralization within the university system have been advanced as important elements stimulating the growth of research. Although both factors greatly influenced the development of research in the universities, neither in itself can fully explain that development.

The history of universities has taken insufficient consideration of a third important factor affecting the professorate: the state. In particular, the government's changing and always-critical role in making university appointments and in setting appointive criteria has been generally neglected. This study examines precisely this role, limiting its scope to the Prussian universities during the *Vormärz* period, and stressing the important, though by no means exclusive, part which the state played in establishing the new professorial values. As a basis for this discussion the study first summarizes the development of the Prussian universities after 1800, emphasizing those developments which particularly influenced science and scholarship. It goes on to critically examine the hypotheses traditionally advanced to explain the growth of research. Finally it discusses the new appointive criteria which promoted the growth in Prussia of a new type of professor committed to critical scholarship and original research.

1. THE DEVELOPMENT OF THE PRUSSIAN UNIVERSITIES, 1806 TO 1848

The nineteenth century opened in Prussia to strident demands for university reform. Throughout the eighteenth century the universities had suffered from poor financing by the state, precipitously declining enrollments, and a widespread public conviction that the universities were obsolete and corrupt. In the 1790's these chronic problems evoked widespread calls for government action.[5] When Prussia's defeat by Napoleon at Jena in 1806 brought reform ministries to power, Prussia embarked upon a thorough reform of its educational structure, including a basic reorganization of its university system. During the era of educational reform, which lasted from 1806 to 1818, Prussia founded new universities at Berlin (1810) and Bonn (1818), consolidated various institutions to form the universities of Halle-Wittenberg and Breslau, abolished several nonviable smaller schools, and provided the University of Königsberg with extensive new institutes and personnel. The reforms channeled new funds into the universities and established them upon a viable economic and geographical basis that was absent during the eighteenth century.[6]

No less important, the state moved to aid its universities through a simultaneous reform of secondary schools. Under Wilhelm von Humboldt's brief but epochal leadership, the Department of Educational Affairs initiated compulsory testing for all students wishing to enter the universities. Control of this examination not only permitted the state to reform and to regulate the secondary school curricula but also to improve the universities by enhancing the quality of their students and by sharply distinguishing secondary from higher education. In July 1810, the state took a similar step in decreeing compulsory examinations for all teachers seeking employment in the secondary schools. This edict initiated the reform of secondary teaching and led to the creation of an entire class of professional gymnasium instructors to be trained by the philosophical faculties of the local universities. Both these edicts produced beneficial, immediately noticeable effects in educational affairs.[7]

5. Though most Prussians favored extensive reform, many advocated outright abolition of the universities. See J. H. Campe, *Allgemeine Revision des gesammten Schul- und Erziehungswesens* (Vienna, 1792), *16*, 145-220, and Adolf Stölzel, "Die Berliner Mittwochsgesellschaft über Aufhebung oder Reform der Universitäten (1795)," *Forschung zur brandenburgischen und preussischen Geschichte, 2* (1889), 201-222. On the condition of the eighteenth-century universities see Karl Biedermann, *Deutschland in achtzehnten Jahrhundert* (Leipzig, 1854), *2*, 513, 678-686, and passim.

6. Eduard Spranger, *Wilhelm von Humboldt und die Reform des Bildungswesen*, 2nd ed. (Tübingen, 1960), esp. pp. 199-234; C. Varrentrapp, *Johannes Schulze und das höhere preussische Unterrichtswesen in seiner Zeit* (Leipzig, 1889), pp. 225-285; Ernst Müsebeck, *Das preussische Kultusministerium vor hundert Jahren* (Berlin, 1918).

7. Friedrich Paulsen, *Geschichte der gelehrten Unterrichts* (Leipzig, 1885), pp. 567-590.

Not all the reforms of the reform era were institutional and financial, nor were all accomplished by the state. Between 1805 and 1810 a number of theorists turned to the spiritual and philosophical rejuvenation of the universities. This group, which included Wilhelm von Humboldt, J. G. Fichte, Friedrich Schleiermacher, Heinrich Steffens, and F. A. Wolf, elaborated and popularized a new concept of the university and its relationship to morality, to the state, and to scholarship.[8] Reacting sharply to the educational theory of the Enlightenment, which stressed the utilitarian or purely propaedeutic function of university education, these men denied that the universities were merely pedagogical institutions or professional schools producing well-trained, docile subjects for the state. They created a new image of the scholar as an individual of moral insight and courage, simultaneously aware of his radical personal freedom and his responsibilities to the state. He was to be, in Fichte's words, the living expression of the divine idea, "morally the best man of his age;—he should exhibit in himself the highest grade of moral culture then possible."[9]

To be sure no consensus emerged on how to educate such a man. Humboldt and Wolf, representatives of the neohumanist movement, looked to philological inculcation in the values of Greece and Rome to cultivate the ethical and aesthetic refinement which they called *Bildung*.[10] Fichte and Schelling—who, with Kant, were the founders of German Idealism—emphasized philosophy as the discipline in whose light all others were to be studied. True education, they argued, could only be insight into *Wissenschaft,* the great organic unity of all knowledge postulated by idealist philosophy. In Schelling's words, "A methodology of university study must be

---

8. The major treatises upon this theme by Schelling, Fichte, Schleiermacher, Steffens, and Humboldt are collected in *Die Idee der deutschen Universität,* ed. Ernst Anrich (Darmstadt, 1964). For F. A. Wolf's ideas see *Über Erziehung, Schule, Universität,* ed. Wilhelm Körte (Leipzig, 1835). The German secondary literature upon this new ideology, here to be called *Wissenschaftsideologie,* is immense. See particularly Helmut Schelsky, *Einsamkeit und Freiheit; Idee und Gestalt der deutschen Universität und ihrer Reformen* (Reinbek, 1963). An excellent English discussion of the basic categories of the new ideology is found in Fritz K. Ringer, *The Decline of the German Mandarins. The German Academic Community, 1890-1933* (Cambridge, 1969), pp. 85-96.
9. Johann Gottlieb Fichte, *The Vocation of the Scholar. A Series of Lectures Given at Jena in 1794,* trans. William Smith (London, 1847), pp. 58-59.
10. Eduard Spranger, *Wilhelm von Humboldt und die Humanitätsidee* (Berlin, 1909), pp. 456-477.

rooted in actual and true knowledge of the living unity of all the sciences, and . . . without such knowledge any guidance can only be lifeless, spiritless, one-sided, limited."[11] In defending the universities the theorists invoked the universities' venerable historical role. They portrayed the universities as hoary symbols of German cultural unity embodying the ideal of *Wissenschaft.*

This radically new concept of the universities' essence and mission allied the new, reformed universities not only to German patriotic sentiment but also to German philosophic and scholarly thought. This *Wissenschaftsideologie* rapidly became the dominant ideology of the universities and was echoed and elaborated in rectorate addresses and academic ceremonies throughout the century.[12]

With the founding of Bonn University in 1818, the period of fervent reform ended abruptly; liberal agitation, centered in the universities, provoked a state policy of official mistrust and initiated two decades of censorship and political persecution of students and teachers all across the German Confederation.[13] Nevertheless, the *Vormärz* period witnessed not only some of the greatest achievements of Prussian scholarship, particularly in law, linguistics, and history, but also the continued prosperity and modest expansion of the universities in students, faculty, and funds.[14]

---

11. F. W. J. Schelling, *On University Study,* trans. E. S. Morgan (Athens, Ohio, 1966), p. 7; also J. G. Fichte, "Deduced Scheme for an Academy to be Established in Berlin," in *The Educational Theory of J. G. Fichte,* ed. and trans. G. H. Turnbull (London, 1926), p. 195.

12. For example, August Boeckh, "Ueber die Pflichten der Männer der Wissenschaft gemäss der bisherigen Entwickelung und dem gegenwärtigen Standpunkt derselben," *Gesammelte kleine Schriften* (Leipzig, 1859), 2, 115-131; also Eduard Zeller, *Über akademisches Lehren und Lernen. Rede zur Gedächtnissfeier der Friedrich-Wilhelms-Universität zu Berlin* . . . (Berlin, 1879).

13. On the political persecutions see Varrentrapp, pp. 287-350 and Max Lenz, *Geschichte der königlichen Friedrich-Wilhelms-Universität zu Berlin,* 4 vols. (Halle, 1910-1911), 2, 34-176. The liberal movement within the universities is treated in a broader political context by Franz Schnabel, *Deutsche geschichte im neunzehnten Jahrhundert,* 4 vols. (Freiburg, 1949), 2, 234-271.

14. The total budget of the University of Berlin, for example, grew from 241,324 marks in 1820 to 529,710 marks in 1850. Her student body numbered 942 in 1817/18 and grew to 1,540 by 1848. (Lenz, *Berlin, 3,* 529, 490-495.) At Halle the number of *Privatdozenten* in the philosophical and medical faculties more than doubled, growing from seven in 1827 to fifteen in 1857; the student enrollment of the philosophical faculty grew from fifty-one to ninety-three during the same period. The higher faculties at Halle, however, declined in enrollment. *(Verzeichniss des Personals der Universität Halle* [Halle, 1827-1857].) For a survey of the other Prussian universities during the *Vormärz* period see Carl

The *Vormärz* era brought significant changes to the internal structure of the universities. These changes represented in part the delayed, often unintended consequences of earlier reforms, in part the universities' response to social and intellectual pressures new to the *Vormärz* era. During the eighteenth century, for example, the philosophical faculty had been merely a poorly attended, penurious preparatory school for the three professional faculties of theology, law, and medicine. After 1820 the philosophical faculty began to expand phenomenally, largely in response to its new role in the education of Prussia's secondary teachers. It attracted new groups of students, and its teaching body swelled with specialists moving into more and more esoteric areas of philology, history, and the sciences.[15]

This growth in personnel occurred not only in the philosophical faculty but also in all branches of the university. The lower ranks of the academic hierarchy, the *Privatdozenten* and junior professors (*Extraordinarien*), expanded most rapidly, and this expansion led to an atmosphere of intense competitiveness, both among younger instructors for promotion and between them and the full professors (*Ordinarien*) for students.[16] Upon beginning lectures at Berlin as a *Privatdozent,* for example, du Bois-Reymond wrote to Karl Ludwig that it was his "single hope to give [Johannes] Müller successful competition," and Ludwig in reply urged him to announce physiology lectures with vivisectional demonstrations, since these innovations would surely reveal the crassness of Müller's own lectures.[17] This sort of academic competition had been induced during the reform era when the universities, following Berlin's lead, had closely regulated the qualifications of *Privatdozenten* and had established them

---

Friedrich Wilhelm Dieterici, *Geschichtliche und statistische Nachrichten über die Universitäten im preussischen Staate* (Berlin, 1836). On Prussian scholarship during the *Vormärz* period see Schnabel, *3*, 36-172.

15. Paulsen, *Universitäten*, pp. 76-77. Counting all ranks, Berlin's philosophical faculty had thirty-two instructors in 1820; this number had grown to ninety-one by 1848. During this period the lower faculty's share of the total teaching body increased from 45 percent to 56 percent. (Lenz, *Berlin, 3*, 490 and 504.) The state exercised much initiative in expanding the philosophical faculty into specialized areas. See Varrentrapp, pp. 444-460.

16. Lenz, *Berlin, 2*, 407-411. The growth of the faculty was accompanied by declining salaries and a sharp decrease in the individual's chances of promotion.

17. Emil du Bois-Reymond and Karl Ludwig, *Zwei grosse Naturforscher des 19. Jahrhunderts. Eine Briefwechsel . . . ,* ed. Estelle du Bois-Reymond (Leipzig, 1927), pp. 132-133, 146-147.

as competitors both for students and for professional chairs.[18] Responding to the enhanced attractiveness of careers in the reformed universities, young men flocked to them all during the *Vormärz* period. As *Privatdozenten* they became the chief agents for specialization and innovation in the curriculum.[19]

In addition to the struggle within their own faculties, universities in the *Vormärz* period began to compete more intensively with each other for students, reputable professors, and learned prestige. During the eighteenth century the Prussian government, like the governments of most German states, had exercised a policy of "academic mercantilism" toward its universities. In the eyes of the state the universities existed to train bureaucrats, ministers, and other professional groups for civic life. By removing from these groups the necessity of studying outside Prussia, the territorial universities prevented the drain of talent and wealth from Prussia and allowed the state more easily to enforce political and religious conformity. To facilitate this primary function of the universities the state maintained monopolistic conditions for its own schools. It depressed inter-university competition by consistently prohibiting student migration between universities and by limiting the professor's right to resign or to change posts.[20] During the reform period, however, the state gradually retreated from its mercantilistic policies and guaranteed both these specific rights.[21] As a result the *Vormärz*

18. Alexander Busch, *Die Geschichte des Privatdozenten* (Stuttgart, 1959), pp. 20-23. The eighteenth-century *Privatdozent* had been literally a "private teacher"; only after the regulation of the *Habilitation* procedure in 1816 did the nineteenth century *Privatdozent* begin to attain his "apprentice" status within the university.

19. During the *Vormärz* period full professors and *Privatdozenten* reached a tacit understanding whereby the former offered only elementary, well-attended, "survey" courses and the latter, more specialized and advanced lectures. The full professors preferred the more elementary courses because attendance, and hence student fees, were correspondingly higher, and they used their power to maintain a monopoly over such courses.

20. Conrad Bornhak, *Geschichte der preussischen Universitätsverwaltung bis 1810* (Berlin, 1900), pp. 118-122.

21. Prussia recognized the professor's right to resign in the *Allgemeines Landrecht für die preussischen Staaten* of 1794, Part 2, Title 10, Paragraph 95. Frederick William III lifted the prohibition on student migration on April 13, 1810. Students were again forbidden to study outside Prussia from 1833 to 1838, at the height of the political reaction. (Johann Friedrich Wilhelm Koch, *Die preussischen Universitäten. Eine Sammlung der Verordnungen . . .* , 2 vols. [Berlin, 1839-1840], *2*, 531-534.)

period witnessed a gradual upswing in professorial mobility and the rise of fervent struggles between universities to woo and win famous professors. The state also abandoned another monopolistic eighteenth-century policy, its outspoken favoritism of Halle University.[22] During the nineteenth century the Prussian government supported its universities as near equals and carefully maintained intense competition both among its own institutions and between them and the universities of neighboring states.

In matters of scholarship the *Vormärz* era witnessed the expansion of activities and institutions devoted to research into all academic fields. Seminars oriented decisively toward research techniques were established in the universities under state auspices after 1810, first predominantly in classical philology and later in history.[23] Their purpose was, as the statutes of August Boeckh's seminar in Berlin insisted,

> to educate by means of all possible diverse exercises leading into the core of learning and by means of literary support of every kind, all those who are properly prepared for [training in] the classics, so that in the future they will be capable of sustaining, propagating, and enlarging these studies. . . . As a rule only those are qualified for acceptance into this institute who dedicate themselves chiefly to philology, not those who expect their future advancement from the exercise of another branch of academic learning.[24]

The statutes of these powerful and influential seminars left no doubt of their "purely scientific" aims, nor of their intention to provide training for specialists, not for the liberally educated.

---

22. Prussia's last consideration of a one-university policy centered upon Berlin seems to have occurred in 1819 after having been advanced by the *Kultusminister* Altenstein. See Lenz, *Berlin*, 2.1, 10-30.

23. These early German "seminars" were programlike institutions in which ten to twenty select students might be enrolled for two or three years, and in which they received intensive practical training under direct supervision of the full professor in charge. Seminars usually possessed their own rooms, libraries, and scientific instruments; their members received subsidies from the state in the form of scholarships, prizes, and exemption from state tests. Only nominally connected with the university faculties, the early seminars constituted an exclusive "inner track" of intensive academic training. The seminars contrasted sharply with the educational experience of the average German student, who merely attended lectures and who rarely had assigned problems and exercises or outside contact with professors.

24. Koch, *2.2*, 560. This and all subsequent translations are my own unless otherwise indicated.

Although many graduates of these early seminars went on to become university professors and although a few went into the bureaucracy, most graduates entered the Prussian gymnasia as advanced instructors. In fact, the need to train a corps of competent teachers for the reformed and rapidly growing gymnasium system had initially motivated the state to found the institutions.[25] Quite early in their history, however, the seminars had asserted their allegiance to philology as a scholarly discipline and their disdain of mere teacher-training. Before 1800 F. A. Wolf, founder of the philological seminar at Halle, pointedly refused to give teaching exercises or instruction in pedagogical theory to his seminar students and insisted upon rigorous philology and critical training as the only proper education for future schoolmen.[26] Prussia rapidly adopted Wolf's views as orthodoxy and embodied them in the statutes of the later seminars of the *Vormärz* period.[27] This approach to the education of teachers not only insured rigorous, advanced instruction in history and classical literature in the gymnasia, but it also created a large reservoir of highly trained scholars capable of independent research on which the universities could draw as they chose.

The development of science and mathematics during the early *Vormärz* period proceeded less rapidly than that of the humanities, yet these disciplines also underwent important changes. During the eighteenth century science and mathematics instruction remained elementary with little practical instruction and only occasional experimental demonstrations in the sciences. The universities taught these disciplines primarily as auxiliary studies to medicine and technology. Until well past 1750 chemistry and many branches of the life sciences were taught exclusively by the medical faculty. Mathematics and the mathematical sciences, on the other hand, were taught largely in the service of technological education, to which the universities increasingly committed themselves after 1760. At

25. Paulsen, *Gelehrten Unterrichts*, pp. 586-589.
26. Wilhelm Schrader, *Geschichte der Friedrichs-Universität zu Halle*, 2 vols. (Berlin, 1894), *1*, 455-456. Also Friedrich August Wolf, *Ein Leben in Briefen*, ed. Siegfried Reiter, 3 vols. (Stuttgart, 1935), *1*, 52-63. The letters include Wolf's correspondence with Freiherr von Zedlitz over the founding of the seminar and his rebellion against responsibility for teaching pedagogy.
27. See Koch, *2.2*, passim.

Göttingen University in 1790, for example, almost three-fourths of the sixty lectures offered in the mathematical sciences stressed surveying, mechanics, machines, perspective, building, or general applied mathematics. Four of the thirteen lectures offered in the experimental and observational sciences stressed technological applications such as the "chemistry of smelting" and the "economics of natural history."[28]

Even though the Prussian universities enjoyed no reputation as centers of research in pure science, scientific activity expanded in Germany as a whole throughout the late eighteenth century. German scholars contributed to European research on particular problems such as electricity and irritability, and by 1790 mature, well-defined German scientific communities had arisen in chemistry, astronomy, and mathematics.[29] New, specialized journals stimulated the consolidation of these communities, journals like C. F. Hindenburg's *Archiv der reinen und angewandten Mathematik* and Lorenz Crell's *Chemisches Journal für die Freunde der Naturlehre*. . . . Although many university professors helped to staff them, these young communities included large numbers of academicians, physicians, and bureaucrats in the technical service.

During the reform period the new *Wissenschaftsideologie*, allied with the related movement of *Naturphilosophie*, had violently attacked the utilitarian approach to science characteristic of the eighteenth century. They condemned utilitarian studies as morally corrosive to the student and as tending to divert his attention from the idealistic ends which they advocated for the universities.[30] Partly as a result of this attack, technological education during the 1820's and 1830's was relegated to other types of institutions, the predeces-

28. "Vorlesungs Verzeichnis der Universität Göttingen," *Göttingische Anzeigen* (1790), pt. 1, 441-456.

29. For chemistry see Karl Hufbauer, *The Formation of the German Chemical Community (1700-1795)*, an unpublished doctoral dissertation from the University of California at Berkeley, 1969. On the community of German mathematicians before 1800 see E. Netto, "Kombinatorik," *Vorlesungen über Geschichte der Mathematik*, ed. Moritz Cantor (Leipzig, 1908), *4*, 201-221 and Wilhelm Lorey, *Das Studium der Mathematik an den deutschen Universitäten seit Anfang des 19. Jahrhunderts*, in *Abhandlungen über den mathematischen Unterricht in Deutschland*, ed. Felix Klein (Berlin, 1916), *3*, pt. 9, 26-29.

30. For example, Lorenz Oken, *Über den Werth der Naturgeschichte, besonders für die Bildung der Deutschen* (Jena, 1809).

sors of the *Technische Hochschulen*.[31] In turn the universities began to teach the sciences for their intrinsic value, and chemistry and the life sciences migrated definitively into the philosophical faculty. Professors began to offer more specialized lectures and to open their own small laboratories to students for occasional practice.

Building upon this basis, Prussian science suddenly began to expand prodigiously on all fronts after 1830. This new development shared traits of previous growth in various fields within the humanities. Young academics cultivated research and publication with intensity; advanced, specialized courses were rapidly introduced; special problems and subfields were quickly developed; and the universities began to establish special institutions for the practical instruction of students.[32] Within two decades Prussian science, like German science in general, rapidly attained the European preeminence which German classical and historical studies had already long enjoyed.[33]

The ethos and the organization of humanistic studies became models for the new academics who helped to develop Prussian science after 1830. The philological seminars insisted upon complete independence from traditional pedagogy and upon their aim of giving students sufficient training in philological techniques to enable them to carry out their own independent investigations. These aims were transferred into the organization of science and mathematics by C. G. J. Jacobi. As a student at Berlin, Jacobi had been a favorite of Boeckh and had attended his seminar, originally

---

31. On German technical education during the nineteenth century see Karl-Heinz Manegold, *Das Verhältnis von Naturwissenschaft und Technik im 19. Jahrhundert im Spiegel der Wissenschaftsorganisation*, in *Technikgeschichte in Einzeldarstellungen*.

32. Note the comparison of the works of Ritschl and Liebig by John Theodore Merz, *A History of European Thought in the Nineteenth Century*, 2nd ed. (New York, 1965), *3*, 145-146.

33. German science lagged behind the humanities in this development for many reasons, not all of which are entirely clear. The achievements of German humanistic scholarship rested largely on methods of textual criticism and reconstruction which German scholars both invented and exploited. The sciences, on the other hand, had to await the importation of mathematical and experimental techniques from France. The repressive role of *Naturphilosophie* has also been frequently invoked to explain the retardation of German science, but in fact we know too little about the real effects of *Naturphilosophie* to assess its role.

intending a career in philology.³⁴ After deserting philology for mathematics, Jacobi was transferred to the University of Königsberg, where he founded a mathematical tradition through his work on elliptical functions and where he introduced the radically new practice of lecturing directly from the material of his research.³⁵ With Franz Neumann in 1835/36 Jacobi founded the Königsberg mathematics-physics seminar modeled directly upon Boeckh's seminar.³⁶ Like the latter the Königsberg seminar insisted upon the use of independent investigation as a pedagogical tool, demanding of students "independent works . . . , which are either purely theoretical or which demand individual observations and measurements on the basis of a mathematical theory."³⁷ Like those of the philology seminars its statutes neglected any mention of teachers or teacher-training, even though its graduates were intended for the secondary schools of East Prussia.

The Königsberg seminar rapidly became the center of German mathematical physics and inspired much imitation. In 1839 Jacobi's student Ludwig Adolph Sohncke founded a general science seminar at Halle modeled upon the Königsberg seminar and the Universities of Göttingen and Berlin founded mathematics-physics seminars in 1850 and 1864 respectively.³⁸ Gustav Robert Kirchoff and Ludwig Otto Hesse exported the Königsberg scientific and pedagogical methods to Heidelberg in the 1850's. Rudolph Clebsch and Paul

34. Lejeune Dirichlet, "Gedächtnissrede auf Carl Gustav Jacob Jacobi," *Abhandlungen der königlichen Akademie der Wissenschaften zu Berlin* (1852), p. 2.
35. Lorey, pp. 59-64.
36. Götz von Selle, *Geschichte der Albertus-Universität zu Königsberg in Preussen* (Würzburg, 1956), pp. 284-285.
37. Koch, 2.2, 859. Jacobi insisted that his students prepare themselves immediately for independent research. He wrote to his brother M. H. Jacobi in 1840: "Ich habe in dieser Beziehung viel an Socoloff gearbeitet, bei dem es mir am meisten zu lohnen schien; er hielt mir immer die gewöhnliche Rede entgegen, wie er denn an eigne Untersuchungen denken könne da ihm noch so viele Kenntnisse fehlen, worauf ich ihm einmal entgegnete, wenn seine Familie von ihm verlangen würde dass er sich verheirathen solle ob er denn sich antworten würde, wie er sich denn verheirathen könne da er noch nicht alle Mädchen kennen gelernt. Erst in der letzten Zeit gelang es mir etwas sie zu eignen Bemühungen zu bringen. . . ." (*Briefwechsel zwischen C. G. J. Jacobi und M. H. Jacobi*, ed. W. Arens, in *Abhandlungen zur Geschichte der mathematischen Wissenschaft*, 22 [1907], 64.)
38. Lorey, pp. 117-120.

Gordan carried them to Giessen in the early 1860's, where Liebig had introduced similar methods in chemistry much earlier.[39] These later seminars retained the preoccupation with research and its use as a pedagogical tool shown by their parent institution. Sohncke's seminar at Halle attempted "to give an introduction to independent study and to the academic lecture, with special emphasis on the education of such teachers . . . who are capable of contributing something not merely to the propagation but also to the expansion of science."[40] These institutions, combined with the growing use of laboratories and field exercises in scientific education, helped to establish the research orientation among the new generation of Prussian science professors.

As Jacobi's innovation at Königsberg demonstrates, the scholarly values and pedagogical institutions developed by the humanities profoundly influenced the later organization and development of the sciences. On the other hand, both sets of disciplines owed their rapid growth within the universities and the quick acceptance of their new, research-directed values to certain institutional factors intrinsic to the university as a whole. These broader factors have, however, received only limited study within university historiography.

2. WISSENSCHAFT AND KONKURRENZ

Not surprisingly, the explanations which German historians have advanced for the growth of research during the *Vormärz* period have corresponded to the stages through which the historiography of the nineteenth century German universities has passed. Before World War I German historians commonly interpreted university history as the institutional realization of—or departure from—the idea of the university laid down by Humboldt, Schleiermacher, and others.[41] Consequently they attributed the growth of research among the professors of the *Vormärz* era to the direct influence of Humboldtian

---

39. Lorey, pp. 71-80. Lorey traces the extension of the Königsberg methods throughout the university system.

40. Koch, 2.2, 839. The discussion is somewhat simplified, since a distinction must be made between the mathematics-physics seminars modeled directly upon the Königsberg seminar and Prussia's general science seminars. The latter retained some pedagogical emphasis and, unlike the former, did not flourish in subsequent decades.

41. Ringer, p. 82. These are the older "Mandarin" historians of whom Ringer speaks.

ideology. After World War I a new historical movement arose which applied sociological and economic analysis to the universities and their history; the work of Helmuth Plessner at Göttingen and of his students and associates is among the most distinguished of this school.[42] Although this approach has rarely dealt with problems of scholarship such as the rise of research, it has directed attention to the role of competition and decentralization among the German universities in stimulating academics to intensive research and specialization. The influence of ideology and competition—*Wissenschaft* and *Konkurrenz*—constitute the two major theories advanced to explain the growth of professorial research.

The influence of ideology upon the subsequent development of professorial scholarship arose primarily from the insistence of Humboldt, Fichte, Schelling, and others that creativity and originality must play a greater role in the universities. As a central part of their new concept of the university, they urged that the creation of knowledge as well as its transmission must be considered the duty of the university.[43] Wilhelm von Humboldt, chief architect of the Prussian reforms, wrote that

> because these institutions can achieve their purpose only if each confronts, insofar as possible, the pure idea of learning, so are solitude and freedom the principles predominating in their circle. But because human intellectual activity thrives only through cooperation, so the inner organization of these institutions must produce and sustain an uninterrupted and self-invigorating, yet unforced and unpredetermined cohesiveness, as well as a similarly unpredetermined cooperation.
>
> It is furthermore a characteristic of these institutions of higher learning that they treat learning [*Wissenschaft*] as a problem ever unsolved, and that they therefore are continually carrying on research. . . .[44]

---

42. Especially Helmuth Plessner, "Zur Soziologie der modernen Forschung und ihrer Organization in der deutschen Universität—Tradition und Ideologie," *Diesseits der Utopie* (Hamburg, 1966), pp. 121-143; Christian von Ferber, *Die Entwicklung des Lehrkörpers der deutschen Universitäten und Hochschulen, 1864-1954*, in *Untersuchungen zur Lage der deutschen Hochschullehrer*, ed. Helmuth Plessner (Göttingen, 1956), vol. 3; and Busch, *Privatdozenten*.

43. Schelling, *On University Studies*, pp. 26-28; Fichte, *The Vocation of the Scholar* (1794), p. 55; Fichte, *On the Nature of the Scholar and Its Manifestations* (Erlangen, 1805), trans. William Smith (London, 1845), p. 208.

44. Wilhelm von Humboldt, "Über die innere und äussere Organisation der höheren wissenschaftlichen Anstalten in Berlin (1810)," *Die Idee der deutschen Universität*, pp. 377-378.

This ideology emerged in the rhetoric and scholarship of Prussian scientists only after the humanists and philosophers had established it as the dominant ideology of the philosophical faculty. During the first few decades of the nineteenth century proponents of the new ideology often scorned the sciences, which still retained their eighteenth-century status as auxiliary studies with utilitarian emphasis. As humanists and philosophers came to dominate university affairs after 1810, they frequently challenged the sciences' status as *Wissenschaft*. They forced the sciences onto the defensive in a continuing struggle for prestige, students, and a voice in the counsels of the university.[45]

In the course of this struggle the scientists began increasingly to couch their defenses in the dominant terminology of *Wissenschaftsideologie*. They argued that the sciences, as well as the humanities, trained the intellect (*Geist*) and led to the refinement of the individual (*Bildung*).[46] Younger scientists emphasized "pure" science and rejected any utilitarian approach, which they, like their colleagues in the humanities, condemned as "bread-study" (*Brotstudium*). Whatever the secondary, material benefits of science, they argued, its cultivation was an end in itself. Carl Jacobi's letters to his brother bristle with defenses of that proposition. He reported to M. H. Jacobi in 1842, "I had the courage there [in Manchester] to assert that it is the honor of science [*Wissenschaft*] to be of no use, which provoked a powerful shaking of heads."[47] Speaking to traditionalists and humanists who scorned science's utilitarian connections and

---

45. General von Müffling wrote: "Ich habe bei der Gelegenheit recht kennen lernen, dass unsere deutschen Philologen eben so intolerant, wie die Jesuiten, sind, und dass eine wahre Verbrüderung Statt findet, die Mathematiker nicht aufkommen zu lassen." This and similar anecdotes upon this theme are collected in Luise Neumann, *Franz Neumann, Erinnerungsblätter von seiner Tochter* (Leipzig, 1904), pp. 242-243. At Berlin the conflict between the humanists and scientists was centered in the Academy of Science. See Adolf Harnack, *Geschichte der königlich preussischen Akademie der Wissenschaften zu Berlin*, 3 vols. (Berlin, 1900), *1*, 680-712.

46. Among many possible examples see G. J. Mulder, *Ueber den Werth und die Bedeutung der Naturwissenschaften für die Medicin*, trans. Jacob Moleschott (Heidelberg, 1844), pp. 31-32; Philipp Phoebus, *Ueber die Naturwissenschaften als Gegenstand des Studiums, des Unterrichts und der Prüfung angehender Aerzte* (Nordhausen, 1849), p. 2; and J. T. C. Ratzeburg, *Die Naturwissenschaften als Gegenstand des Unterrichts . . .* (Berlin, 1849), pp. 6-8.

47. Jacobi, *Briefwechsel*, p. 90; also p. 115.

impugned its place in the university, Franz Neumann answered in 1844: "True, its heritage is not old; true, it was born in the modern state, in the service of those arts and trades which attend only to the requirements and conveniences of the external life; that it does not deny. But through great, unceasing rigors it has emancipated itself and made itself a free man, has created for itself a realm in which reigns only the free force of intellect, and where independent thought and research alone obtain."[48]

The new scientists adopted the belief that independent research served society not only by adding to the sum of learning but also by contributing to the moral and ethical development of the individual carrying on that research. Like their humanist colleagues before them, they employed that faith not only to justify intensive research activity in itself but also to elevate research to a pedagogical tool, an activity to be demanded even of men destined for strictly practical careers. Such training, du Bois-Reymond argued, offers a benefit "which accrues even to the mediocre mind, that, at least once in his life before the overwhelming attraction of practical studies seizes him, he has been compelled to one step over the threshold of pure learning and has felt the breath of its spirit; that at least once he has seen the truth sought, found, and cherished for its own sake."[49]

The role of *Wissenschaftsideologie* within the sciences reflected its role in encouraging intensive professorial research throughout the university. It promoted a lofty, idealistic concept of the universities and also that absolute devotion to learning stereotypical of the German scholar ever since.[50] It lent historical and moral sanction to research activity and justified its employment as a pedagogical tool. Historians of the universities, intensely aware of this role and permeated by the Humboldtian categories, have gone further to maintain that this ideology transformed the approach to knowledge

48. Luise Neumann, p. 360.
49. Emil Heinrich du Bois-Reymond, *Reden,* ed. Estelle du Bois-Reymond (Leipzig, 1912), *1,* 356. A humanistic expression of this faith in *Wissenschaft* and research as the proper foundation for a practical career is found in Friedrich Bülau, "Wirken und Schicksale der deutschen Universitäten . . . ," *Akademische Monatschrift, 2* (1850), 15-16.
50. Fichte called the university "the most holy thing which the human race possesses . . . ; the University is the visible representation of the immortality of our race. . . ." *(Concerning the only Possible Disturbance of Academic Freedom. Rector address at Berlin, Oct. 19, 1811,* in Turnbull, pp. 262-265.)

within the universities. Friedrich Paulsen in his *Die deutschen Universitäten und das Universitätsstudium* wrote:

> The 19th century first introduced the requirement of independent learned research: only he is effective as a teacher of science [*Wissenschaft*] who is himself actively productive in science. . . .
>
> It was the era of the highest intellectual productivity that the German people have ever known, the era of Kant and Goethe, which found the courage to advance this view. Fichte and Schleiermacher first enunciated it decisively in their treatises written upon the occasion of the founding of Berlin University: upon whomever wishes to enter upon a scholarly career is the demand to be placed that he not merely have learned the knowledge at hand, but rather that he also be capable of producing knowledge out of his own independent activity. . . .
>
> Under the dominion of this idea the German universities of the nineteenth century have developed into what they are today: the workshops and the forges of the intellectual life of our people.[51]

The rise of Humboldtian ideology obviously influenced profoundly not only the development of the universities, but also German culture as a whole. Its importance notwithstanding, however, the new ideology alone does not fully explain the growing stress upon professorial research during the *Vormärz* period. The genius of Germany's academic system did not lie in its ideology of research and originality alone. In its universities Germany created an institutional system in which the extrinsic rewards of honor, salary, and career reinforced ideological precepts. Several factors—the criteria of academic appointments, the seminars and laboratories, the rewards accorded to successful professors—gave institutional expression to the philosophy of *Wissenschaft* and creativity. The ideological explanation in its traditional form says little about how such a system was established or how it functioned. Without an adequate account of these factors, no historical explanation can be entirely satisfactory.

Furthermore, factors other than ideological ones largely determined how Germany's ideological precepts manifested themselves in the work of academics. The actual development of the universities

---

51. Paulsen, *Universitäten,* pp. 204-205.

often contradicted rather than corresponded to the precepts and directives of the Humboldtian program. The founders of that program envisioned a type of academic originality and research which served the ends of synthesis, not analysis. Research was to employ philosophic insight in order to return to a postulated grand unity of knowledge, the vision of which had long been lost by fragmented, empirical learning.[52] In Humboldt's words the university was to strive, "first to derive everything from one original principle . . . , further to mold everything to one ideal . . . ," and "finally to unite this principle and that ideal into one idea."[53]

Although it never rejected this scholarly ideal in theory, the tradition of research which actually developed in the universities stressed the elaboration of critical and analytic methods and resulted in the endless fragmentation of learning into disconnected specialties and subfields. German academics remained intensely aware of the contradiction between theory and practice, for well before midcentury it had become embarrassingly obvious that their ideology as practiced was leading further and further from the unity of knowledge which it postulated. By 1845 purists moaned that "the universities have disintegrated into as many institutes as there are disciplines taught, whose [mutual] ties are only of the most superficial kind. They are no longer universities, but only aggregates of special institutes. From this results the one-sidedness of the education which they afford to academic youth."[54] While regretting the fragmentation of learning, August Boeckh in 1850 noted how beneficial this process had been to scholarship: "This division and splintering has incontestably taken a decisive upperhand in our age, in which the celebrated principle of the division of labor has come into widespread currency in science. This has given rise to a mass, indeed we could say a flood, of monographic treatises, to be acquainted with all of which is difficult, but which certainly has contributed very much to the broadening of our knowledge."[55] The conflict between the historical ideal of philosophical synthesis and the historical reality of methodological analy-

52. Schelling, *On University Studies*, pp. 17-32.
53. Wilhelm von Humboldt, "Über die innere und äussere Organisation . . . ," *Idee*, p. 379.
54. Luise Neumann, pp. 360-361.
55. *Gesammelte kleine Schriften*, 2, 190-191.

sis continued to trouble academic theorists throughout the century.[56]

Unlike the explanation based upon the influence of *Wissenschaftsideologie,* the institutional factors of competition and decentralization among the universities have never been advanced within a detailed historical context to explain the emergence of research. They do, however, underlie several recent historical and sociological studies of the universities and constitute an important, suggestive hypothesis for explaining the development of these institutions.[57]

In the sciences during the *Vormärz* era, competition between universities served mainly to propagate new research techniques within existing disciplines. After 1840, however, the same pressures initiated an accelerating process of subject fission as the universities elevated expanding subfields to full disciplines entitled to full professors. Joseph Ben-David has examined this process and its limitations in some detail, and his work has been convincingly confirmed by Awraham Zloczower's study of careers in physiology in the universities.[58] They have described how competitive pressures encouraged young academics to enter new, specialized areas and to develop them rapidly by means of intensive research. These academics then stood first in line for a full professorship in the eventual recognition of the subfield as a full discipline.[59] Through this process the competitive

---

56. Eduard Spranger, *Wandlungen im Wesen der Universität seit 100 Jahren* (Leipzig, 1913), pp. 23-24; Max Weber, "Wissenschaft als Beruf (1919)," *Gesammelte Aufsätze zur Wissenschaftslehre,* ed. Johannes Winckelmann, 3rd ed. (Tübingen, 1968), pp. 584-585, 589.

57. Warren O. Hagstrom, *The Scientific Community* (New York, 1965), pp. 36-40; Joseph Ben-David, "Scientific Productivity and Academic Organization in Nineteenth Century Medicine," *American Sociological Review,* 25 (1960), 828-843; Joseph Ben-David, "Scientific Growth: A Sociological View," *Minerva,* 2 (1963), 455-476.

58. Joseph Ben-David and Awraham Zloczower, "Universities and Academic Systems in Modern Societies," *European Journal of Sociology,* 3 (1962), 45-84; Awraham Zloczower, *Career Opportunities and the Growth of Scientific Discovery in 19th Century Germany with Special Reference to Physiology,* an unpublished master's thesis from the Hebrew University of Jerusalem.

59. The Ben-David-Zloczower thesis is concerned less with the origins of this dynamic of subject-fission than with its limitations. The thesis argues that each newly created discipline experienced a decline in vigor and scientific creativity as its limited number of chairs was filled and the front of specialization moved on to other subfields. Consequently the incentives which the universities could offer young scientists and the creativity of German science as a whole became heavily dependent upon the process of subject fission. That process slowed down after 1880, the thesis argues, largely because the new institute system tended to "bureaucratize" science instruction and research.

university system not only encouraged but obliged ambitious young academics to devote themselves to the search for knowledge. Competition and decentralization accelerated the growth of research among professors during the *Vormärz* period; yet these factors, even in conjunction with the influence of Humboldtian ideology, do not fully explain that development. Far more important than the heightened intensity of competition during the *Vormärz* period was a notable shift from eighteenth- to nineteenth-century standards of competition. Gerlach Adolf von Münchhausen, chief founder of Göttingen University, identified the eighteenth-century standards: in establishing the university, he wrote in 1733, "it appears that the two most important considerations are: 1) to choose capable people who are able to attract a great number of students and 2) to persuade them to accept nominations at Göttingen. . . . It is important above all else that the juridical faculty be staffed with famous, excellent men, for this must lead to many rich, well-born people studying in Göttingen."[60] Münchhausen's precepts, as well as other lengthy, eighteenth-century studies on the management of universities, suggest the prevailing mercantilistic and pedagogical standards of competition.[61] Universities competed over the numbers, wealth, and social position of their students; over the efficiency of the bureaucrats which they trained for the state; over the skill of their teachers; and over the reputation of their faculties, not in the eyes of a small community of specialists but of the general educated public.

After the reform period, as Prussia gradually retreated from her mercantilistic policy toward the universities, more esoteric or scientific criteria began to replace the older standards of competition. The university's fame as a center of learning and basic research came more and more to outweigh its fame as a pedagogical center or as a professional school. The new intensity of competition between universities accelerated the victory of these new competitive standards, but it did not initiate them or the new academic values on which they were founded.

60. "Nachträgliches Votum Münchhausens . . . ," *Die Gründung der Universität Göttingen,* ed. Emil F. Rössler (Göttingen, 1855), pp. 33-34.
61. See Anon. [J. D. Michaelis], *Raisonnement über die protestantschen Universitäten in Deutschland,* 4 vols. (Frankfurt, 1768), *1,* 1-97.

The new ideology and the new competitiveness within the universities went far to transform the professorate into a position based upon critical scholarship and individual research. But there is another major factor that influenced the transformation. It is the mechanism by which the new competitive-ideological values of research and specialization so quickly became the basis of appointments and promotions within the academic system.

3. THE APPOINTMENT OF PROFESSORS IN PRUSSIA

In any established university system professorial duties and values are defined and sustained largely through the criteria imposed upon young academics seeking appointment or promotion. In considering the important changes which occurred in professorial duties and values during the *Vormärz* period, it will, therefore, be useful to study changes in the procedures for professorial appointments and, in turn, the academic qualities which these procedures rewarded. Before examining Prussian university appointments, however, a consideration of the modern professorial post will provide a new framework for investigating the historical situation.

In the modern German university or in any university system significantly influenced by the German model, the professorship is characterized by its peculiar "dualistic" nature. The professor is a man of two loyalties, each with its corresponding activities and academic values. One loyalty looks toward the institution, and the values corresponding to it are pedagogical and collegiate. They idealize the professor who is a stimulating teacher, who fits well socially and intellectually with his colleagues, and who identifies with his institution and accepts his share of its tasks. The second loyalty, however, looks to the profession and primarily values research and other contributions to the academic's specialized, discipline-community beyond the university. The discipline as a whole and the community in particular establish the academic values connected with this professorial role. These values center around the struggle for reputation and recognition within the discipline-community, a struggle central not only to science but to all established academic fields.

Never entirely disjoint, these two academic roles interact most closely in matters of salary, appointment, and promotion. The modern academic system expressly subordinates university-centered values to disciplinary criteria in assessing the individual's fitness for advancement. Although the professor ostensibly is paid for teaching and other university-centered functions, in practice the prestige which he holds (or promises to attain) within his professional community usually determines his "success" within the academic world. Caplow and McGee in *The Academic Marketplace* describe this peculiar arrangement:

> For most members of the profession, the real strain in the academic role arises from the fact that they are, in essence, paid to do one job, whereas the worth of their services is evaluated on the basis of how well they do another. . . . Most professors contract to perform teaching services for their universities and are hired to perform those services. When they are evaluated, however, either as candidates for a vacant position, or as candidates for promotion, the evaluation is made principally in terms of their research contribution to their discipline.[62]

One oversimplifies the dual role of the professor, however, in considering it to be merely the conflict of teaching and research which Caplow and McGee imply. It arises rather from conflicting loyalties, to institution and to discipline, and from the diverse, often incompatible academic values they entail. Nor is this description of American academia to be facilely generalized to other university systems; the modern German universities, for example, emphasize disciplinary criteria almost to the exclusion of collegiate ones. It does, however, describe that universal characteristic of the modern professorate which underlies its unparalleled efficiency as an instrument of science and scholarship: appointive procedures which subordinate institutional to disciplinary values. Through these procedures the university career provides material incentive to disciplinary attainment, just as through its teaching function it provides for the continuity of the discipline as well.

The historical development of the professor's post in Prussia can be considered in terms of this dual role characteristic of the modern professorship. Since appointive procedures which stress research and

---

62. Theodore Caplow and Reece J. McGee, *The Academic Marketplace* (New York, 1958), pp. 82-83.

disciplinary reputation sustain that dual role today, the historical origin of that dual role ought to be found at the point in time when disciplinary criteria begin to supplant collegiate ones in determining academic promotions. This section examines how professors were appointed in eighteenth-century Prussia and how the focus of authority for such appointments changed during the *Vormärz* period. The final section shows how the criteria imposed upon academics in the selection of professors were largely dictated by the group which controlled appointments. It demonstrates how the changes in appointive authority brought about corresponding changes in appointive criteria and how these ultimately affected the professorate in Prussia.

During the eighteenth century there existed no clear distribution of authority in filling vacancies within the German universities. By 1700 the territorial princes had generally usurped the universities' ancient corporate privilege of self-recruitment, though a few institutions and individual faculties retained that right throughout the eighteenth century. Many universities enjoyed the *Vorschlagsrecht,* the right to propose a list of candidates for a vacant post from which the prince or his delegate might make the final choice. In other cases the state might fail to consult the universities in any formal way at all. But even though the territorial princes imposed their favorites upon the eighteenth-century universities frequently and imperiously, on the whole the universities succeeded either in filling their professorial ranks with candidates of their own choosing or in implicitly defining those criteria which the appointees of others were obliged to satisfy.[63]

The Prussian universities enjoyed somewhat less autonomy than other German institutions, and in particular they possessed no statutory right to nominate candidates; nevertheless, they retained considerable influence over appointments into their professorial ranks. Conrad Bornhak's study of the Prussian university administration before 1810 cites numerous cases preserved in ministerial records in which Prussian universities were called upon to propose candidates for vacant chairs. In the case of Königsberg University, which Berlin administered indirectly through the provincial government, the

63. Paulsen, *Universitäten,* pp. 101-106.

state left appointments almost entirely in local hands.[64] Bornhak concludes that "the participation of the university in the filling of vacant chairs was in no way extinguished and can be demonstrated during the whole century."[65]

The full professors of the local faculties maintained their influence over appointments and other local affairs partly by default of the state. Before 1810 the Berlin ministry never employed its power of appointment in a consistent policy to shape the universities, and it lacked the bureaucratic surveillance and control over the universities necessary to enforce such a policy had it existed. Throughout the eighteenth century, state power in the local university was vested in an official called the *Direktor*. This director, through whom the authorities maintained surveillance over local conditions and saw that orders from Berlin were carried out, was invariably a full professor of the faculty.[66] As such he possessed certain vested interests in university affairs and rarely scrupled to use his state influence to protect them. By often subordinating state power to one of several quarreling factions among the faculty, the director system only imbittered faculty feuds and acerbated intrigues for vacant posts.[67]

Legal authority for appointments lay with the administrative heads of the universities called *Kuratoren*, men usually appointed by the king from the ranks of prominent Berlin bureaucrats. In their Berlin posts these men possessed only two sources of information about conditions in the local universities: correspondence with the director and other favorites and large-scale inspection tours or visitations. Both proved ineffectual; the curators remained generally out of touch with the local situation and intervened only occasionally in university affairs. The uniting of the various curators into a single council called the *Oberkuratorium* in 1747 remedied this situation somewhat. But although the *Oberkuratorium* functioned ostensibly as a centralized administrative body, in practice it was merely a subordinate council deep in the bureaucracy of the Justice Depart-

---

64. Von Selle, *Königsberg*, pp. 158-161; Bornhak, p. 180.
65. Bornhak, pp. 99-100.
66. Bornhak, pp. 176-178.
67. J. J. Moser's brief directorate at Frankfurt-an-der-Oder from 1736 to 1739 offered the most scandalous example of this effect. A prejudiced account is Johann Jacob Moser, *Lebensgeschichte* (1768), pp. 60-83.

ment. The real interests and responsibilities of its members lay elsewhere, a condition which contributed to the lack of initiative characteristic of Frederician university policy.[68]

These inadequacies of the bureaucratic structure and of the personnel who staffed it vitiated Prussia's occasional attempts to regulate its universities. The curators rarely exercised personal initiative or judgment in selecting professors. In many appointments free of theological or legal controversy they simply acquiesced to the wishes of the local faculty or its dominant group, which then imposed its own appointive criteria upon candidates. When the curator did not acquiesce, the appointment was thrown open to intrigue, lobbying, and favoritism. Under these conditions no consistent criteria were employed.[69] By 1805 these administrative inadequacies in all areas of university affairs had become patently obvious and had evoked demands for reform. Ludwig Heinrich Jacob wrote in 1798:

> The top university administrators know too little about the condition of the institutions they are to govern. They are usually ministers or *Geheimräte* who have had little opportunity in their lives to acquaint themselves with the universities in detail, men who lack a general overview of the sciences, their systematic interconnections, and a correct judgment about the importance of each individual part. They feel no special need to acquire such knowledge at their age. Usually the supervision of the universities is only a secondary responsibility *(Nebenfach)*; another department where his colleagues can better judge him, where he has worked out a better routine and where his influence is greater interests and occupies the minister more. Hence he worries little about

---

68. Bornhak, pp. 179-188. Freiherr von Zedlitz' leadership of the *Oberkuratorium* proved a partial exception to this judgement, for Zedlitz exercised vigor and imagination both in controlling the universities and in appointing new professors. Most of his activities, however, were devoted to the secondary schools. See Friedrich Adolf Trendelenburg, "Friedrich der Grosse und sein Staatsminister Freiherr von Zedlitz," *Kleine Schriften Trendelenburgs* (Leipzig, 1871), *1,* 127-158.

69. University factions lobbied through the provincial assemblies in which they were represented; the church authorities, through the *Oberkonsistorium*. Under Frederick William I the theological faculty at Halle influenced most appointments of theological and philosophical faculty professors throughout the state, as did the prince's personal physician Eller for the medical faculties. There were even cases in which the size of a scholar's voluntary contribution to the *Rekrutenkasse* exercised a definite influence upon his subsequent career. (Bornhak, pp. 101-105.)

the universities and has no knowledge of the local situation. From this circumstance results the fact that the regulations and orders which are sent out from the residence to the academic senate are seldom carried out. Counterproposals are made, difficulties on the side of the universities pile up, and in the end everything usually remains the same.[70]

In response to these criticisms and to its growing commitment to the universities, the Prussian government moved during the reform era to create a new bureaucratic structure through which to administer its universities. The reforms had brought successive reorganizations of the entire bureaucracy, and in each reorganization university administration received higher rank within the bureaucratic hierarchy. Finally in 1817 von Hardenberg accorded cabinet status to educational affairs as part of the newly created *Ministerium der geistlichen, Unterrichts-, und Medizinalangelegenheiten*, commonly called the *Kultusministerium*. The new minister, Karl Freiherr vom Stein zum Altenstein, assumed direct control of university affairs, thus centralizing university administration under one responsible official at the highest level of the Berlin government.[71]

"Reform" on the local level occurred as a result of the political reaction which marked the close of the Humboldtian era in Prussian education. A royal decree of 18 November 1819 established at each Prussian university a salaried official to censor publications and lectures, to prevent student associations, and to persecute liberal sentiment in accordance with the Karlsbad Decrees. Although conceived for a purpose antithetical to the principles of the earlier reforms, this bureaucratic innovation actually continued their thrust by vastly extending the state's administrative authority over the universities. The powers of these *Regierungsbevollmächtigte* went far beyond their original political purposes, and they replaced the curators and directors as the supreme representatives of the state in the local universities. Intensely resented by the faculties, these officials nevertheless played crucial parts in the development of the

---

70. Ludwig Heinrich Jacob, *Ueber die Universitäten in Deutschland, besonders in den königl. preussischen Staaten* (Berlin, 1798), pp. 22-24.
71. See Müsebeck, *Das preussische Kultusministerium*, passim. Also see Karl-Heinz Manegold, "Das 'Ministerium des Geistes'; Zur Organisation des ehemaligen preussischen Kultusministeriums," *Die deutsche Berufs- und Fachschule, 63* (July, 1967), 512-524.

universities.⁷² Responsible only to the Minister of Education in Berlin, they constituted an important part of the powerful and efficient bureaucratic mechanism created during the reform period.

Prussia immediately began to utilize this new bureaucratic machinery in an explicit new policy aimed at reforming and molding the universities through aggressive, centralized control of all professorial appointments. The ministry in Berlin began to play a much more active role in evaluating and selecting professorial candidates than it had done in previous decades, and with growing frequency it refused to consult the local corporate faculties either directly or indirectly. As a result of this policy the authority to make or to influence appointments shifted gradually away from the corporate body of professors and the various lobbying groups toward the top figures of the Berlin ministry.

The actual origins of this policy lay in the various treatises on university reform written during the eighteenth century, all of which had insisted that the Prussian government, in practice as well as in theory, exercise a closer and more consistent control over university appointments. Only later during Wilhelm von Humboldt's fourteen-month leadership of Prussia's Department of Educational Affairs from February 1809 to June 1810 was this policy officially formulated. Despite the theories of academic freedom and cultural liberalism which he espoused, Humboldt centralized and extended state power over all aspects of Prussia's educational affairs. For the universities he insisted upon strict government control of appointments and government maintenance of competition.

> The appointment of university professors must be reserved exclusively to the state, and it is certainly no good arrangement to allow the faculties any more influence in these matters than a reasonable, fair curatorial council will do of its own accord. For in the university antagonism and friction are wholesome and necessary, but the collisions that naturally arise between instructors in the course of their affairs can unintentionally distort their point of view. Furthermore the condi-

---

72. For the role of *Regierungsbevollmächtigte* at Berlin see Lenz, *Berlin, 2.1*, 101-116; at Bonn, Friedrich von Bezold, *Geschichte der rheinischen Friedrich-Wilhelms-Universität* (Bonn, 1920), pp. 127-137.

tion of the universities is too closely bound up with the direct interests of the state.[73]

In the founding of Berlin University, Humboldt personally led the recruitment of professors for the new institution, thus setting a precedent for his followers.

Although his successor, Friedrich von Schuckmann, upheld Humboldt's insistence upon government control of all appointments, only under the long rule of Stein zum Altenstein over the Ministry of Education did professorial appointments become a virtual monopoly of the minister. Max Lenz, considering Altenstein's relationship with Berlin University, wrote:

> The faculties had hardly anything to say about appointments; only very few instructors were appointed with their collaboration. Most were procured directly by the minister. . . . This practice corresponded to Humboldt's precepts, who, however, usually consulted with professors whom he knew well, like Schleiermacher, Savigny, and Reil. Schuckmann followed Humboldt in this respect. . . . Under Altenstein such consultations took place no more. Among the doctors he soon gained Rust as an acceptable adviser, and among the theologians he called upon Eylert or Strauss—all men who stood in direct official connection to him. Consultation with the philosophical faculty, on the other hand, was almost forgotten by the ministry after 1820. The minister alone became the source of all grace; whoever desired advancement was forced to turn to him.[74]

The Prussian bureaucracy had many reasons for seizing control of university appointments in the early nineteenth century. It had always possessed the legal authority for such control, and after 1806 the administrative centralization made such control practically feasible as well. To exercise it seemed to be a natural extension of the bureaucratizing thrust of the whole reform movement. Perhaps more important, the Prussian bureaucracy emerged from the eighteenth century with a profound distrust of the corporate

---

73. Humboldt, "Über die innere und äussere Organisation . . . ," *Idee,* p. 385. Numerous historians have written on the apparent paradox in Humboldt's liberal philosophy and his authoritarian approach to educational affairs. See René König, *Vom Wesen der deutschen Universität* (Berlin, 1935), pp. 162-166, 171-177; Spranger, *Reform des Bildungswesen,* pp. 86 ff., 107 ff., Lenz, *Berlin, 1,* 195.
74. Lenz, *Berlin, 2.1,* 407.

faculties. The eighteenth-century universities abounded in minor scandals: professorial laziness, nepotism, academic monopolies, interminable faculty feuds, and teachers who winked at student violence and debauchery. The new bureaucrats who came to the fore during the reform era blamed the abuses on the universities' corporate intransigence. They were determined, as Humboldt indicated, to root out such abuses by controlling university affairs and appointments and by allowing the corporate faculties a minimal role at best.

Practiced in all the universities across Prussia, this policy of rigid ministerial control produced some unfortunate results. The utter dependence of academics upon the ministry often encouraged subservience and flattery. Also the ministry never hesitated to use its power in order to further controversial intellectual movements which enjoyed its support. Hegelianism became a virtual state philosophy in Prussia before 1830, and the ministry guaranteed its dominance by assuring Hegel's students a near monopoly over chairs of philosophy in Prussia.[75] On the whole, however, Altenstein's appointment policy proved extremely beneficial to scholarship within the universities. Particularly among the emerging philological, historical, and scientific disciplines of the philosophical faculty, Altenstein's ministry brought some of the most noted scholars of the century into Prussian chairs.

The ministry attained its monopoly over appointments not through any new legal powers but rather through the vigor of its personnel and the efficiency of the bureaucracy. Ostensibly the *Vorschlagsrecht* of both Berlin and Bonn, legally recognized in Prussia through the respective statutes of these new institutions, provided these universities a voice in appointments by guaranteeing their right to propose candidates. In practice, however, the ministry still frequently failed to consult the universities about vacancies; or, when it did receive their nominations, it often disregarded them and appointed its own candidate. After the death of Altenstein in 1840 the Bonn faculty angrily protested to the new minister Eichhorn "that many things would have developed differently or more favorably at our university if, in earlier times, the faculties' right of

75. Varrentrapp provides a lengthy discussion of the ministry's relationship to Hegel, pp. 433-443.

nomination, guaranteed by the statutes, had been less frequently ignored."[76] Eichhorn, however, quickly adopted a policy of appointments more imperious than Altenstein's, especially in the particularly sensitive theological and juridical faculties.[77] The controversies which Eichhorn's appointments provoked raged until the Revolution of 1848, in which recognition of their *Vorschlagsrecht* constituted a major university demand.[78]

4. THE NEW PROFESSORS AND THE OLD

Both before and after the Prussian reforms many factors helped to determine the academic criteria imposed upon professorial candidates, but the direct needs and interests of the group controlling these appointments played the major role. During the eighteenth century the Prussian universities themselves defined the general criteria used for appointments, either through direct control of promotions or through their influence and advice in these matters. Not surprisingly, they insisted upon qualities in academics directly beneficial to their local corporate interests; unlike modern appointive procedures, eighteenth-century ones valued institutional over disciplinary criteria in assessing potential academics.

Critics and reformers of the eighteenth-century universities—and there were many—have left thorough, if occasionally biased, descriptions of these procedures and their results. Although generally admitting that faculty-controlled appointments promoted solidarity within the university corporation, reformers invariably condemned these appointments as damaging to scholarship. Critics insisted that they encouraged professorial monopolies and restricted the healthy competition necessary to vigorous intellectual life. J. C. Hoffbauer, in writing of the Prussian universities in 1800, urged that

> every instructor ought to enjoy the fullest independence from every other. . . . In my opinion all relationships which make an instructor dependent on the interests of others in any manner must be banned. . . . I know of cases in which younger instructors have oriented their choice

76. Quoted in Bezold, *Bonn,* p. 295.
77. Lenz, *Berlin, 2.2,* 12-20.
78. Lenz, *Berlin, 2.2,* 163-164; Karl Griewank, *Deutsche Studenten und Universitäten in der Revolution von 1848* (Weimar, 1949), pp. 52-53 ff.

of lectures, however unwillingly, in accordance with the wishes of their seniors in order not to displease them, because they hoped either for further advancement through their recommendation or for other sorts of advantages arising from their favor. . . . Everyone who seeks advancement in the university knows that it depends upon whether the faculty will recommend him or not. . . . Often everything hangs upon the will of one individual, to whose vote the other members of the faculty conform more than they should.[79]

Reformers observed that institutional concerns dominated appointive procedures. They argued that the criteria upon which professors evaluated candidates looked more to their social and corporate acceptability than to their skill in their particular discipline. Christoph Martin Wieland complained bitterly of the University of Erfurt that

all along the philosophical faculty, instead of concentrating at all times and to the best of its ability on the best possible choice, has let itself be led by completely false premises; it has notoriously concerned itself more with its relatives and personal friends, more with religious, fraternal, or collegiate relationships and the like in the selection of its new members than with true learned capability. Out of this practice has arisen not only a mass of quarrels, but also—understandably—the circumstance that it was only a fortunate coincidence when a really skillful man ever found his way to a teaching post.[80]

In addition to these university-centered criteria based upon "religious, fraternal, or collegiate relationships," the universities demanded candidates with sufficient learning in their field to teach successfully. Usually they hoped for the author of a treatise or text with some literary reputation. Beyond this, however, disciplinary criteria played little role in appointments. Already mature and well-defined discipline-communities had arisen in branches of science and in classical philology by 1790. But although a few Prussian professors worked actively in extending the sciences and participated in these young communities, no such activity had yet become incumbent upon the academic. While this activity was honored and

---

79. *Über die Perioden der Erziehung* (Leipzig, 1800), pp. 182-184. Hoffbauer was professor of philosophy at Halle.
80. Wilhelm Stieda, *Erfurter Universitätsreformpläne im 18. Jahrhundert* (Erfurt, 1934), p. 227.

occasionally rewarded, few men considered it an intrinsic part of the professor's duty and few argued that it could or ought to be required of all academics.[81] Indeed the full faculty, dominated by professors of theology and law, usually lacked the ability to judge candidates in specific fields upon disciplinary grounds, and was certainly unprepared to judge specialized research within these fields. A contemporary noted that

> because a professor does not teach all the sciences and consequently does not need to understand them, so-called scholars can be guilty of still greater misjudgments about professors. Let us assume, for example, that a university has only one professor of mathematics and that this chair is to be filled. Then among the men who will make the appointment there are no real professional mathematicians; what, then, makes their judgement particularly accurate in comparison with that of others? The same case can occur in many other fields. [Such circumstances promote] . . . only too often the most common personal considerations which in no way further learning.[82]

This emphasis upon institutional criteria by no means contradicted the university's yearning for famous professors. On the contrary reform treatises usually urged professors to publish in order to attain a literary reputation and to enhance the reputation and attractiveness of their universities. The Prussian government also made a few, largely ineffective attempts to require publication of its professors. J. J. Moser, while director of Frankfurt-an-Oder, reported to Berlin in 1737 that the local professors were unknown in the scholarly world because they published nothing. His report

---

81. Eighteenth-century ideas about the duties of the professor and the relationship of the university to the advancement of learning spanned a wide range of opinion, the disagreements having been caused partly by the criticism of the universities late in the century. Spokesmen for the universities at Göttingen, who led the counterattack against their critics, began after 1790 to recognize the general responsibility of the universities to expand learning whenever possible as well as to transmit it. See Ernst Brandes, *Ueber den gegenwärtigen Zustand der Universität Göttingen* (Göttingen, 1802), pp. 16-20, 159-161. In Prussia, on the other hand, the climate of opinion seemed to have favored just the opposite view: that the shortcomings of the universities arose from their neglect of pedagogy and that reform must be in the direction of a more *schulmässig* institution. See Jacob, *Ueber die Universitäten*, p. 254. The Prussian government shared this preference for pedagogy and a suspicion of professorial scholarship up to the beginning of the Humboldt ministry. See Bornhak, p. 147.

82. Wieland, quoted in Stieda, pp. 153-154.

resulted in a command to the collective faculty to begin writing. Similar reports came from Halle during the 1768 visitation of Geheimer Tribunalrat Steck, and the state in the same year commanded the Halle faculty to begin to publish in order to ensure the reputation of that institution.[83]

Such emphasis upon publication, however, must be distinguished from the nineteenth-century movement toward a "publish or perish" policy. Not only was the pressure to publish much less, but also the aim and nature of publication was quite different. Eighteenth-century publication aimed at establishing literary reputation within the eyes of the general learned public, not literary reputation within a small group of specialists gained through original contributions from one's research. The Prussian government made this quite clear to its professors. Frederick William I ordered the Halle professoriate to neglect esoteric treatises and to concentrate on more widely available works of practical interest to the common man.[84] Geheimer Tribunalrat Steck, who criticised the Halle professors in 1768 for their failure to publish, asserted in his visitation to Frankfurt-an-der-Oder in 1770 that the business of the universities was not discovery but the "service of the state and the enlightenment of the nation." As late as 1802 a decree of the government to Halle University argued that the purpose of the university was not the "Erweiterung der Wissenschaft," but rather teaching, which, the decree asserted, would lead indirectly to discovery.[85] In keeping with such a conception of professorial publication and with the often-cited encyclopedic and pedagogical tendencies of the age, professors devoted much of their efforts to textbooks, handbooks, and compendia. Unlike the nineteenth century, the eighteenth regarded such works as creative in their own right, as central to the professor's literary activity.[86] The eighteenth-century universities differed from the

83. Bornhak, p. 148.
84. Bornhak, p. 129.
85. Bornhak, p. 147.
86. Hoffbauer, for example, defended the proliferation of textbooks in Prussia. He argued that such literary work encouraged professors to organize their learning and offered the public a basis on which to judge them. Writing even a mediocre text, he claimed, demands more knowledge and diligence than the "anderweitige Behandlung eines wissenschaftlichen Gegenstandes, im mehrern Bänden. . . . Aus diesen Gründen wäre es vielleicht zu wünschen, dass jeder Docent nur über eigne Lehrbücher lese." (Hoffbauer, p. 180.)

nineteenth not so much in that they emphasized scholarship and publication less, but that they emphasized a very different kind of scholarship pursued for distinctly different reasons.

The eighteenth-century stress upon institutional criteria in appointments resulted largely from the fact that these criteria were defined primarily by the local universities. During the early nineteenth century, as the previous discussion showed, the state began to exercise more consistent control over the universities and their professorial appointments. By 1825 the minister of education had attained a virtual monopoly over such appointments, and he rarely deigned to consult the local corporate faculties. This policy began as part of a general program of administrative centralization and was intended to correct certain corporate abuses; the policy was certainly not addressed, at least initially, to the reform of scholarship.

The decisive shift in appointive power in the reform era brought about a corresponding shift in appointive criteria which profoundly affected Prussian scholarship and ultimately the nature of the professorial post. The officials of the Ministry of Education, Altenstein and his vigorous advisor Johannes Schulze, paid little attention to the fraternal, collegiate, and pedagogical values which eighteenth-century appointments had stressed. Indeed, they had no way of judging these factors, for the ministry usually knew its appointees only through their publications. Instead ministerial decisions laid greatest stress upon the originality and the depth of a candidate's scholarship and publications. The ministry demanded that the candidate possess or promise quickly to attain a reputation within the relevant discipline-community in Germany. It relied almost exclusively upon the specialists within those communities for advice about candidates and almost totally neglected the full faculty involved.

By deciding appointments in these ways, the ministry encouraged disciplinary criteria of value to the broader community of scientists over institutional criteria of value to the local community of professors. In place of literature addressed to the general learned public, the state promoted literature bearing the results of research and addressed to other contributors in the field. This insistence upon disciplinary criteria gradually established these as decisive in professorial appointments. The change established the dualistic professorate in Prussia and with it research as an integral part of the academic's life and work.

The gradual shift in appointive power toward the state naturally brought new academic criteria in its wake; various historical factors influenced the state's decision to emphasize rigid disciplinary criteria rather than pedagogical or utilitarian ones. On the one hand the chief figures of the Ministry of Education felt deep personal allegiance to the *Wissenschaftsideologie,* which unequivocally rejected pedagogical or utilitarian views of the professor's role. Although a retiring and indecisive man, Altenstein had nevertheless been a fervent admirer of Fichte, and he shared Fichte's desire for a new national education based upon creativity and philosophic synthesis.[87] In philosophical appointments he therefore favored idealists like Hegel and his disciples; in philology and history he favored men who had envinced their creativity in their judgment and their critical method. Schulze had been trained in F. A. Wolf's philological seminar, and he retained the neohellenist, rather elitist outlook of German philology with its emphasis upon rigor and critical method.[88] Wolf's techniques had been rapidly changing the nature and standards of Prussian classical scholarship since 1795, and Schulze used his office to further that scholarly movement in which he himself had been trained. The critical, analytic tendencies of the new philology clashed sharply with the philosophic program of a grand synthesis of learning. Nevertheless, the state sponsored both. After 1830 the critical outlook of the new philology largely replaced the philosophical tradition; before 1830 both coexisted in a fruitful if uneasy equilibrium.

In part the ministry insisted upon disciplinary distinction among its professors because the relationship of the universities to the state had changed. Before 1806 the universities served Prussia in two ways: through training bureaucrats and professional men and through their mercantilistic function of bringing money into the state. After 1806 the universities certainly continued to train bureau-

---

87. Varrentrapp, pp. 272-275; Lenz, *Berlin*, 2.1, 3-10; Schulze's personal assessment of Altenstein's personality and accomplishments is in Johannes Schulze, "Beiträge zur Geschichte des Ministeriums der Unterrichtsangelegenheiten von 1818 bis 1840 und zur Charakteristik des verewigten Ministers Freiherrn v. Altenstein." (Müsebeck, pp. 293-307.)

88. Varrentrapp, pp. 26-44, 71-79.

crats and professional men; indeed, Altenstein expected his new appointive policy to improve professional education in the higher faculties. The universities' mercantilistic function gradually disappeared, however, to be replaced by another, more fundamental one. The universities came to be regarded by the state as showplaces of Prussian intellect and the chief foci of German culture. As such they were to be groomed and maintained as national symbols. This new role for the universities can be dated almost precisely from the founding of Berlin University, when that institution became the phoenixlike symbol of Prussia's resistance to Napoleon. Prussians patriotically contrasted their universities to the French system of schools and academies which had rejected the university model. As national showplaces the universities required scholars of European reputation, not teachers or bureaucratic favorites. Altenstein and Schulze set out to recruit such scholars whenever possible.

In carrying out this policy Altenstein and Schulze employed a candidate's publications as the basis of his promotion. Professors submitted their new publications to the ministry for review, and these treatises received careful attention despite their increasingly esoteric natures.[89] Schulze defended this close attention to professors' publications, noting that these practices "have contributed not a little to establishing and maintaining a beneficial relationship between the minister and the authors in question. At home and abroad they have brought the ministry the reputation of honoring every learned endeavor in accordance with its merit, of recognizing and encouraging subordinate talent, and of knowing how to find the correct criteria for judging literary works."[90] Schulze used publications as the basis of an outspoken "publish or perish" policy. He would make no one a professor, he insisted, until "he has written a solid book, a work which one can display and reap honor from, a work one can stand on."[91]

Still another aspect of the ministry's direct, personal approach to

---

89. Altenstein's letters to Johannes Müller, for example, are intelligent and perceptive, and they reveal a close acquaintance both with Müller's publications and with the state of Prussian physiology. Manfred Stürzbecher, "Aus dem Briefwechsel des Physiologen Johannes Müller mit dem Preussischen Kultusministerium," *Janus, 49* (1960), 273-284.
90. Quoted in Varrentrapp, pp. 488-489.
91. Quoted in Varrentrapp, p. 488.

university administration was its pronounced tendency to favor with financial support and quick promotion young scholars whose early work had particularly impressed it. Although this favoritism led often to inequities, it did encourage many young men to enter and to pursue scientific and professorial careers. Among the scientists whose early labors the ministry financed were Eilhard Mitscherlich, Ludwig Ferdinand Moser, Georg Friedrich Pohl, Franz Neumann, Heinrich Dove, Enno Dirksen, C. G. J. Jacobi, Friedrich J. Richelot, Julius Plücker, Ludwig Adolph Sohnke, Peter Gustav Lejeune-Dirichlet, Jacob Steiner, and Martin Ohm.[92] Schulze's greatest find proved to be Johannes Müller, who had attended the Koblenz gymnasium from 1816 to 1818 while Schulze was director there, and who considered Schulze his "long time patron."[93]

Continuing Wilhelm von Humboldt's policy, the Altenstein ministry regarded an intelligent policy of appointments as "the first and most difficult task in the administration of the German university."[94] The ministry went beyond Humboldt, however, to employ its appointive power as a tool to shape the universities as it thought best. Although the ministry used this tool most effectively in the philological and historical disciplines, it also brought significant results in the sciences, medicine, and mathematics. In 1826, for example, the ministry moved to counteract the provincial isolation of Königsberg University and to encourage the sciences there. It took the unprecedented step of offering two hundred thaler each to three *Privatdozenten*—Franz Neumann, Heinrich Dove, and C. G. J. Jacobi—as incentive to go to Königsberg to teach.[95] The appointments established Königsberg as the center of Prussian science and mathematics throughout the *Vormärz* era.

Very often the appointments of the ministry met much opposition, especially when they opposed vested interests and aimed clearly at altering the scientific makeup of the university. In the 1830's the

---

92. Schulze himself provides this list in his "Beiträge zur Geschichte des Ministeriums . . . ," p. 302.
93. See the various letters of Müller to Schulze from 1830 to 1831 printed in Lenz, *Berlin, 4,* 529-531.
94. Schulze, in Varrentrapp, p. 486.
95. On the growth of science at Königsberg and the 1826 appointments see Hans Prutz, *Die königliche Albertus-Universität zu Königsberg i. Pr. im neunzehnten Jahrhundert* (Königsberg, 1894), pp. 152-173.

ministry set out to reform the Berlin medical faculty, which had become dominated by clinicians and practitioners.[96] The ministry desired to introduce a more scientific and theoretical approach to medicine, and it seized its first opportunity to do so after the death of the anatomist-physiologist Rudolphi. Although the medical faculty urged that no successor to Rudolphi's post be appointed and that his teaching duties (and his salary) be divided among the other full professors, the ministry, supported by the philosophical faculty, summoned Johannes Müller from Bonn in 1833 to fill the vacancy.[97] Müller's student Henle wrote enthusiastically of his call: "The high officials hope from him a violent shake-up in academic life, particularly in the study of medicine, formerly so indolent and mechanical here. His colleagues and especially his rivals feel his superiority and resign themselves to it, at least outwardly. His subordinates and the young instructors, who depend on him in part, cannot sufficiently praise his modesty, friendliness, and civility."[98] Because Altenstein had ignored the wishes of the medical faculty, Müller's appointment provoked indignation in Berlin, particularly among the numerous professors in the conservative party who held simultaneous, influential posts within the state medical administration.

Following his appointment Müller introduced a vigorous experimental activity in physiology and became the leader of the faculty's reform party which opposed the empirical, clinical approach to medicine. The final victory of the reform party occurred in 1839 when Müller and his followers compelled the faculty to nominate the clinical theoretician Johann Lucas Schönlein.[99] Although Altenstein hesitated to propose Schönlein to Frederick William III because of his suspect political background, Schulze persuaded both king and minister to call Schönlein to Berlin.[100] The reform of the Berlin medical faculty, accomplished largely through the controversial appointments of Müller and Schönlein, opened the way for the scientific work of Müller's students during the 1840's.

96. Lenz, *Berlin, 2.1,* 462-465.
97. Gottfried Koller, *Das Leben des Biologen Johannes Müller, 1801-1858* (Stuttgart, 1958), pp. 96-104.
98. Friedrich Merkel, *Jacob Henle. Ein deutsches Gelehrtenleben* (Braunschweig, 1891), pp. 102-103.
99. Lenz, *Berlin, 2.1,* 470-472.
100. Varrentrapp, pp. 469 ff.

As discipline-centered criteria replaced institution-centered criteria in university appointments, the various specialist-communities determined precisely what disciplinary standards would be imposed. Reluctant to consult the local faculties, the ministry looked more and more to the circle of specialists within a candidate's field for evaluation of his work and advice on the appointment. Such discipline communities had existed in the eighteenth century. Within their own circle they had developed well before 1800 new academic values which stressed intensive, specialized research and which existed alongside the older and more encyclopedic concept of scholarship prevailing in the universities. As these small communities became the chief advisers to the ministry, they were able, through the state, to impose their methods and standards of scholarship upon the entire university. By 1830 a candidate's reputation within his circle of specialists largely determined his academic success. In this way the extrinsic rewards of career and salary became harnessed to the intrinsic rewards of scholarly recognition and interaction, providing an additional powerful incentive to careers in scholarship and research.

Especially in mathematics and physics, areas in which they distrusted their personal competency, Altenstein and Schulze both looked to the respective scientific groups for advice.[101] Jacobi's promotion to professor at Königsberg suggests the growing dependence of scientists upon the scholarly community. Having been appointed to Königsberg as a salaried *Privatdozent* in 1826, Jacobi wrote to Altenstein on 27 October 1827, requesting promotion and announcing that since arriving in Königsberg he had "won through brilliant

---

101. During the *Vormärz* period the ministry gradually established regular channels of contact with the various scientific communities. One such channel was Alexander von Humboldt, who after his return from Paris in 1828 became the chief mediating link between the state and its scientists. See Kurt-R. Biermann, "Über die Forderung deutscher Mathematiker durch Alexander von Humboldt," *Alexander von Humboldt, Gedenkschrift zur 100. Wiederkehr seines Todestages* (Berlin, 1959), pp. 81-159; Gerhard Dunken, "Alexander von Humboldt und der Plan der Gründung einer höheren technischen Lehranstalt in Berlin," *Wissenschaftliche Zeitschrift der Humboldt-Universität zu Berlin. Mathematisch-Naturwissenschaftliche Reihe, 8* (1858/59), 131-133. Another important mediator between the ministry and the Prussian scientific communities in mathematics and physics was A. L. Crelle, editor of the *Journal für die reine und angewandte Mathematik*. See Kurt-R. Biermann, "Urteile A. L. Crelles über seine Autoren," *Journal für die reine und angewandte Mathematik, 203* (1959), 216-220.

discoveries a not insignificant reputation in the scholarly world, as Herr Hofrat Gauss in Göttingen and Herr Professor Bessel here will gladly attest to the ministry. . . ."[102] After making inquiries with these mathematicians, the ministry granted the promotion on 16 January 1828.

Jacobi's promotion also illustrates the particular interest of the ministry in a candidate's Parisian reputation. Jacobi wrote later that his promotion had been decided by the recognition of his work by Legendre in Paris.[103] Jacobi had been in contact with the French mathematician over his elliptical functions, and Legendre had praised his work before the Parisian Academy in November 1827. Reprinted in various German journals and newspapers, Legendre's remarks had come to the attention of the ministry and had ensured Jacobi's advancement.[104] Concerning the ministry's respect for a foreign reputation, Alexander von Humboldt remarked that "in Germany one impresses a few great personages only through the reflection of one's reputation abroad."[105] Humboldt himself became the ministry's principal tool in assessing the Parisian reputation of scientists. In this manner the ministry's appointment policy stimulated the importation of French scientific methods and theories into northern Germany.

Only after bitter, repeated conflicts all during the *Vormärz* period did disciplinary values gradually replace institutional concerns as the principal criteria in professorial appointments. Many of Altenstein's most famous and most controversial appointments resulted in the clash of disciplinary and collegiate issues in the local universities. Shortly after Jacobi's arrival at Königsberg, Bessel had written to Gauss that "he has made almost all here his enemy because upon arriving he said something disagreeable to each one: he assured the born Königsbergers that he considered his present location an exile; to the philosophers he praised Hegel, to the philologists, Boeckh, everything in a manner for which no one will

---

102. Leo Koenigsberger, *Carl Gustav Jacob Jacobi* (Leipzig, 1904), pp. 55-56.
103. Koenigsberger, *Jacobi,* p. 57.
104. Koenigsberger, *Jacobi,* p. 44; Jacobi, *Briefwechsel,* pp. 5-6.
105. Kurt-R. Biermann, "Alexander von Humboldts wissenschaftsorganisatorisches Programm bei der Übersiedlung nach Berlin," *Monatsberichte der deutschen Akademie der Wissenschaften zu Berlin. Mitteilungen aus Mathematik, Naturwissenschaft, Medizin, und Technik,* 10 (1968), 144.

pardon him."¹⁰⁶ Consequently when Altenstein called upon the Königsberg faculty to advise upon Jacobi's promotion, it freely admitted the distinction of his mathematical achievements, yet recommended that the promotion be denied because of the "unsuitable manner in which he expressed himself concerning the university and its teachers." The curator also noted Jacobi's unpopularity resulting from his refusal to associate with any instructors except Bessel, Lobeck, and the younger *Privatdozenten*.¹⁰⁷

This clash between Jacobi and the Königsberg faculty had no basis in scientific issues, although by influencing Jacobi's future its outcome might well have affected the development of mathematics in Prussia. Collegiate, not disciplinary, issues underlay the hostility, for whatever the validity of its sentiments, the faculty considered Jacobi personally unsatisfactory as a colleague with whom to live and work. Had the ministry valued collegiate criteria, or had the faculty's influence been as strong as it had been during the previous century, Jacobi would certainly have remained a *Privatdozent*. In reality such concerns, however important to the faculty, meant nothing to Altenstein when compared to Jacobi's burgeoning reputation abroad, and he quickly promoted the irascible mathematician. Such victories of disciplinary over collegiate criteria gradually established the former as the undisputed bases of professorial appointments.

The same confrontation of values can be seen in another of Altenstein's famous appointments made, as was frequently the case, against the wishes of the faculty involved. Around 1820 the young physiologist Evangelista Purkinje attracted the notice of Goethe, Alexander von Humboldt, and Schulze for his treatise, *Beiträge zur Kenntniss des Sehens in subjectiver Hinsicht*. In 1821 the chair of physiology at Breslau became vacant, and the faculty nominated Gruithuisen for the post with the recommendation of the curator Neumann. The ministry, however, ignored the nomination and appointed the little-known Purkinje.¹⁰⁸

Purkinje's subsequent career at Breslau was marked by continued

---

106. Koenigsberger, *Jacobi*, p. 27.
107. Koenigsberger, *Jacobi*, p. 56.
108. "Johannes Evangelista P. Purkinje," *Allgemeine Deutsche Biographie* (Leipzig, 1888), *26*, 717-718.

state support for the purchase of instruments, the prosecution of his researches into microscopic anatomy, and finally the founding of the Breslau physiological institute. Purkinje's relationship with the local faculty, however, began quite poorly. As in Jacobi's case, Purkinje's difficulties with the faculty reflected no confrontation over scientific issues. His colleagues generally applauded and supported his personal research as well as his pioneering efforts to teach physiology through experiments and demonstrations.[109] They objected, however, to Purkinje's inadequacy as a lecturer. The anatomist Otto complained in a report to the ministry in 1825 that Purkinje's courses attracted only handfuls of students, that his German was faulty and difficult to understand, and that his lectures confused students through their abstractness. In order to remedy these problems, Otto recommended that Purkinje be compelled to lecture in Latin and from a relatively elementary textbook. The curator Neumann suggested that Purkinje's efforts be supplemented by appointing the botanist Treviranus to give additional physiology lectures.[110]

The ministry, however, did not share the pedagogical concern foremost in the minds of the faculty. Schulze's reply to Otto's report not only refused to appoint a second physiologist but also instructed the curator to express to Purkinje the minister's special satisfaction with his research and experimental demonstrations. The minister felt sure, Schulze said, that Purkinje would eventually accustom himself to lecturing.[111] Through its decision the ministry again subordinated institutional to disciplinary values, even though in Purkinje's case the former involved the sensitive issue of effective teaching.

By the days of Helmholtz and Clausius in the 1840's the "dual nature" of the professorate had become a fact of academic life in Prussia; disciplinary reputation had become not only the accepted

---

109. When Purkinje began to accompany his physiology lectures with experimental demonstrations in 1824, the anatomist Otto, later Purkinje's opponent, provided laboratory space in the anatomy building. The *Prosektor* in anatomy Sehrig assisted Purkinje. The curator Neumann helped secure funds from the state to finance the demonstrations. (*ADB*, 26, 718-719.)

110. Hürthle, "Die Gründung des physiologischen Instituts in Breslau durch Joh. Ev. Purkinje . . . ," *Allgemeine Medicinische Central-Zeitung*, 77 (2 January 1908), 72-73; Henry J. John, *Jan Evangelista Purkyne, Czech Scientist and Patriot, 1787-1869* (Philadelphia, 1959), p. 21.

111. *ADB*, 26, 719-720.

basis of promotion but also the standard of university competition. The letters of du Bois-Reymond and Karl Ludwig from midcentury illustrate the new professional outlook of that self-conscious generation of German scientists. Research and the publication of results dominated the activities of these men, relegating to insignificance the writing of textbooks, the publishing of compendia, and even the teaching of more elementary topics. Although much of even Johannes Müller's fame had been based upon his *Handbuch der Physiologie des Menschen,* in 1849 Karl Ludwig dismissed the writing of textbooks as "Dilettantenarbeit," and du Bois-Reymond sympathised with his "reluctance to give up your own research for the sake of a mere compilation (*Sammelwerk*)."[112] Not only did these men look upon achievement through specialized research as the single proper road to the professorship, but they also regarded it as prerequisite to effective teaching as well. Du Bois-Reymond wrote to Ludwig in 1849:

> I would unconditionally prefer as a teacher the one-sided scholar who is nevertheless outstanding in his subject to the roundly educated man who has never really achieved anything. For the first—and this appears to me to be the main thing in teaching—will be permeated more deeply by the spirit (*Geist*) and the method of science (*Wissenschaft*) than the second, and will be better able to communicate them.[113]

The acceptance of these new professorial values was attested to most eloquently by the protests they provoked. They were met by bitter, sometimes pathetic pleas for a return to a simpler university oriented toward teaching and professional training in which rank had privilege and *Privatdozenten* kept their place. Old Ernst Bischoff, who had been a professor of Johannes Müller's at Bonn, angrily wrote in 1842: "The purpose of the university on the whole, however, is: teaching—the intellectual equipping of sages and leaders in the life of the state . . . in the three academic branches of the civil service. . . . According to such a concept teaching, not research into all higher knowledge, is recognized as the purpose and task of the German university."[114] With passion and hyperbole

112. Du Bois-Reymond and Ludwig, *Briefwechsel,* pp. 36, 43.
113. Du Bois-Reymond and Ludwig, *Briefwechsel,* p. 42.
114. C. H. Ernst Bischoff, *Einiges, was den deutschen Universitäten Noth thut,* 2 vols. (Bonn, 1842, 1848), *1,* 2.

Bischoff attacked appointment procedures which based promotions upon the results of a candidate's research rather than upon his teaching skill, his maturity, or his experience. No one, he claimed, could indulge in research without neglecting his pedagogical obligations. The "whirl of a premature, muddled literary officiousness" leads the professor "to deprive his teaching post of the best part of his intellectual means and achievement and to devote it to those confused, multifarious strivings after research and literary production characteristic of our time."[115]

His greatest scorn Bischoff reserved for the state-imposed principle of *Lehrfreiheit*—the right of every *Privatdozent* and junior professor to offer lectures in competition with full professors. This pernicious practice, Bischoff claimed, had not only deprived professors of their livelihoods and subverted every pedagogical value but had also introduced into the universities great numbers of arrogant, incompetent young men who justified their shallow pretensions and impotence as teachers with high-sounding but empty appeals to *Wissenschaft* and *Forschung*. The growth of their numbers was "one main source—if not the first and most important—of the deep confusion in academic studies today, as indeed of the entire university life."[116]

Such protests could not stem the new movement within the universities, however; they could only attest to its pervasiveness. By 1840 the new conception of the professorate which regarded research and disciplinary prestige as the supreme professorial values had permeated all the universities of Prussia. Viewed in historical perspective, obviously many factors, most of them introduced during the Prussian reform period, caused the spread and final victory of the movement. The *Wissenschaftsideologie* with its dynamic concept of learning and its emphasis upon creativity gave the new movement an ethical and epistemological basis. The triumphs of critical scholarship in classical and Germanic philology provided a disciplinary model. The critical, analytic tendencies inherent in this scholarly tradition largely dissipated the emphasis upon philosophical synthesis characteristic of the early *Wissenschaftsideologie*. Finally competition within and among the universities ensured that the new

115. Bischoff, *1*, 3-5.
116. Bischoff, *2*, 19.

professorial values, once established, would rapidly be propagated throughout the university system.

As this study has emphasized, however, the state's new, aggressive control of professorial appointments acted as a third, equally important stimulus to the rise of the new professors. It became the specific institutional mechanism through which the new scholarly values—induced partly by ideological and competitive pressures—came so quickly to dominate the universities. More important, the state's appointment policy promoted these values directly. By stressing disciplinary criteria over the collegiate and pedagogical concerns of direct interest to the local, corporate university, the state gradually established the disciplinary criteria as accepted professorial norms. These scholarly values introduced and sustained the dualistic professorial role in Prussia and with it the emergence of the Prussian university as a center of research.

Individualism and the Structure of British Science in 1830

BY J. B. MORRELL*

Over half a century ago the polymathic J. T. Merz produced his monumental classic on the intellectual history of nineteenth-century Europe. One of its most welcome features remains his account of the different ways in which science had been institutionalized in France, Germany, and England.[1] Merz saw, inter alia, that science is not merely positive knowledge which has been rigorously acquired and then patiently accumulated in textbooks and journals, but also that it is a socially organized intellectual activity. Accordingly he concerned himself with scientific institutions as well as scientific ideas. Particularly since 1960 this aspect of Merz's approach has enjoyed a belated revival. In their different ways, Cannon's exploration of the intellectual network centred on the University of Cambridge during the 1830's, Mendelsohn's survey of the professionalization of science in Europe, Ben-David's insistence on the competitive nature of the German university system, Crosland's examination of Napoleonic patronage of science, and Ackernecht's appraisal of the Paris clinical school have all drawn attention to and illuminated key aspects of

---

* Department of Social Sciences, University of Bradford, Bradford, 7, Yorkshire, England.

I am grateful to the University of Pennsylvania for the tenure of a visiting professorship in spring 1970 when this paper was written. For help and encouragement I am indebted to the faculty and students of its Department of History and Philosophy of Science, particularly to Dr. R. McCormmach and Dr. A. W. Thackray, to Professor H. Guerlac, Dr. W. H. Brock, Dr. W. F. Cannon, Dr. D. Fraser, and to the members of seminars held at Johns Hopkins University on 6 March 1970 and the University of North Carolina on 13 March 1970 to whom preliminary versions of this paper were read.

1. J. T. Merz, *A History of European Thought in the Nineteenth Century* (London, 1904; reprint, New York, 1965), *1*, 89-301.

the changing social structure of nineteenth-century science.[2] In the previous decade, too, Cardwell had usefully scrutinized the organization of science in England in a pioneering and indispensable book published (one notes) in a series dedicated to sociology.[3]

It appears that much recent social history of nineteenth-century science has been focused on the European continent, if one excludes quite arbitrarily the genre of heroic biography.[4] Perhaps, therefore, it would be advantageous to turn to nineteenth-century British science, and to survey its organization not over the whole century but in and around a particular year. I have chosen 1830 because crucial changes in both British science and British society took place then or thereabout. In that climacteric year Charles Babbage vehemently denounced the Royal Society of London, which was suffering traumatic internal dissent apropos its future development and the related question of the election of its next president. In the following year the British Association for the Advancement of Science, the first national pressure group for professionalizing science, was established at York. Shortly afterward the Reverend William Whewell obligingly coined the term "scientist" to acknowledge those changing or novel characteristics and aims which were inadequately conveyed by the term "cultivator of science." The year also saw the accession of William IV to the British throne and the first gleams of the Age of Reform. My aim in this paper is therefore to analyze the leitmotiv and the related institutional structure of British science, not merely English science, in 1830 and immediately afterward. To clarify the discussion, my paper is divided into four related sections. First I draw attention to the prevalence of individualism as a social and political ideology in Britain around 1830. Second I argue that the conditions under which the British govern-

2. W. F. Cannon, "History in Depth: The Early Victorian Period," *History of Science, 3* (1964), 20-38; *idem,* "Scientists and Broad Churchmen: An Early Victorian Intellectual Network," *Journal of British Studies, 4* (1964), 65-88; E. Mendelsohn, "The Emergence of Science as a Profession in Nineteenth-Century Europe" in K. Hill, ed., *The Management of Scientists* (Boston, 1964), 3-48; M. Crosland, *The Society of Arcueil, A View of French Science at the Time of Napoleon I* (London, 1967); J. Ben-David, "Scientific Productivity and Academic Organization," *American Sociological Review, 25* (1960), 328-343; E. H. Ackernecht, *Medicine at the Paris Hospital, 1794-1848* (Baltimore, 1967).

3. D. S. L. Cardwell, *The Organization of Science in England* (London, 1957).

4. The outstanding example from the last decade is L. P. Williams, *Michael Faraday* (London, 1965).

ment was likely to disturb the dominant individualism by state intervention were palpably not fulfilled with regard to the alleged decline of British science. Third I describe the institutional structure of British science, which reflected the individualism prevalent if not pervasive in 1830. In this third section I try to give a synoptic anatomical view, illustrated with apposite but not exhaustive detail; and I take the liberty of stressing those provincial and Scottish institutions and influences whose importance seems to me to be frequently underestimated. Lastly, as a coda, I emphasize the provincial origins of the British Association for the Advancement of Science; and I analyze the apparent paradox that the Association generally shunned the question of state support for science.

I

Writers from Babbage onward, including the redoubtable Merz, seem dazzled by the successful efforts made by the French governments between 1794 and 1808 to create de novo and sometimes to refurbish an impressive set of scientific institutions of which the Ecole Polytechnique was the enviable pinnacle.[5] Certainly the professionalization of French science was achieved through centralized bureaucratic innovation and revival which was organized and supported by the state, a tradition previously characteristic of enlightened despotism. In contrast with French science, that of Germany became professionalized chiefly through the decentralized University system operated by the various states in a spirit of cultural competition. It was after all in the small provincial University of Giessen in Hesse-Darmstadt that the messianic Justus von Liebig started and developed those two basic aspects of professionalized science, laboratory teaching and the research school. Though the German universities were largely autonomous intellectually, they were administratively state institutions of which the capital and running expenses were borne by the appropriate government. We must realize that in both France and the German states the professionalization of science occurred mainly in organizations which were both sponsored and managed by the state: patronage by the relevant

5. C. Babbage, *Reflections on the Decline of Science in England* (London, 1830), passim; Merz, *op. cit.,* pp. 243 and 245.

government maintained both the Ecole Polytechnique and the University of Giessen. Clearly if one measures British science against this criterion culled from the French and German modes of practice, the results will be hopelessly inappropriate.

Again, from Babbage onward one senses that the diversity of the organization of British science and the motive behind it have been persistently underestimated.[6] Babbage himself gave scant attention to University College London, the mechanics' institutes, popular scientific education, and the provincial scientific societies. For polemical reasons he simplified the structure of British science and exaggerated some of its alleged defects. Yet his dissatisfaction with the intellectual and ethical standards adopted by the Royal Society of London and his protest about lack of public recognition for science both show the self-awareness that was increasingly displayed by British scientists from about 1830. It must not be forgotten that the growth of self-consciousness—the realization of collective identity —is probably the most important element in the professionalization of an activity.[7] At the personal level, Babbage's specialism was higher mathematics and its applications, a field which featured prominently in his book. As his difficulties with the Treasury in connection with his calculating machine were increasing in 1830, his plea for munificent state patronage of higher mathematics was scarcely concealed self-interest. Furthermore his envy of Edward Sabine, named in 1828 as one of the Admiralty's three advisers along with Thomas Young and Michael Faraday, was shown in his attacks on Sabine's competence and character. It is clear that Babbage's book confirms the old adage that personal testimony is not necessarily reliable evidence. One of the most serious defects of his book was his exclusion of the Scottish universities, among which Edinburgh and Glasgow were distinguished in scientific research and teaching, embarrassingly so to Babbage. Perhaps he had a valid excuse: until about 1850 obtuse writers used "England" to include "Scotland," and "English" to subsume "Scottish," heinous practices which linguistic Scottish nationalism eradicated in mid-century. For polemical and personal

---

6. Merz appreciated individualism in the English character, but failed to see the rich variety of institutions it produced: Merz, *op. cit.*, pp. 264-275 and 279.
7. G. Millerson, *The Qualifying Associations: A Study in Professionalization* (London, 1964), pp. 10-12.

reasons Babbage revealed a limited awareness of the sheer variety of ways in which science was organized in England, not to mention those frequently ignored Celtic fringes called Scotland and Ireland.[8]

A necessary antidote to the narrow perspective dispensed by Babbage et alii is the recognition that basic traits in British society in 1830 were self-help, voluntarism, individualism, and libertarianism. All these characteristics implied the tenet and belief that individuals should initiate and support their own activities at the grass-roots level by exercising their own powers of will, choice, and action, as opposed to looking to state intervention or control. The persistence of this voluntarist tradition is shown by the wide span of time through which it was enunciated and elaborated. From Adam Smith in 1776 to John Stuart Mill and Samuel Smiles in 1859, libertarianism was noisily lauded, and the domestic function of government was generally seen as negative restriction and not positive action. Indeed, the opening pages of Smiles's famous homily *Self-Help* unambiguously identified the chief source of British power:

> "Heaven helps those who help themselves" is a well-tried maxim, embodying in a small compass the results of vast human experience. The spirit of self-help is the root of all genuine growth in the individual; and, exhibited in the lives of many, it constitutes the true source of national vigor and strength. Help from without is often enfeebling in its effects, but help from within invariably invigorates. Whatever is done *for* man or classes, to a certain extent takes away the stimulus and necessity of doing for themselves; and where men are subjected to over-guidance and over-government, the inevitable tendency is to render them comparatively helpless. . . . The solid foundations of liberty must rest upon individual character; which is also the only sure guarantee for social security and national progress. John Stuart Mill truly observes that "even despotism does not produce its worst effects so long as individuality exists under it; and whatever crushes individuality *is* despotism, by whatever name it be called."[9]

---

8. As early as the 1830's Whewell's wholesale disregard of Scottish achievements in science was angrily corrected by David Brewster, "Whewell's History of the Inductive Sciences," *The Edinburgh Review*, 66 (1837), 110-151 (147-149). It would be invidious to mention recent writers who, failing to distinguish between Scotland and England, have blithely described Brewster as English.

9. See particularly A. Smith, *The Nature and Causes of the Wealth of Nations* in *The Works of Adam Smith* (London, 1811), 4, 41-43; J. S. Mill, *On Liberty* (London, 1859), pp. 21-22; and S. Smiles, *Self-Help* (London, 1859), pp. 1-3.

II

Adoration of British liberty had been markedly strengthened by the fervid reaction against the excesses committed by the Jacobins during the Terror.[10] It shone in contrast with the intolerance and despotism which allegedly were bringing ruin upon France, so that the deep-seated British suspicion of the arbitrary prerogative of the state was conspicuously reinforced. By 1830, at the dawn of the Age of Reform, it was generally if not totally accepted that the state would be imprudent or iniquitous to undertake actions and duties which were already adequately performed by individuals free from state control. When Lord Melbourne, who was twice British prime minister during the 1830's, put the question "Why can't you let it alone?" he was not for once exposing his incontrovertible indolence but expressing trust in individualism. This emphasis on personal responsibility was clearly not the sole property of one political group such as the Whigs or Benthamites; and it was consolidated by the corresponding and widespread appeal to personal religion made by the thriving Evangelical movement.

Yet owing to the rapid pace of social change, state intervention and collectivism had begun inevitably to gather force in the mid-1820's and within ten years were well launched. It seems that the British government disturbed the prevalent spirit of individualism when at least one of two not totally separable conditions was fulfilled: first, when intolerably cruel or unsafe situations outraged humanitarian consciences; second, when apposite opportunities arose for the deployment of Benthamite ideology, which was deeply motivated by the desire of its middle-class adherents to gain power. In opposing deliberate and wanton cruelty, Benthamite utilitarians, Whig philanthropists, Tory humanitarians, and bristling Evangelicals willingly joined forces during the 1820's when inter alia the

---

10. The parts of this section which deal with the general conditions for state intervention have been synthesized from such standard sources as: P. A. Brown, *The French Revolution in English History* (London, 1918); L. Woodward, *The Age of Reform 1815-1870* (Oxford, 1964); A. Briggs, *The Age of Improvement 1783-1867* (London, 1959); E. Halévy, *The Growth of Philosophic Radicalism* (London, 1949); *idem, The Triumph of Reform 1830-1841* (London, 1961); A. V. Dicey, *Lectures on the Relation between Law and Public Opinion in England during the Nineteenth Century*, 2nd ed. (London, 1914); J. B. Brebner, "Laissez Faire and State Intervention in Nineteenth-Century Britain," *Journal of Economic History,* Supplement to 8 (1948), 59-73.

public whipping of women was abolished and measures for the protection of children were introduced. It was, of course, a basic element in the Benthamite creed that pain should be minimized and if possible eliminated. The success enjoyed by various juntos in initiating the Age of Reform, aided by the persistently enunciated doctrines associated with utilitarianism, shows that pressure-groups used state intervention for their own not disreputable ends. In short, state action in Britain around 1830 was likely to occur only when intolerable conditions or opportunities for Benthamite legislation could be exploited by powerful and united pressure-groups.

If we consider British science at this time, it hardly provided examples of outrageous circumstances likely to arouse shame, wrath, and action from conscience-stricken humanitarians. Compared with large imminent issues such as slavery, working conditions in factories, the deficiencies of the Poor Law, the inadequacy of municipal administration, sanitary conditions in towns, and other problems concerning the patent misery and discontent of the working classes, the allegations made by Babbage about the decline of science in England were insultingly trivial and eminently dispensable. Nor did the organization of British science provide signally favorable opportunities which benevolent Benthamite legislators could appropriate and use for political ends. Their chosen targets were usually obvious abuses out of which widespread and energetic concern could be generated. In any event, except when humanitarian questions arose, the utilitarians generally tried by legislation to extend and not to curtail individual liberty and rights. Furthermore, that minority among scientists which urged either occasional or persistent state intervention in British scientific activity did not form a pressure group sufficiently well represented or active in government to achieve it.[11] In any event it is doubtful whether the condition of scientific instruction and institutions was at this time a party question. With respect to science there is no doubt that the two prerequisites for state intervention rarely occurred. It should be remembered, too, that such an important national matter as the education of English children was organized exclusively on the voluntarist system until 1833. Even a reforming Lord Chancellor,

11. Among politicians important in 1830 only Sir Robert Peel, Henry Brougham, and maybe Lord Althorp were strongly interested in science.

Henry Brougham, argued strongly in that year that "The efforts of the people are still wanting for the purpose of promoting Education; and Parliament will render no substantial assistance, until the people themselves take the matter in hand with energy and spirit, and the determination to do something."[12] Indeed when in 1833 the government granted £20,000 to two religious groups— the National Society and the British and Foreign School Society —for the building of new schools, the grant was available only if voluntary contributions met half their cost. If the state was slow and cautious in entering the field of national education, it is no wonder then that the state generally left British science to run itself in a voluntarist way.

The British government did, however, support scientific activity of various kinds when national security, internal need, and intermittently, sheer prestige were felt to be either directly or obliquely at stake. For decades the government's concern with navigation and cartography had led it to patronize observational astronomy and voyages of discovery. Quite traditionally, therefore, a number of the Admiralty's enterprises abroad, such as giving a five-year honorary research fellowship to Charles Darwin in 1831, occasionally patronized scientists at a time when the British were seeking to extend their naval and colonial dominion.[13] Less directly the state patronized science abroad through the East India Company, whose president had been a member of the British cabinet from 1812. For instance, the Company not only maintained observatories at Madras, Bombay, and St. Helena, but also supported the Calcutta Botanical Gardens of which the botanist Nathaniel Wallich was the superintendent. On the domestic front security and need jointly motivated the government's continuing creation of Regius chairs of medicine at the Universities of Edinburgh and Glasgow during the opening decades of the nineteenth century. It wanted trained and competent medical manpower. Yet quite characteristically these Regius professors received a low salary or none at all from the Crown, which relied on the ability of the professors and their universities to meet

12. Cited by G. Combe, *Lectures on Popular Education* (Edinburgh, 1833), p. 1.
13. Cannon, "Scientists and Broad Churchmen," pp. 82-83.

the heavy expenses associated with their activities.[14] Sometimes government departments, such as the Boards of Excise and Ordnance, turned to chemists for data which would serve as the basis of legislation or employed them for reasons of safety. The military schools, too, employed scientists: we find that in 1830 James Inman was resident professor of mathematics at the Royal Naval College, Portsmouth, and Michael Faraday was a visiting lecturer at the Royal Military Academy, Woolwich. Finally, less irregular patronage was dispensed to major national institutions for reasons associated with prestige and sometimes utility. Establishments such as the Royal Observatory, the British Museum, the Mint, the Botanic Gardens at Kew, Edinburgh and Dublin, the Natural History Museums at Edinburgh and Dublin, and the Royal Dublin Society were partly or totally financed by the government.[15] Yet this state patronage only nibbled incidentally and innocuously at the total body of scientific activity in Britain. Compared with the amounts which the governments of France and the German states were willing to spend on scientists and scientific institutions, the British government's contribution in 1830 was characteristically erratic and possibly niggardly: with few exceptions it expected people and organizations to look after themselves, as its recent closing of the Board of Longitude in 1828 had conspicuously shown.

The result of putting what may be called Smilesian virtues to work has been readily appreciated by general historians of Britain during the first industrial revolution. For example, Pollard and recently Musson and Robinson have forcibly argued that the

---

14. A. Duncan, *Memorials of the Faculty of Physicians and Surgeons of Glasgow, 1599-1850* (Glasgow, 1896), pp. 162-186; J. Coutts, *A History of the University of Glasgow* (Glasgow, 1909), pp. 512-569; A. Grant, *The Story of the University of Edinburgh during Its First Three Hundred Years* (London, 1884), *1*, 321-328.

15. The financing of science by the British government remains to be investigated, though H. F. Berry, *History of the Royal Dublin Society* (Dublin, 1915) contains useful information on Ireland. The variability of government patronage may be judged from the annual grants given in 1830 to the Botanic Gardens in Edinburgh and the Natural History Museum, both of which were associated with the University of Edinburgh: the former received £819; the latter received £100. In both cases Professors Graham and Jameson met the annual deficits as best they could. See: *Report of the Royal Commission of Inquiry into the State of the Universities of Scotland, Parliamentary Papers, 12* (1831), 151 and 177.

entrepreneurs of the industrial revolution, while not enjoying state sponsored and formalized technical training, frequently availed themselves of local opportunities which were usually sufficient for their immediate purposes.[16] If this sort of thesis is extended to the organization of science in Britain in 1830, then the effects derived from voluntarism and individualism were understandably great variety and apparent confusion. This diversity frequently emanated from local initiative which was exercised in response to local conditions and pressures. Such bewildering disorder should not be regarded as a regrettable deviation from the French and German patterns of professionalization. On the contrary it represented a distinctly national style of science of which individualism was the chief leitmotiv. Nor was the disorder really haphazard and bewildering. If differences of function, location, and social class are made, the variegated structure of institutionalized British science at the accession of William IV seems almost tailor-made to satisfy the needs and aims of professionals, devotees, and amateurs alike. Correspondingly the commitment to science shown by institutions and the degree of professionalization of the sciences cultivated in them varied widely, being frequently dependent on the extent to which private initiative could create or exploit local opportunities, traditions, and contingencies. What then was this structure and how did it reflect the dominant individualism?

III

First, at the national level, each of the three metropolitan societies—the Royal Society of London, the Royal Society of Edinburgh (f. 1783), and the Royal Irish Academy (f. 1786)—possessed its own characteristic membership, degree of professionalization, and favored fields of interest.[17] For example, the Edinburgh Society gained intellectually and socially from the contribution and expertise

16. S. Pollard, *The Genesis of Modern Management: A Study of the Industrial Revolution in Great Britain* (London, 1965); A. E. Musson and E. Robinson, *Science and Technology in the Industrial Revolution* (Toronto, 1969).

17. The Royal Society of London has been served by C. R. Weld, *A History of the Royal Society* (London, 1848) and H. Lyons, *The Royal Society 1660-1940* (Cambridge, 1944). Serious studies devoted to the remaining metropolitan societies are still required.

readily given by the city's leading cultural and professional groups, the University professoriate, the lawyers, and the medicals. It contrasted sharply with the Royal Society of London where the gentry, clergy, and military officers were trying to ensure that the dominant ethos remained dilettante and amateur: a greater proportion of its members published, and the cohesion provided by the unique combination of its three leading groups seems to have prevented the sort of internal dissension which so troubled its London counterpart in 1830. Second, as a result of group initiative a galaxy of specialized societies flourished chiefly in two of the three metropolitan centers. One can instance the following London societies: the Medical (f. 1773), Linnean (f. 1788), Mineralogical (f. 1799), Medical and Chirurgical (f. 1805), Geological (f. 1807), Astronomical (f. 1820), Zoological (f. 1826) and Geographical (f. 1830), not forgetting the Society of Arts (f. 1754) and the Institute of Civil Engineers (f. 1818).[18] At their best the style of work and composition of membership of these specialized societies subversively threatened the hegemony relished by the Royal Society of London, whose acute premonition had led it to try to suppress the embryonic Geological and Astronomical Societies in 1807 and 1820 respectively.[19] Elsewhere in England the dominance of natural history and allied fields, which every man could cultivate, is seen in the Geological Society of Cornwall (f. 1814) and the Natural Historical Society of Manchester (f. 1821). In Scotland, however, more emphasis was placed on medicine and related subjects, many of the societies being predictably associated with universities often in the form of student societies whose activities were nurtured by the professoriate. At Edinburgh alone one can point to the Royal Medical Society (f. 1734), the Harveian Society (f. 1752), the Royal Physical Society (f. 1771), the Wernerian Natural History Society (f. 1808), the Scottish Society of Arts (f. 1821), the Plinian Society (f. 1823), and the Hunterian

---

don, 1907); J. L. E. Dreyer and H. H. Turner, *History of the Royal Astronomical Society 1820-1920* (London, 1923); D. Hudson and K. Luckhurst, *The Royal Society of Arts, 1754-1954* (London, 1954).

19. See, e.g., M. J. S. Rudwick, "The Foundation of the Geological Society of London: Its Scheme for Co-operative Research and Its Struggle for Independence," *The British Journal for the History of Science*, 1 (1963), 326-355.

18. Though indispensable, histories of these societies are usually narrow and hagiographical: A. T. Gage, *History of the Linnean Society of London* (London, 1938); H. B. Woodward, *The History of the Geological Society of London* (Lon-

Medical Society (f. 1824).[20] By contrast, for reasons which remain obscure, specialist societies in Ireland were rare.

Third, again as the result of group initiative, a multitude of provincial general scientific societies, often carrying the ennobling title "Literary and Philosophical Society," had burgeoned in the English manufacturing towns and also in traditional regional centers. It is probable that Charles Lyell did not exaggerate when he asserted that compared with the specialist London societies the provincial ones constituted "a still more novel and characteristic feature of the times."[21] A sample list culled from the North of England is impressive even allowing for qualitative differences: Manchester (f. 1781), Liverpool (f. 1812), Newcastle (f. 1813), Leeds (f. 1818), Yorkshire, Sheffield, Hull, Whitby (all f. 1822), and Scarborough (f. 1830). Scotland and Ireland could each boast at least a pair: Perth (1784) and Glasgow (1802), Cork (f. 1819) and Belfast (f. 1821).[22] Naturally these societies varied widely in membership and technical competence: for instance, in 1830 the Philosophical Society of Glasgow was still dominated in its membership by intellectually isolated craftsmen and manufacturers who generally neither attended meetings nor offered papers. Indeed the Society slid from crisis to crisis until in 1834 its renovation began with the election to membership and immediate office of distinguished professors from the University of Glasgow and Anderson's University who had previously ignored it.[23] Yet the provincial societies provided a local focus of activity where meetings, lectures, a library, sometimes apparatus, an opportunity for publication, social legitimation, and the satisfactions of power, vanity, and emulation were all to be savored.

20. C. P. Finlayson, "Records of Scientific and Medical Societies preserved in the University Library, Edinburgh," *The Bibliotheck: A Journal of Bibliographical Notes and Queries Mainly of Scottish Interest, 1* (1958), 14-19.
21. C. Lyell, "Scientific Institutions," *The Quarterly Review, 34* (1826), 153-179 (163).
22. The list is taken from A. Hume and A. I. Evans, *The Learned Societies and Printing Clubs of the United Kingdom* (London, 1853). It would appear that the call issued by Schofield for more social history of nineteenth-century British science, based on these societies, has remained largely unheeded: R. E. Schofield, "Histories of Scientific Societies: Needs and Opportunities for Research," *History of Science, 2* (1963), 70-83. But see two studies of ephemeral societies: W. H. Brock, "The London Chemical Society 1824," *Ambix, 14* (1967), 133-139; and N. G. Coley, "The Animal Chemistry Club," *Notes and Records of the Royal Society, 22* (1967), 173-185.
23. A. Kent, "The Royal Philosophical Society of Glasgow," *The Philosophical Journal, 4* (1967), 43-50.

A fourth category was constituted by those institutions, devoted equally to teaching and research, which were created and maintained by individual initiative or jeu d'esprit. Founded as a philanthropic establishment for artisans by Count Rumford in 1799, the Royal Institution in 1830 combined popular teaching and specialized research. Under Sir Humphry Davy and then Michael Faraday it had capitalized on fashionable London's interest in physical and chemical science; in its laboratory brilliant individual work flourished, but Faraday led no research school.[24] Its frequently ignored Irish equivalent, the Royal Dublin Society (f. 1731), was furnished with a devoted staff, led by Charles Giesecke, which specialized in natural history. Its English imitators and descendants, such as the London, Surrey, Liverpool, and Birmingham Institutions, served a mainly middle-class audience with the lectures and libraries which were such characteristic features of an age of self-improvement.

When one turns to organizations which were avowedly concerned with teaching, those that taught some science are best understood in terms of the class needs they fulfilled, as Cardwell has clearly demonstrated with respect to England alone.[25] Though a variety of establishments diffused science, it should be noted as an indicator of the degree of professionalization of British science in 1830 that not one of them gave a qualification solely in science. Nor was the context of science teaching uniform: some subjects such as natural philosophy appeared in the arts curriculum, some like botany in the medical program, and flexible ones like chemistry in both. Again in some institutions a scientific subject could be irrelevant to the process by which a degree was acquired: the geology class at the University of Cambridge was run by Sedgwick as a nongraduating one. In other establishments such as the Scottish universities the students enjoyed the privilege of *Lernfreiheit* and attended classes in science, whether graduating or nongraduating ones, according to their intellectual preferences and pockets.

Within these diverse patterns, the constituent colleges of Oxford and Cambridge Universities, Trinity College Dublin, and King's College London provided some education in science largely for those

---

24. T. Martin, *The Royal Institution* (London, 1961).
25. Cardwell, *Organization of Science in England*, pp. 28-29.

aspiring to the status of the cultivated English gentleman.[26] There were, however, sharp differences between these four institutions. By 1830 an increasingly powerful group of research-oriented teachers at Cambridge had not only modernized the mathematics tripos, then the dominant degree, but were also rejuvenating natural historical fields. In the former activity George Biddell Airy (Plumian professor of mathematics and natural philosophy) and Babbage (sinecurist Lucas professor of mathematics) were prominently supported by William Hopkins, a private tutor, and George Peacock and the Reverend James Challis, both of whom were Fellows of Trinity College Cambridge. In the latter activity those two key patrons of Darwin, the Reverend Adam Sedgwick (professor of geology) and the Reverend John Stevens Henslow (professor of botany), scorned their sinecures by developing field methods of instruction. The link between the mathematicians and the natural historians was provided by the voraciously versatile Whewell, then professor of mineralogy. With the exception of the Observatory no great institutional demands for accommodation or apparatus were found to be necessary. At Oxford, which was oriented toward classics and about to be racked by tractarianism, only two men, the Reverend William Buckland (professor of mineralogy and geology) and Charles Daubeny (professor of chemistry), ranked equally with the Cambridge group. Both the charismatic Buckland in field and cave and Daubeny in his laboratory at Magdalen College worked hard and to little long-term effect in a deteriorating situation. In the Emerald Isle the solitary William Rowan Hamilton (professor of astronomy) saved Trinity College Dublin from total scientific obscurity. Lastly, King's College London, founded in 1828 by Anglican and Tory groups as an immaculate rival to University College London, functioned mainly

26. This paragraph is based on relevant works previously cited and others such as: C. E. Mallet, *A History of the University of Oxford* (Oxford, 1924-1927), 3; D. A. Winstanley, *Unreformed Cambridge* (Cambridge, 1935); C. Maxwell, *A History of Trinity College, Dublin* (Dublin, 1946); J. F. C. Hearnshaw, *The Centenary History of King's College, London, 1828-1928* (London, 1929); N. Barlow, ed., *The Autobiography of Charles Darwin* (London, 1958); N. Barlow, ed., *Darwin and Henslow* (London, 1967); J. W. Clark and T. M. Hughes, eds., *The Life and Letters of the Reverend Adam Sedgwick* (Cambridge, 1890); E. O. Gordon, *Life and Correspondence of William Buckland* (New York, 1894); R. T. Gunther, *A History of the Daubeny Laboratory Magdalen College Oxford* (London, 1904); F. S. Taylor, "The Teaching of Science at Oxford in the Nineteenth Century," *Annals of Science*, 8 (1952), 82-112; and the indispensable portraits provided by the *Dictionary of National Biography*.

as a theological and classical preparatory department for the Universities of Oxford and Cambridge. Accordingly it placed less emphasis on science than its London competitor, though in 1830 the recruitment of Charles Lyell and John Frederic Daniell to the chairs of geology and chemistry was imminent.

Science for the middle classes and above was more widely available at the six Scottish universities, at University College London, and tenuously at their Ulster offspring, the Belfast Academical Institution's collegiate department.[27] Great qualitative differences existed among the three pre-Reformation universities of Glasgow, St. Andrews, and King's College Aberdeen, the late sixteenth-century pair of Edinburgh and Marischal College Aberdeen, and that late eighteenth-century monument to private enterprise, Anderson's University in Glasgow. Of these, Edinburgh and Glasgow were preeminent in offering inexpensive and secular studies which spanned liberal, useful, and vocational fields. Though in 1830 the University of Edinburgh was losing its reputation as the outstanding English-speaking University in the world of science, its medical school remained internationally important. Of its scientific professoriate, John Leslie (natural philosophy), Thomas Charles Hope (medicine and chemistry), Robert Jameson (natural history), and Robert Graham (medicine and botany) were outstanding professors who had initiated or expanded effective use of lecture demonstrations or of field work in spite of financial and other constraints.[28] Laboratory

27. Histories of these universities vary widely in quality. For the eight institutions in question, H. H. Bellot, *University College, London 1826-1926* (London, 1929) is exemplary. Also useful are: D. B. Horn, *A Short History of the University of Edinburgh 1556-1889* (Edinburgh, 1967); Grant, *op. cit.*; J. D. Mackie, *The University of Glasgow 1451-1951* (Glasgow, 1954); Coutts, *op. cit.*; T. W. Moody and J. C. Becket, *Queen's, Belfast 1845-1949: The History of a University* (London, 1959). R. G. Cant, *The University of St. Andrews, A Short History* (Edinburgh, 1946) contains little about science. R. S. Rait, *The Universities of Aberdeen: A History* (Aberdeen, 1895); J. Muir, *John Anderson, Pioneer of Technical Education and the College He Founded* (Glasgow, 1950); and A. H. Sexton, *The First Technical College: A Sketch of the History of "The Andersonian"* (Glasgow, 1894) are now dated.

28. See: *Evidence, Oral and Documentary, Taken and Received by the Commissioners for Visiting the Universities of Scotland: The University of Edinburgh, Parliamentary Papers, 35* (1837), passim; M. Napier, *Biographical Notice of Sir John Leslie* (Edinburgh, 1836); J. B. Morrell, "Practical Chemistry in the University of Edinburgh, 1799-1843," *Ambix, 16* (1969), 66-80; L. Jameson, "Biographical Memoir of the late Professor Jameson," *The Edinburgh New Philosophical Journal, 57* (1854), 1-49; and C. Ransford, *Biographical Sketch of the Late Robert Graham* (Edinburgh, 1846).

work was available only in chemistry, but Hope led no research team. However, the much maligned Robert Jameson was remarkably successful in producing a notable group of students, probably as a combined result of his rigorous training in the field and of his seminars held in the Natural History Museum of which he was Keeper. Westward at Glasgow, the scientific reputation of the medical school had risen dramatically through the ambitious efforts of Thomas Thomson (professor of chemistry) and William Jackson Hooker (professor of botany).[29] The former's valiant attempt to establish a research school had foundered on political rocks and on his inability to devise basic techniques which his students could systematically apply to a new field; but the practical training in chemical analysis which he had given for twelve years to his laboratory class was unique in a British university. At Anderson's University, founded in 1796 by John Anderson to remedy the defects of the University of Glasgow and to extend the availability of useful knowledge, the outrageously spectacular Andrew Ure (professor of chemistry and natural philosophy) was about to be replaced by the broodingly introspective Thomas Graham, whose new laboratory was soon to be a mecca for keen young Northern chemists.[30]

South of the border in 1826 the ideological drive and political agility of dissenters, chiefly Whig alumni of the University of Edinburgh, had disturbed Oxbridge's dominance by their private enterprise in founding University College London as a joint-stock company which offered an attractive rate of interest.[31] Its organizational characteristics and intellectual emphases showed its Scottish and particularly Edinburgh origin: a wide range of modern subjects, liberal and professional, was taught in a secular context by lectures to nonresidential students who had free access to single courses; and professors were remunerated by small salaries plus student fees. The debt to Edinburgh was particularly strong in the professoriate,

---

29. J. B. Morrell, "Thomas Thomson: Professor of Chemistry and University Reformer," *The British Journal for the History of Science*, 4 (1969), 245-265.

30. R. A. Smith, *The Life and Works of Thomas Graham* (Glasgow, 1884) is disappointing on Graham's Glasgow period; but W. S. C. Copeman, "Andrew Ure," *Proceedings of the Royal Society of Medicine*, 44 (1951), 655-662, is enticing.

31. This paragraph is largely based on: Bellot, *op. cit.*; and C. W. New, *The Life of Lord Brougham to 1830* (Oxford, 1961).

in the stress placed on medicine and science, and not least in the internecine quarrels which disturbed its opening years. Indeed in 1830 University College London was only four years from building, on the Edinburgh model, the first hospital attached to an English university to be used for clinical teaching of medical students. Just four years after its foundation, though smarting from the autocratic wardenship of the Whig geologist Leonard Horner, the first scientist to head a British university, it was furnished with seven professors of science. The spectrum was wide: it ranged from their doyen Charles Bell (physiology) through the painstaking Edward Turner (chemistry) and John Lindley (botany) to the intellectually stagnant Robert Edmond Grant (comparative anatomy and zoology), the flamboyantly fashionable Dionysius Lardner (natural philosophy), and the execrably incompetent Granville Sharp Pattison (morbid anatomy). None of these led a research school, but it was symptomatic that, for example, Turner, a master of exact analysis, had just opened a practical laboratory class.[32]

The artisan classes were also catered for by 1830.[33] Inspired by the success of George Birkbeck's popular lectures to working men at Anderson's University between 1800 and 1804 and by the enterprise of Leonard Horner in founding the Edinburgh School of Arts in 1821, the London Mechanics' Institute had been created largely by Birkbeck himself, Thomas Hodgskin, and Lord Henry Brougham in 1823. Supported by dissenters and radicals, and supervised by Whigs who feared the rebellious propensities of the uneducated urban masses, the mechanics' institute movement had reached its peak in 1830 shortly before it was appropriated by the bourgeoisie. Aided by the popular education groups, such as Brougham's Society for the Diffusion of Useful Knowledge, and by encyclopedias ranging from the *Britannica* to the *Metropolitan*, the movement constituted a remarkable national attempt on a self-help basis to educate the populace at a time when the state was manifestly indifferent.

32. H. Terrey, "Edward Turner, M.D., F.R.S. (1798-1837)," *Annals of Science*, 2 (1937), 137-152.
33. My interpretation is indebted to: New, *op. cit.*; T. Kelly, *George Birkbeck: Pioneer of Adult Education* (Liverpool, 1957); M. Tylecote, *The Mechanics' Institutes of Lancashire and Yorkshire before 1851* (Manchester, 1957); and J. N. Hays, "Science and Brougham's Society," *Annals of Science*, 20 (1964), 227-241.

Finally two types of teaching institution, the national medical colleges and the private tutor, complete the broad picture. Wide differences of social class and professional aims existed among the six metropolitan Colleges of Physicians and of Surgeons which flourished in London, Edinburgh, and Dublin.[34] Yet their concern with teaching preclinical medical subjects and the continuing importance in science of the medical profession of which they were the pinnacles are frequently ignored. For example, the first Edinburgh medical institution to make a laboratory course of practical chemistry compulsory for its students—in 1829—was significantly the Edinburgh College of Surgeons and not the prestigious University with its distinguished chemical lineage.[35] Again, in 1830 the Hunterian Museum of the Royal College of Surgeons in London employed and gave facilities to William Clift and the young Richard Owen who acted as its chief and assistant curators. Of the increasing body of private teachers of science performing in 1830, only that arch entrepreneur John Dalton was a nationally known figure.[36] Generally second-order men prospered in metropolitan or regional centers where a fashionable or interested audience could be attracted, and they flourished as private pre-clinical medical teachers.

IV

The tradition of self-help did not die immediately after 1830, but on the contrary persisted with characteristic vigor. It is true that the 1830's witnessed an acceleration in the status and recognition afforded to science, a movement in which the British Association for the Advancement of Science (f. 1831) was conspicuously active.

---

34. Competent accounts are given by: G. N. Clark, *A History of the Royal College of Physicians of London* (London, 1964-1966), 2; J. D. H. Widdess, *A History of the Royal College of Physicians of Ireland 1654-1963* (Edinburgh and London, 1963); idem, *The Royal College of Surgeons in Ireland and its Medical School 1784-1966* (Edinburgh and London, 1967). Z. Cope, *The Royal College of Surgeons of England: A History* (London, 1959) is less satisfactory. C. H. Creswell, *The Royal College of Surgeons of Edinburgh: Historical Notes from 1505-1905* (Edinburgh, 1926) is fragmentary. Both Edinburgh Colleges would repay detailed study.

35. J. B. Morrell, "Practical Chemistry," p. 72.

36. A. W. Thackray, "Fragmentary Remains of John Dalton, Part I: Letters," *Annals of Science*, 22 (1966), 145-174.

Yet the British Association itself, in spite of being a body devoted inter alia to the professionalization of science, strikingly shows the continuing vitality of local private enterprise in its origins and crucially successful first meeting. Its foundation primarily reflected provincial Northern initiative and not the struggle for power which split the Royal Society of London during 1830.[37]

Several elements in its creation and early history may be easily distinguished. First, up to and including 1830 but especially from 1827, such diverse figures as John Playfair, John Frederick William Herschel, Davy, David Brewster, Babbage, and Daubeny had expressed concern about the status of scientists and the concomitant public support of some kinds of science. Brewster in particular repeatedly pressed hard for increased recognition for scientists and science through the columns of his periodical, the *Edinburgh Journal of Science*.[38] The deaths of William Hyde Wollaston, Thomas Young, and Davy within the space of six months between December 1828 and May 1829 not only staggered him but confirmed his view that an era in British science had ended.[39]

Second, in his famous critique of Babbage's book Brewster proposed in October 1830 that an "association of our nobility, clergy, gentry and philosophers" should be formed to remedy the depressed state of science.[40] Third, by spring 1831 Brewster, who at that time

---

37. My interpretation is merely an extension of W. F. Cannon, "History in Depth," pp. 24-25. It is implicit in O. J. R. Howarth, *The British Association for the Advancement of Science: A Retrospect 1831-1931*, 2nd ed. (London, 1931), and the early *Reports of the British Association for the Advancement of Science*. Cf. L. P. Williams, "The Royal Society and the Founding of the British Association for the Advancement of Science," *Notes and Records of the Royal Society*, 16 (1961), 221-233. The foundation of the British Association in 1831 was paralleled by the establishment by Charles Hastings in 1832 of the Provincial Medical and Surgical Association which was later renamed the British Medical Association. Both these groups aimed to improve professional status, and both were founded in the provinces by the exertions of alumni of the University of Edinburgh.

38. See: Brewster, "Memoir of the Life of M. Le Chevalier Fraunhofer," *Edinburgh Journal of Science*, 7 (1827), 1-11; Brewster, "Exhibition of the National Industry of France," *ibid.*, 8 (1828), 344-346; Brewster, "Mr. Dalton's System of Chemical Philosophy," *ibid.*, 8 (1828), 346-355; G. Harvey, "On the Science of Ship-building," *ibid.*, 9 (1828), 298-300; Babbage, "Great Congress of Philosophers at Berlin," *ibid.*, 10 (1829), 225-234.

39. Brewster to Örsted, 24 June 1829, M. C. Harding, ed., *Correspondance de H. C. Örsted avec divers savants* (Copenhagen, 1920), 2, 282.

40. In his "Decline of Science in England," *The Quarterly Review*, 43 (1830), 305-342 (341), Brewster reached a large Anglican and Tory audience.

lacked means, prospects, and a permanent post commensurate with his achievements, had gained for his project the support of John Phillips, Secretary to the Yorkshire Philosophical Society, and of Roderick Murchison, the Scottish President of the Geological Society of London. Largely through the efforts made by Phillips and his colleague, the Reverend William Vernon Harcourt, Vice-President of the Yorkshire Philosophical Society and conveniently a son of the then Archbishop of York, ably supported by three Scots, John Robison (jnr.), James Finlay Weir Johnston, and James David Forbes, the first meeting at York was arranged and social ostracization avoided.[41] Not unexpectedly that first meeting was strongly Northern and Scottish in its inspiration, composition, and initiative: Harcourt, "the soul of our meeting" as Brewster called him, laid down the objects and rules of the association whose title he invented.[42] As Murchison recalled: "It was then and there resolved that we were ever to be *Provincials*. Old Dalton insisted on this —saying that we should lose all the object of diffusing knowledge if we ever met in the Metropolis."[43] Indeed, London scientists with the exception of the geologists were absent; the University of Oxford was represented by the solitary Daubeny; and the Universities of Cambridge, Edinburgh, Glasgow, and London held themselves totally aloof. Clearly many of the savants associated with the old and new British universities, keen to maintain their aristocracy, were doing more than weighing the upstart provincial association; they were also holding back through resentment of embarrassing allegations about the decline of British science and the accompanying call for

---

41. Brewster, "Great Scientific Meeting to be Held in York," *Edinburgh Journal of Science*, 4 (April 1831), 374; Brewster, "Observations on the Decline of Science in England," *ibid.*, 5 (July 1831), 1-16; Brewster, "Proposed Scientific Meeting at York," *ibid.*, 5 (July 1831), 180-182; Brewster, "Decline of Science in England," *ibid.*, 5 (October 1831), 334-358; J. W. F. Johnston, "Scientific Meeting at York," *ibid.*, 6 (January 1832), 1-33; A. Giekie, *Life of Sir Roderick I. Murchison* (London, 1875), *1*, 184-189; J. C. Shairp, P. G. Tait, and A. Adams Reilly, *Life and Letters of James David Forbes* (London, 1873), pp. 75-79; M. M. Gordon, *The Home Life of Sir David Brewster* (Edinburgh, 1869), pp. 141-149; I. Todhunter, *William Whewell, D.D.: An Account of His Writings with Selections from His Literary and Scientific Correspondence* (London, 1876), 2, 126-132, and 237-238.

42. Gordon, *Brewster*, p. 144.

43. Giekie, *op. cit. 1*, 187-188. The italics are Murchison's.

state attention to be given to science, which to them constituted explicit criticism of what they regarded as their own adequacy and success. It must be stressed that the London institutions and the University of Cambridge (including Babbage) supported the British Association from 1832 only after it had become apparent that the fledgling group was sufficiently important to be invaded and remodeled on less contentious lines.[44] Accordingly direct national encouragement of science, one of the Association's original aims, was quietly and quickly dropped in spite of Brewster's public opposition.[45] During the 1830's, therefore, the Association rarely approached the government, the chief exception being the successful recommendation it made in 1838 to Lord Melbourne's administration that an Antarctic expedition be mounted. Not surprisingly the first British national pressure-group for science and scientists relied extensively on voluntary support and individual zeal; and its membership embraced savants and sciolists alike. Quite simply the very diverse local and personal interests in the Association wanted to preserve opportunities in which their own initiative could be exercised; any profound surrender of autonomy was fundamentally anathema to their own concerns, however badly coordinated at a national level their activities might have been.

Self-help as a motive inherent in the organization of British science persisted well into the Age of Reform, its inability to cope with the problems of scale and expense of professionalized science becoming slowly visible. Yet it had hardly sunk into its geriatric phase in 1851 when the Great Exhibition dazzlingly confirmed the superiority of British manufacturing enterprise. In that same year the senior British scientific civil servant, the Astronomer Royal, was

---

44. Giekie, *ibid.*, p. 188. Buckland, President of the British Association's Oxford meeting in 1832, ensured that its social tone was effortlessly superior: see, e.g., E. C. Curwen, *The Journal of Gideon Mantell, Surgeon and Geologist* (Oxford, 1940), pp. 102-104. It is apparent from Clark and Hughes, *op. cit.*, *1*, 390 and 407, that Sedgwick reluctantly attended the 1832 meeting because of the pressure put on him by his Oxford equivalent, Buckland, and by Murchison. The next year as president of the Association, which met at Cambridge, he was its animating spirit.

45. Brewster, "The British Scientific Association," *Edinburgh Review*, 60 (1835), 363-394; idem, "British Association for the Advancement of Science," *North British Review*, 14 (1850), 235-287; idem, "Presidential Address," *Report of the British Association for the Advancement of Science, 19* (1850), xxxi-xliv.

far from singing the requiem of the libertarian tradition in his presidential address to the British Association. On the contrary Airy's words epitomized a feeling almost as prevalent then as in 1830: "this absence of Government-Science harmonizes well with the peculiarities of our social institutions. In Science, as well as in almost everything else, our national genius inclines us to prefer voluntary associations of private persons to organizations of any kind dependent on the State."[46]

---

46. Airy, "Presidential Address," *Report of the B.A.A.S., 20* (1851), xxxix-liii (li).

Social Support for Chemistry in Germany during the Eighteenth Century: How and Why Did It Change?[1]

BY KARL HUFBAUER*

Chemistry, J. F. Gmelin reminded the reader of his massive *Geschichte der Chemie* in 1797, used to be misunderstood, contemned, and ridiculed. In his day, however, the science had become "the idol before which all peoples and all orders, princes and subjects, clergy and laymen, the educated and uneducated, high and low, bend their knees; the favorite science of the great . . . who reward it with royal generosity. . . ."[2] Though Gmelin exaggerated, esteem and support for chemistry did improve immensely during the century of Enlightenment. H. Guerlac has illuminated the crucial stage in the development of such support in France.[3]

---

* Department of History, University of California, Irvine, California 92664.
1. This essay pulls together ideas which were necessarily diffused in my study, "The Formation of the German Chemical Community (1700-1795)," unpublished dissertation (University of California at Berkeley, 1970). My aim here is to expose an explanation for changes in the social support for chemistry in eighteenth-century Germany. I seek to make this explanation plausible without marshalling all available evidence. The reader who desires further information should see the parent study and the host of publications cited there. In thinking about this essay's topic, I have been fortunate to receive comments and criticisms from R. Hahn, H. Rosenberg, P. Forman, A. Thackray, R. Multhauf, H. Snelders, E. Bowden, S. Turner, and R. McCormmach.
2. J. F. Gmelin, *Geschichte der Chemie seit dem Wiederaufleben der Wissenschaften bis an das Ende des achtzehenden Jahrhunderts,* photographic reprint (Hildesheim, 1965), *1,* 2.
3. H. Guerlac, "Some French Antecedents of the Chemical Revolution," *Chymia,* 5 (1959), 73-112.

Unfortunately, comparable studies are lacking for other countries.[4] This paper seeks to correct this deficiency for Germany[5] by first tracing, then suggesting an explanation for, the development of social support for chemistry there between 1700, when German chemists were beginning to see their subject as an independent science, and 1800, when the independence of their science was beginning to be seriously questioned.

Social support for a scholarly discipline comes from two main sources—patrons and participants.[6] The support of the patrons is expressed in salaried positions, facilities, or budgets to promote teaching and pursuit of the discipline. The support of the participants is expressed in their number, their intelligence, their dedication, their productivity, and, if patronage is inadequate, their willingness to use personal resources. Since social support takes so many forms, there are many conceivable quantitative indices for tracing changes in its level. Few indices, however, can be used for looking at developments in the eighteenth century because of the difficulty or impossibility of obtaining the requisite data. For example, in tracing how social support for chemistry changed in Germany between 1700 and 1800, one must rest content with three rough indices—(1) the number of salaried positions in schools and academies whose occupants were partly or wholly responsible for chemistry, (2) the number of learned institutions with functioning

---

4. However, many of the materials needed for an examination of changing social support for chemistry in Britain are readily available in J. R. Partington, *A History of Chemistry* (London, 1961-1962), 2-3; Archibald and Nan L. Clow, *The Chemical Revolution: A Contribution to Social Technology* (London, 1952); A. E. Musson and Eric Robinson, *Science and Technology in the Industrial Revolution* (Toronto, 1969); Robert E. Schofield, *Mechanism and Materialism: British Natural Philosophy in An Age of Reason* (Princeton, 1970); and Arnold Thackray, *Atoms and Powers: An Essay on Newtonian Matter-Theory and the Development of Chemistry* (Cambridge, Mass., 1970).

5. Since "Germany" was not a political entity in the eighteenth century, it is necessary to define the term. For the purposes of this essay, "Germany" means those parts of Europe where German was the predominant language—the Holy Roman Empire (less the Hapsburg possessions in The Netherlands and Italy) plus the German-speaking parts of Alsace, Switzerland, and Eastern Europe (e.g., old Prussia and western Hungary).

6. For an illuminating discussion of contemporary social support for science, see Alvin M. Weinberg, *Reflections on Big Science* (Cambridge, Mass., 1967), pp. 65-114.

laboratories, and (3) the number of fairly productive and renowned chemists.[7]

At the beginning of the eighteenth century, a few medical schools were the only learned institutions in Germany with salaried positions for chemistry.[8] In the 1720's the number of medical professors who were at least partly responsible for chemistry began to grow at a rapid rate. From the 1740's this growth was complemented by the rise of chemical positions in scientific academies[10] and several nonmedical schools.[11] Throughout the remainder of the century,

Salaried Positions for Chemistry in German Schools and Academies, 1700–1800[9]

	1700	1710	1720	1730	1740	1750	1760	1770	1780	1790	1800
Number new during preceding decade	1	2	0	7	5	6	5	9	9	8	14
Total number on given date	5	5	4	10	13	18	21	28	33	38	48

7. The evidence presently available is either too fragmentary or too unwieldy to permit one to follow quantitatively the development of support for chemistry outside of learned institutions, changes in overall financial support for the science, the growth in the number of chemical authors, variations in overall or average productivity, etc.

8. Thanks to the iatrochemists, who campaigned vigorously for the use of chemical products and concepts in medicine, courses on chemistry were offered at one time or another during the seventeenth century in most German universities with medical faculties. The term "medical schools" will be used in this essay for medical faculties in universities and upper schools and for medical-surgical colleges.

9. For information on which schools and academies had salaried chemical positions on the given dates, see Table I. In calculating the number of new positions during the preceding decade, I have counted those few which were established and disestablished in the same decade (the positions in this category are not listed in Table I) but excluded those which were merely transferred from a closing institution to a successor institution.

10. In 1744 the Chemists J. H. Pott and A. S. Marggraf began receiving salaries from the Berlin Academy *qua* academicians. However, a decade later, when Marggraf became Director of the Academy's laboratory, the number of salaried positions for chemistry in the Academy was reduced to one. I am indebted to Dr. Christa Kirsten, Director of the Academy's archives, for this information. Late in the century the Munich Academy charged one of its salaried members with chemistry as well as physics.

11. The first nonmedical schools with salaried chemical positions were the new schools for prospective mining administrators in Schemnitz (1763), Idria (1763), Prague University's Law Faculty (1763), Freiberg (1765), and Berlin (1770). Their example was soon followed by a few schools whose primary purpose was to prepare the sons of nobles and high officials for bureaucratic or military careers. Toward the end of the century, a few universities established salaried chemical positions in economics and philosophical faculties.

Table I
Salaried Chemical Positions and Laboratories in German Schools and Academies, 1700-1800*

	1700	1710	1720	1730	1740	1750	1760	1770	1780	1790	1800
1. MEDICAL SCHOOLS											
Altdorf U. Med. Fac.	1,L	1,L	1,L	1,L	1,L	1,L	1,L	1,L	1,L	1,L	1,L
Bamberg U. Med. Fac.	X	X	X	X	1	X	X	0	1	0	1
Berlin Med. Surg. College	X	X	X	2	1	1	1(2)	1(2)	1	1	1
Bonn U. Med. Fac.	X	X	X	X	X	X	X	X	X	X	X
Bonn Central Sch. Med. Fac.	X	X	X	X	X	X	X	1	1	1	1
Budapest U. Med. Fac.	0	0	0	0	0	0	0	0	1	1	2
Cologne U. Med. Fac.	0	0	0	0	0	0	0	0	1,L	1,L	1
Duisburg U. Med. Fac.	0	0	0	0	0(1)	0(1)	0(1)	0(1)	0(1)	0(1)	X
Erfurt U. Med. Fac.	0	0	0	1,L	0	0	2	1	1,L	0(L)	0
Erlangen U. Med. Fac.	X	X	X	X	X	1,L	1	1	1	1	1
Frankfurt a.d.O. U. Med. Fac.	0	0	0	0	1	1	1	1	1	1	1,L
Freiburg i. Br. U. Med. Fac.	0	0	0	0	0	0	0	1	1	1	1
Fulda U. Med. Fac.	X	X	X	X	X	X	X	X	X	X	X
Giessen U. Med. Fac.	0	0	0	0	0	0	0	1(0)	1(0)	1,L	1(L)
Göttingen U. Med. Fac.	X	X	X	X	X	X	X	1	1	1,L	1,L
Graz Lyceum Med. Fac.	X	X	X	X	X	X	X	X	X	X	X
Greifswald U. Med. Fac.	0	0	0	0	0	0	0	0	0	1	1
Halle U. Med. Fac.	1	1	1	0	0	0	0	0	1	1	0
Heidelberg U. Med. Fac.	X	0	0	1(0)	1	1	1	1	1	1	1
Helmstedt U. Med. Fac.	1	1	1	1	1	1	1	1	1	1	1
Herborn High Sch. Med. Fac.	0	0	0	0	0(1)(L)	0(1)(L)	0	0	0	0	1,L
Ingolstadt U. Med. Fac.	0	0	0	0	1	1	1	2(L)	1,L	1,L	1,L
Innsbruck U. Med. Fac.	X	X	X	X	X	X	X	1	1	X	1(L)
Innsbruck Lyceum Med. Fac.	X	X	X	X	X	X	X	X	X	1(L)	X
Jena U. Med. Fac.	1	1	1	1	1	1	1	1	1,L	1	1
Kassel Collegium Med. Fac.	X	X	X	X	X	X	X	X	X	X	X
Kiel U. Med. Fac.	0(1)	0	0	0	0	0	0	0	1,L	0(1)	1
Königsberg U. Med. Fac.	0	0	0	1	1	1	1	1	1	1	1
Leipzig U. Med. Fac.	0	0	0	0	0	1	0	0	0	1,L	1,L
Mainz U. Med. Fac.	X	X	X	X	X	X	X	X	X	X	X
Mainz Central Sch. Med. Fac.	0(1)(L)	0(L)	0(L)	0	0	0	0	0	0	0	1,L
Marburg U. Med. Fac.											1,L

Münster U. Med. Fac.	X	X	X	X	X	X	X	X	X	X	1
Olmutz Lyceum Med. Fac.	X	X	X	X	X	X	X	X	X	1	1
Prague U. Med. Fac.	0	0	0	0	0	1	1	1	1	1,L	2(1),L
Rostock U. Med. Fac.	0	0	0	0	1	1	0	0	0	0	0
Strasbourg U. Med. Fac.	1	1	1	1	1	1	1	1	1	1	X
Stuttgart U. Med. Fac.	X	X	X	X	X	X	X	X	1	1,L	X
Tübingen U. Med. Fac.	0	1	1	1	1(0)	1	1,L	1,L	1,L	1,L	1,L
Vienna U. Med. Fac.	0	0	0	0	0	1	1,L	1,L	1,L	1,L	1,L
Vienna Med. Surg. Acad.	X	X	X	X	X	X	X	X	X	1	1
Würzburg U. Med. Fac.	0	0	0	0	1	1	1	1	1	1,L	1,L
TOTALS	5,1L	5,1L	4,1L	10,2L	13,1L	16,2L	20,3L	22,3L	26,9L	30,12L	34,10L
2. OTHER SCHOOLS											
Berlin Artillery School	X	X	X	X	X	X	X	X	X	X	1
Berlin Mining School	X	X	X	X	X	X	X	1	1	1	1
Clausthal Mining School	X	X	X	X	X	X	X	X	X	1	1
Erlangen U. Phil. Fac.	X	X	X	X	X	0	0	0	0	0	1,L
Freiberg Mining Academy	X	X	X	X	X	X	X	1	1	1	1,L
Giessen U. Econ. Fac.	X	X	X	X	X	X	X	X	1	X	X
Halle U. Phil. Fac.	0	0	0	0	0	0	0	0	0	0	1(L)
Heidelberg U. State Econ. Fac.	X	X	X	X	X	X	X	X	X	1,L	1,L
Jena U. Phil. Fac.	0	0	0	0	0	0	0	0	0	1,L	1,L
Kaiserslautern Cameral High School	X	X	X	X	X	X	X	X	1,L	X	X
Kassel Collegium	X	0	0	0	0	0	0	1,L	0	X	X
Marburg U. State Econ. Fac.	X	X	X	X	X	X	X	X	X	1	1
Prague U. Law Fac.	0	0	0	0	0	0	0	1	0	0	0
Rostock U. Phil. Fac.	0	0	0	0	0	0	0	0	0	0	1
Schemnitz Mining School	X	X	X	X	X	X	X	1,L	1,L	1,L	1,L
Vienna Ritterakademie	X	X	X	X	X	0	0	0	1	X	1
TOTALS	0	0	0	0	0	0	0	5,2L	6,2L	7,3L	12,5L
3. ACADEMIES											
Berlin	0	0	0	0	0	2	1,L	1,L	1,L	1,L	1,L
Munich	X	X	X	X	X	X	0(1)	0	0(1)	0(1)(L)	1(L)
TOTALS	0	0	0	0	0	2	1,L	1,L	1,L	1,L	2,L

* This table is a revised and improved version of Appendix V-A in K. Hufbauer, *op. cit.*, pp. 364-367. Only those learned institutions are listed which had salaried chemical positions on one of the dates given. X = the institution was not in existence. 0,1,2 = the number of salaried chemical positions (laboratory assistants have not been included). (0),(1),(2) = a possible alternative number of salaried chemical positions. L = institutional laboratory probably in use. (L) = institutional laboratory existed but was probably not in use for one reason or another. The totals for salaried chemical positions equal the sums of the numbers not in parentheses. The totals for laboratories equal sums of those probably in use.

the pace of this combined growth never slackened. Consequently, the number of salaried chemical positions in German learned institutions had climbed to forty-eight by 1800, nearly ten times higher than in 1700.[12]

Just as remarkable was the growth in the number of schools and academies with functioning laboratories.[13] Before 1750 the men who controlled these institutions displayed little interest in providing adequate facilities for chemistry. But then the number of laboratories began to climb, reaching six by 1770 and sixteen by 1790. During

Functioning Laboratories in German Schools and Academies, 1700–1800[14]

	1700	1710	1720	1730	1740	1750	1760	1770	1780	1790	1800
Number new during preceding decade	0	0	0	1	1	1	3	2	6	8	4
Total number on given date	1	1	1	2	1	2	4	6	12	16	16

the 1790's growth ceased as the rate at which new laboratories were constructed fell to that at which old ones were abandoned. Still, as the century closed, the number of German schools and academies with laboratories was many times higher than in 1700.

Counting "chemists" is not such a straightforward task. Procedures are needed for identifying them and specifying their periods of activity. Using a variety of approaches, I have developed a list of seventy-one chemists who were active in Germany during the eighteenth century and who enjoyed at least a modicum of con-

---

12. The occupants of many of these positions, it must be emphasized, encountered such difficulties as very low salaries, responsibilities that went far afield of chemistry, and a lack of facilities. Still, over half these men were provided some opportunities and incentives to participate in the development of chemistry. As of 1800, around two-thirds of the chemical representatives had at least one publication on a chemical topic to their credit.

13. Most of these laboratories were used by chemical representatives for their research and preparation of classroom experiments. Only a few were teaching laboratories open to all serious students—e.g., those at the Schemnitz and Freiberg mining schools.

14. For details, see Table I.

temporary renown.[15] I have established their periods of activity by assuming (1) that a man became an "active chemist" by publishing three short essays or a book on chemistry and (2) that he ceased being an active chemist three years after his last publication on a chemical topic, unless, of course, he died first.[16] The upshot is that

	"Active Chemists" in Germany, 1700–1800[17]										
	1700	1710	1720	1730	1740	1750	1760	1770	1780	1790	1800
Number new during preceding decade	2	0	1	6	2	5	5	5	16	13	11
Total number on given date	7	5	4	9	8	8	12	12	23	31	30

the number of chemists, after declining in the first two decades, jumped somewhat beyond its starting level in the 1720's and then remained essentially stable until the 1750's. The number then began to climb, reaching twelve by 1760 and, thanks to the influx of well over one new chemist per year during the 1770's and 1780's, thirty-one by 1790. During the 1790's, however, the number crested at thirty-five and then, as the rate of influx slowed, fell somewhat.[18]

15. In developing this list of chemists, I started with a list consisting of all those who received more than half a column in Gesellschaft für Geschichte der Pharmazie, *Chemisch-Pharmazeutisches Bio- und Bibliographikon,* ed. Fritz Ferchl (Mittenwald, 1937). Then, on the basis of more than two dozen contemporary lists of notable chemists, I removed and added several names. Finally, on the basis of my own knowledge of eighteenth-century German chemistry, I made a few more adjustments. The basic trends for the numbers of chemists are essentially the same, regardless of whether one uses the fairly "arbitrary" list based on Ferchl's work or this more "natural" list. The trends might well be altered, however, if one only counted those few chemists who made *significant* contributions to the development of chemical theory or technique.

16. These criteria for establishing a chemist's years of activity do something to improve on the arbitrary procedure of saying that a man was active from age twenty-five or thirty until his death.

17. For a chart which reveals the names and active periods of the chemists, see Table II.

18. Additional evidence of declining activity in chemistry during the 1790's is provided by the fact that the number of contributors to Germany's chief chemistry journal, L. Crell's *Chemische Annalen,* fell from over forty in 1789 to under twenty in 1797. In 1800 J. B. Trommsdorff complained that "the spirit of experimental investigation is beginning to get drowsy in Germany"—*Journal der Pharmacie,* 7:2 (1800), 246.

TABLE II

"Active Chemists" in Germany, 1700-1800

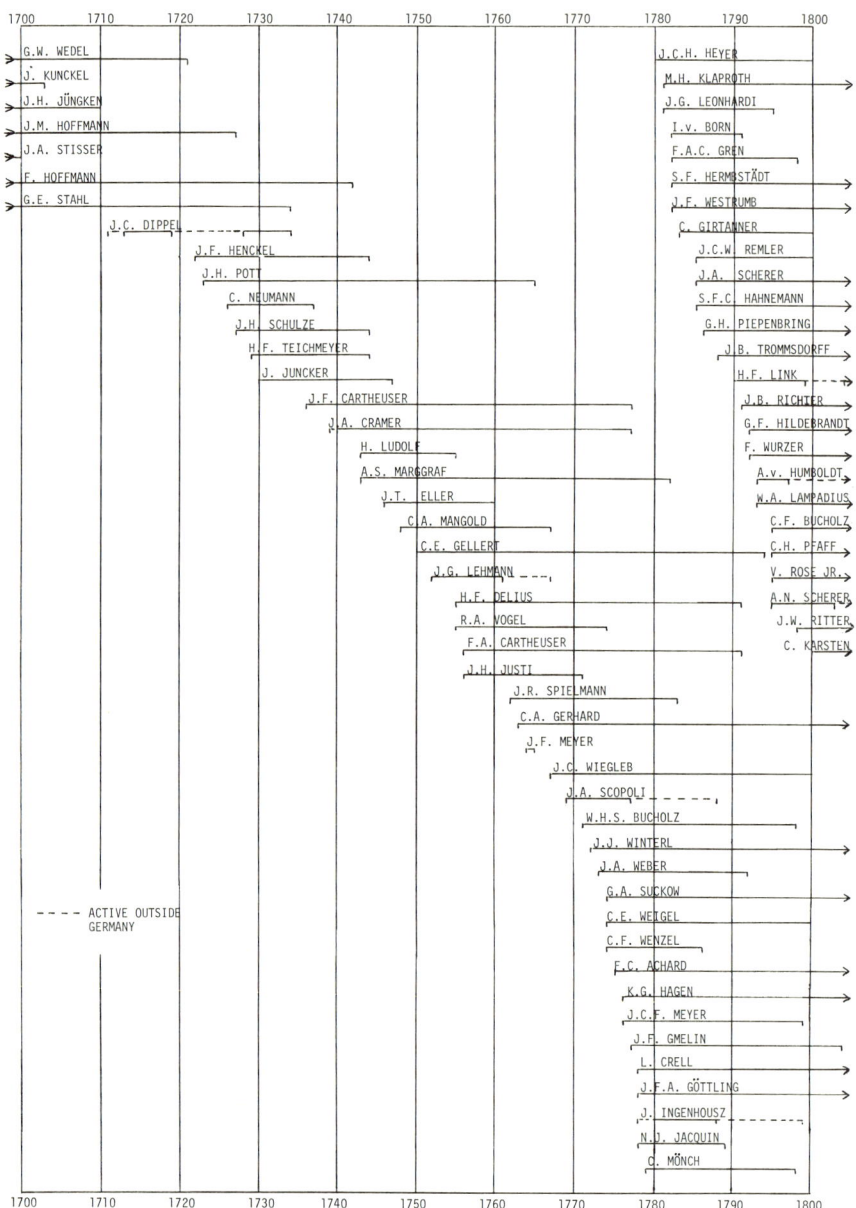

Nevertheless, there were thirty chemists in Germany at the end of the century, more than four times as many as at the start.

To sum up, during the first five decades of the century, the number of salaried chemical positions climbed significantly, but the numbers of laboratories and chemists registered only slight gains. By contrast, all three indices made quite impressive gains between 1750 and 1790, especially between 1770 and 1790. Finally, during the 1790's, while the number of positions continued to rise, the number of laboratories leveled off and the number of chemists declined slightly.

Social support for a scholarly discipline depends, I have suggested, upon the decisions of the discipline's potential patrons and participants. The higher the prestige (status, standing, esteem) of the subject in the eyes of such men, the greater the possibility that their decisions will be supportive. But what determines the prestige that a social group (e.g., the potential patrons) assigns to a discipline?[19] Apparently, two factors are crucial—(1) the group's image of the subject, which need not be and often is not accurate, and (2) the group's values. If the perceived attributes of the discipline are incompatible with or irrelevant to the group's goals, the group will have little esteem for the subject. On the other hand, if the perceived attributes harmonize with the group's values and aspirations, the group will grant the subject high status. Clearly, a change in either or both factors can affect the group's ranking of and inclination to provide support for the discipline. Indeed, I wish to suggest that the development of social support for chemistry in eighteenth-century Germany depended upon coinciding changes in *both* the image of chemistry *and* the values held by its potential patrons and participants.[20]

19. For a useful discussion of the prestige of scholarly disciplines, see Warren O. Hagstrom, *The Scientific Community* (New York, 1965), pp. 167-176. He examines how contemporary scientists assign prestige to "disciplines" (e.g., physics, chemistry, biology) and "specialties" (e.g., elementary particle physics, nuclear physics, solid-state physics), finding that they tend to esteem those fields which are fundamental (the net-flow of information is outwards) and theoretical. He does not, however, discuss how nonscientists assign prestige to disciplines or why the prestige of disciplines changes with time.

20. To establish this, or any other explanation of the development of social support for chemistry in eighteenth-century Germany would be an immense undertaking. The potential patrons—the professors, academicians, high officials and rulers who, during the course of the century, controlled Germany's many

1. 1700–1750

The prevailing image of chemistry during the first half of the eighteenth century is reflected in three important lexicons published in this period. The first was J. Hübner's *Curieuses und reales Natur-Kunst-Berg-Gewerck- und Handlungs-Lexicon* (1712). Its main article on chemistry appeared under the heading of *"Alchymia,"* to which readers who looked up either *"Goldmacherey"* or *"Chemia"* were referred. Alchemy was described as

> an art by virtue of which the pure is separated from the impure, or an art [for acquiring] an active and working knowledge (*Wissenschaft*) of natural things. It may be called practical physics (*Physica Practica*) because it resolves and dissolves all sublunary bodies into their first seeds or prime matter and recoagulates these into their former bodies so that the medicines which can be made from these [bodies] will be safer and healthier. . . . One can divide the chemical art into the vulgar or common [chemistry] known by apothecaries and physicians and the secret [chemistry] called alchemy which concerns itself with the preparation of the stone of wisdom upon which the transmutation of metals depends.

Elsewhere in the lexicon, the article on medicine identified chemistry as one of five branches of medicine, defining it as "the art which teaches how the best force can be drawn from natural things by means of fire." And under *"Lapis Philosophorum,"* it was observed that "most" of those who devoted themselves to "chemistry" believed in the stone, even though it could not be produced according to existing recipes which resulted from either "vain hope" or "plain fraudulence."[21]

A similar picture of chemistry emerges from J. T. Jablonsky's

---

schools and academies—probably numbered over two thousand. The potential participants—physicians and, as the century progressed, pharmacists and others who learned something of the science as youths—were far more numerous. Even if one focused on the actual patrons (roughly 350 men for the salaried chemical positions and 100 men for the laboratories) and the actual participants (71 men according to my criteria), the task of relating their images of chemistry and their values to their patronage and their careers would be formidable.

21. J. Hübner, *Curieuses . . . Lexicon . . .* (Leipzig, 1717), pp. 47-49, 383, 723, 931, 1039. I have not seen the first edition but there is no reason to presume that it was very different. This popular lexicon went through many editions, the last appearing in 1792.

Allgemeines Lexicon der Künste und Wissenschaften (1721). He described *"Chymie, Alchymie, Chymia,"* which he regarded as one of the six branches of medicine, as

> an art which teaches how one should dissolve natural substances, separate them from one another, combine them, and prepare wholesome medicines out of them; or [how one should] analyze mixed, compound, or aggregate bodies into their fundamental parts or synthesize the same bodies out of such fundamental parts. . . . Chemistry, insofar as it investigates the powers and properties of nature and teaches the preparation of drugs is very useful and necessary to the doctor. However, one must not go further, nor allow oneself to be seduced by it [chemistry], to try to make the notorious philosophers' stone and thereby gold and silver.

Anyone who did so, Jablonsky warned his readers, would soon be just as impoverished as "many poor chemists."[22]

The third lexicon was J. H. Zedler's *Grosses vollständiges Universal-Lexicon aller Wissenschaften und Künste* (64 vols., 1732–1750). Its main article on chemistry appeared in 1742 under the heading *"Scheidekunst."* After giving all the synonyms for chemistry, this article, in contrast to those in the earlier lexicons, clearly pointed out that one should distinguish *"Chymie"* from *"Alchymie"* since the one belonged to medicine, the other to gold-making. The article went on to give a brief history of chemistry, then defined it as the

> art which uses fire and adept manipulations to decompose substances that are precisely and firmly united by nature whether they belong to the metal, mineral, animal, or vegetable realms; which recombines the separated substances into a single substance; and which prepares wholesome medicines from these substances to preserve and restore the health of mankind.

The article then gave all the operations of chemistry, described the traditional artistic symbol for alchemy and chemistry—an old woman whose age represented experience—and closed with an etymology of the word.[23]

---

22. J. T. Jablonsky, *Allgemeines Lexicon* . . . (Leipzig, 1721), pp. 50, 137-138.
23. J. H. Zedler, *Grosses vollständiges Universal-Lexicon* . . . , photographic reprint (Graz, 1961), *34*, 1110-1111.

The treatment of chemistry in these three lexicons indicates that during the first half of the eighteenth century it was seen as being immediately concerned with the analysis and synthesis of natural substances. In addition, chemistry's popular image had three other prominent features. First, the subject was viewed as a manipulative art, not an experimental science. Second, it was linked with alchemy, quite closely in the early decades of the century and then, judging from Zedler's lexicon, less closely. And third, it was associated with medicine, especially pharmacy.

These three attributes of chemistry's image, much more than its immediate subject-matter, influenced how educated Germans rated the discipline. To an age that looked down upon manual labor, chemistry's association with craft-like empiricism was demeaning. Indeed, many thought of chemists as "charcoal boys," as "filthy cooks, laboratory helpers, and generally useless riff-raff,"[24] or as mere "artisans."[25] To an age that condemned swindling, chemistry's association with alchemy was dubious. This was small wonder, for alchemists, both fraudulent and sincere, had long been arousing hopes for great riches, hopes which they failed to fulfill. To an age that was beginning to abandon resignation to sickness and early death, chemistry's association with medicine gave rise to mixed feelings. On the one hand, powerful chemical drugs (usually metallic compounds) were known to bring about remarkable cures. On the other hand, these powerful drugs were unreliable, sometimes exacerbating the illness they were supposed to cure. And the claims for drugs were so exaggerated that many people were suspicious. How could they wholeheartedly accept the word of physicians as to the value of drugs when so many doctors boasted about "their fountains of youth, their incombustible oils, their hermetical anti-

24. These terms are from a debate about the merits and faults of chemistry in a rambling novel of the late seventeenth century—J. C. Ettner, *Des getreuen Eckharts entlauffener Chymicus, in welchem vornemlich der Laboranten und Process-Krämer Bosheit und Betrügerey, wie dieselben zu erkennen und zu fliehen; hernach bewärteste Artzney-Mittel in allerhand Kranckheiten und Zufällen menschlichen Leibes zu gebrauchen; dann sonderliche, philosophische, politische, medicinische am meisten aber chymische Anmerckung und Process* . . . (Augsburg and Leipzig, 1697), pp. 48-54.

25. The influential philosopher C. Wolff lumped "chemists, alchemists, and artisans" together as men whose unexplained experiments needed investigation. See his *Vernünfftige Gedancken von den Würckungen der Natur*, 3rd ed. (Halle, 1734), p. 567.

dotes, their elixirs of gold, their snake-powders, and precious stones, their remedies for snake bites, their tinctures and panaceas, and their six hundred other remedies out of Arabia. . . ."[26] Consequently, even chemistry's association with drugs was not regarded as an entirely positive one.

In sum, the educated public's attitude toward chemistry was strongly colored with contempt and suspicion in the early decades of the eighteenth century. Though the fairly favorable treatment of the subject in Zedler's lexicon suggests that this attitude was dissipating by the 1740's, the layman did not yet have much reason to esteem chemistry and its devotees. This, it would seem, was why the numbers of laboratories and productive chemists in Germany failed to increase appreciably during the first half of the century.

The only sizable group that did not share the prevailing attitude toward chemistry was the physicians.[27] They had a different estimate of the subject because both their background and values were different. Most doctors had learned enough about chemistry as students to avoid confusing it with alchemy. Moreover, most placed a much higher value on empiricism and good health than their contemporaries. Some doctors, still enamored with the ambitious claims of the iatrochemists, regarded chemistry as the "principal and most noble part of medical study."[28] A growing majority, though they held a more moderate estimate of chemistry's merits, esteemed it

26. J. B. Mencke, *The Charlatanry of the Learned,* trans. by F. E. Litz and notes by H. L. Mencken (New York, 1937), p. 59. Mencke, a distant relative of the American pundit, also commented in his essay of 1715 that "although they [the doctors] are profoundly ignorant about the real effect of their medicines, they administer pills, syrups, drops and I cannot tell what other panaceas to their patients with an assurance so overwhelming that sometimes they even promise to restore the dead to life" (p. 165).

27. Some of the early cameralists recognized the great technical potential of chemistry; see U. Troitzsch, "Ansätze technologischen Denkens bei den Kameralisten des 17. and 18. Jahrhunderts," *Schriften zur Wirtschafts- und Sozialgeschichte,* 5 (1966), *passim.* So did some men associated with the mining industry; see, for instance, W. Herrmann, "Bergrat Henckel. Ein Wegbereiter der Bergakademie," *Freiberger Forschungshefte: Kultur und Technik, D37* (1962), *passim.* And some natural philosophers and natural historians recognized chemistry's relevance to their fields; see, for example, A. Harnack, *Geschichte der königlich preussischen Akademie der Wissenschaften zu Berlin* (Berlin, 1900), *passim.* These men, however, played but a secondary role in the transformation of chemistry's public image which is discussed below.

28. R. Klüpfel, *Geschichte und Beschreibung der Universität Tübingen* (Tübingen, 1849), p. 163.

as an important auxiliary discipline because of its relevance to pharmacy. Indeed, as doctors, especially professors of medicine and physicians to rulers, became increasingly committed to public health, they took steps to insure that chemistry, which had been haphazardly taught in most medical schools, occupied a regular place in the curriculum. In most cases, their goal was to prepare the physician for the important task of testing prospective apothecaries and inspecting pharmaceutical shops. The result of these efforts was, as we have seen, that the number of medical schools with salaried positions for chemistry grew from four in 1720 to sixteen in 1750. Another consequence of the doctors' favorable attitude toward chemistry and their increasing contact with the subject was that most of the men who became chemists before 1750 were trained physicians.

Those doctors and occasional others who became chemists must have found the prevailing lack of respect for their discipline quite discouraging. They could not deny that most chemists of the past had relied upon crude empiricism, pursued alchemical fantasies, and tried to prepare miracle drugs. But, influenced by Pietism and Western European rationalism, they came to think that their subject could be a great deal more than it had been. That is, in the early decades of the eighteenth century, they came to entertain and to want the public to entertain a new image of chemistry.

First, they had to dissociate themselves and their subject from simple artisans and avaricious alchemists. G. E. Stahl, the most influential proponent of the chemists' new position, complained in 1726 that the writers of dictionaries, unaware that alchemy and chemistry had parted company, still equated the two. He then characterized alchemy as a "bewildering," "incomprehensible," and "futile" undertaking to make gold. Chemistry, by contrast, was "a rational, deliberate, and comprehensible investigation and processing [of substances which] leads to fundamental knowledge."[29] On another

29. G. E. Stahl, "Bedencken von der Gold-Macherey," in J. J. Becher, *Chymischer Glücks-Hafen* . . . , 3rd ed. (Leipzig, 1755). For Stahl's influence in Germany, see W. Strube, "Die Auswirkung der neuen Auffassung von der Chemie in Deutschland in der Zeit von 1745 bis 1785," unpublished dissertation (Karl-Marx-University in Leipzig, 1961) and "Die Ausbreitung der Naturanschauung G. E. Stahls unter den deutschen Chemikern des 18. Jahrhunderts," *Zeitschrift für Geschichte der Naturwissenschaften, Technik und Medizin*, 1:2 (1964), 52-61.

occasion, he indignantly insisted that "bungling is characteristic of day-laborers, not chemists."[30] One of Stahl's disciples grumbled that "people generally call those men chemists who know how to carry out a task with fire according to a prescribed recipe. . . ." He also warned that anyone who approached Stahl's work as "a mere recipe-monger or greedy goldgrub" would find nothing of interest.[31] Another of Stahl's followers, J. Juncker, urged the reader of his text (1730) not to confuse "true" or "philosophical" chemistry with the "common chemistry" of pharmacists and artisans which was "purely empirical or mechanical." He also insisted that a recommendation to pursue chemistry was not the same as advising someone to embark "on the wild sea and open ocean of alchemy or goldmaking where . . . many a man has already been shipwrecked and lost his capital, his profession and standing, his health, honor, and life, yes, his temporal and probably even his eternal well-being." True chemistry did not deserve, therefore, the "great fear and contempt" with which many regarded it.[32]

In attempting to dissociate chemistry from the crafts, however, chemists did not deny the necessity of work in the laboratory. Rather, they portrayed experimentation as the means, not the end, of chemistry. Stahl went so far as to claim that "chemistry is a science, a science of causes in which the laboratory operations are the least part. . . ."[33] Likewise, in their denunciations of alchemy, the chemists did not deny the possibility of transmuting metals. Rather, they maintained that transmutation must be approached with the hope of enriching knowledge, not making a quick fortune. They stressed the importance of first determining the composition of gold and silver before trying to produce these metals.[34] In brief, though chemists sought to set themselves apart from artisans and alchemists,

---

30. Quoted in Partington, *op. cit.*, *2*, 664.
31. Translator's preface to G. E. Stahl, *Chymia rationalis et experimentalis; oder gründlicher, der Natur und Vernufft gemässe und mit Experimenten erwiesene Einleitung zur Chymie*, 2nd ed. (Leipzig, 1729).
32. J. Juncker, *Conspectus Chemiae Theoretico-Practicae* . . . , 3rd ed. (Halle, 1749-1753), *1*, 1-3, 45, 52. I have not seen the first edition, but the third is probably not greatly changed.
33. Quoted in Partington, *op. cit.*, *2*, 664.
34. For Stahl's views on how the transmutation of metals should be approached, see his "Bedencken . . . ," *op. cit.*, and Partington, *op. cit.*, *2*, 685-686.

they continued to affirm the importance of experimentation and the possibility of transmutation.

Chemistry's association with medicine, it will be remembered, was the most favorable feature of the subject's popular image in the early eighteenth century. Throughout the century, chemists continued to call attention to their discipline's importance to medicine. However, Stahl and an increasing number of chemists argued that the subject should be more than a handmaiden to medicine. They came to see that chemistry's "true nature, effectiveness, attractiveness, and usefulness"[35] qualified it to be an independent and broadly applicable science.

For one thing, a growing number regarded chemistry as a penetrating tool of the natural philosopher. Some, like F. Hoffmann and H. F. Teichmeyer, put the chemical approach to nature on a par with the mechanical approach.[36] Others went further. Stahl, for example, maintained that chemistry penetrated much deeper than the highly touted mechanical philosophy which "scratches the shell and surface of things and leaves the kernel untouched."[37] In a similar vein, Juncker insisted that chemistry, not mathematics, was the "true key for penetrating the deepest secrets of natural objects." Only chemistry reveals their "essential, real, and original constituents."[38]

Not only did more and more German chemists regard their science as a fundamental part of natural philosophy, but they also came to believe in its *general* usefulness. In recognition of its many uses, both real and potential, they divided chemistry into various branches. One of the first to do so was Teichmeyer who, in his chemistry text of 1729, partitioned the field into "physical chemistry" (concerned with understanding nature), "medical chemistry" (concerned with therapy and drugs), "metallurgical chemistry" (concerned with assaying), "transmutational chemistry" (concerned with transmuting metals), "mechanical chemistry" (concerned with crafts

---

35. Translator's preface to G. E. Stahl, *Zymotechnia Fundamentalis* ... (Frankfurt and Leipzig, 1734). The translator gave Stahl credit for bringing chemists to adopt this new attitude toward chemistry.

36. F. Hoffmann, *Opuscula Pathologico-Practica* ... (Venice, 1739), p. 275, and H. F. Teichmeyer, *Elementa Philosophiae Naturalis Experimentalis* ..., 3rd ed. (Jena, 1733), preface to first ed. (1716).

37. Quoted in Partington, *op. cit.*, 2, 665.

38. Juncker, *op. cit.*, p. 10.

such as glass-making, dyeing, printing, and salt-making), and "economical chemistry" (concerned with agriculture).[39]

During the first half of the eighteenth century, therefore, most German chemists came to regard their discipline as an experimental science which was both fundamental and generally useful. They disseminated this view primarily as teachers and as text-writers. Though their courses and texts were oriented toward pharmaceutical chemistry, they often opened with general observations about the science's great potential. Some students most likely embraced these general notions about chemistry, passing them on to others after leaving the universities. Even before 1750, therefore, the chemists must have made some headway in their campaign to replace the prevailing image of chemistry as a menial, alchemical, and pharmaceutical subject with their image of it as a fundamental and useful experimental science.[40] However, it was after 1750, as educated and powerful Germans embraced new values and aspirations, that the science's prestige really climbed.

2. 1750–1790

During the 1750's and 1760's, the Enlightenment triumphed in Germany. This triumph had long been in the making. Early in the century many Germans had been persuaded by the Pietists

---

39. H. F. Teichmeyer, *Institutiones Chemiae Dogmaticae et Experimentalis*... (Jena, 1729), p. 4.

40. Indeed, in 1746 the Berlin chemist J. H. Pott wrote that "rational chemistry and its further investigation presently awakens such general approval from most of the learned and reasonable men within not only the cultured nations but probably also the barbaric peoples that they always receive its experiments and endeavors with gratitude." See his *Chymische Untersuchungen welche fürnehmlich von der Lithogeognosia oder Erkäntniss und Bearbeitung der gemeinen einfacheren Steine und Erden ingleichen von Feuer und Licht handeln* (Potsdam, 1746), preface. So far as I know, this is the first time that a member of the Stahlian school spoke favorably of the educated public's opinion toward chemistry. Pott was not without reason for this high appraisal, since patronage for his work had been increasing steadily since the 1720's. Nevertheless, I believe Pott's appraisal was too optimistic. For instance, in 1755 J. C. Zimmermann, a chemist-physician in Schneeberg, wrote that "very few people know what chemistry actually is. In general either they equate chemistry and alchemy, calling the chemist an alchemist, goldcook, or swindler, or they assume that chemistry deals only with smelting and assaying, only with the mineral and not the vegetable and animal realms." See his *Allgemeine Grundsätze der Theoretisch-Practischen Chemie* . . . (Dresden, 1755), p. 2.

and the "moral weeklies" to abandon traditional otherworldly Christianity with its pessimistic acceptance of the evil *status quo*. They were to seek, instead, the improvement of private and public morality by living as productive and socially useful Christians. Somewhat later many Germans were won to the gospel of rationalism by C. Wolff and his disciples. In the 1740's these two currents coalesced into a single movement dedicated to achieving social betterment through economic growth, medical progress, educational reform, and the advance of knowledge. During the 1750's and 1760's, thanks to state encouragement and, after the Seven Years' War (1756–1763), unprecedented prosperity, this popular Enlightenment gained the allegiance of most educated Germans.

Meanwhile, the ideal of "enlightened" despotism was gaining the allegiance of ruling elites in Germany. Rulers and their close advisors had long been striving to consolidate and strengthen state power. Prussia's Frederick II and Empress Maria Theresa, both of whom came to the throne in 1740, were among the first to recognize that the growing commitment of educated Germans to the Enlightenment could be used to this end. Thus, though the ultimate goal of these rulers and a growing number of emulators was usually greater power rather than social meliorism, they too began to promote material and intellectual progress. At first they proceeded at a leisurely pace. But after the Seven Years' War, with its legacy of immense debts and uneasy peace, ruling groups approached the task of creating "enlightened" despotisms with urgency and dedication.

Thus, enthusiasm for material and intellectual progress came to the fore in the 1750's and 1760's. This enthusiasm—it remained important through the 1770's and 1780's—created a favorable climate of opinion for all the natural sciences. Enlightened men looked to these sciences to promote economic growth by increasing the efficiency of existing productive activities and suggesting new ones. Moreover, they counted on the natural sciences to improve medicine. Last, and probably least, they saw the natural sciences as an important source of profound and exciting insights about the world.

Chemistry was one of the natural sciences which attracted the attention of enlightened men. Physicians, because of their contact with the subject as students, were among the first to broadcast its virtues. In 1752, for instance, F. Börner introduced a biography of the chemist J. H. Pott with the following remarks:

> ... it cannot be denied that there has never been a lack of persons who conduct themselves as the declared and sworn enemies of chemistry and its admirers and who have banned it as completely useless or unnecessary from the republic of letters. On the other hand, there has never been a lack of men who, as experienced judges of useful knowledge, have counted it among the sciences which are indispensable for a true doctor and perform the most beneficial services for people of nearly all ranks. For it is chemistry alone which unlocks the secrets of nature and leads us to the very heart of things, discovering the most wonderful contrivances which otherwise would be eternally hidden from us . . . ; it contributes much to the commonweal; it delights, entertains, and benefits; it is indispensable.[41]

Though this paean indicates that many people still entertained serious doubts about chemistry, it also suggests that esteem for the subject was beginning to climb.

Nine years later, the chemist J. G. Lehmann introduced a collection of A. S. Marggraf's treatises with a preface which further illuminates public attitudes toward chemistry. He opened by remarking that

> There has probably never been a century richer in chemical authors than the present and it goes with chemistry as with medicine, *Fingunt se Chymicos omnes.* The state official, the financier, the barber and surgeon, the brewer and distiller, the dyer, the tanner, the old woman, the charcoal-carrier and woodcutter, and, yes, the project-maker (oh, what a deplorable name), all are clever enough to count themselves among the chemists. Thousands, who have been ruined by such people, are the sad witnesses to this fact.

He went on to insist that Marggraf had nothing in common with such frauds, warning alchemists that they would find nothing of interest in the book. Then, he condescendingly furnished those "narrow" men who were only interested in chemistry's practical applications with some ideas for using Marggraf's results. Finally, he declared his pride in being a midwife for "such a useful and impor-

---

41. F. Börner, *Nachrichten von den vornehmsten Lebensumständen und Schriften jeztlebender berühmter Aerzte und Naturforscher in und um Deutschland* (Wolfenbüttel, 1752), 2, 485-487.

tant work," one that he believed needed no recommendation for true scholars.[42] While Lehmann's preface reveals that chemistry still retained something of its old reputation as a risky subject, it also indicates that the science had numerous followers. Moreover, both his haughty treatment of the cameralists—an important group of allies—and his confidence in the general scholar's interest suggest a new self-assurance on the part of chemists.

Lehmann and his fellow chemists had good reason to be more confident than their predecessors. Not only were physicians like Börner portraying chemistry as an important subject, but so were influential proponents of economic development. In 1763 the cameralist D. G. Schreber gave chemistry a prominent place in an article recommending the establishment of scientific-technical schools for prospective administrators. He thought such schools should have five professors, one of whom would be responsible for mineralogy and chemistry. This man would teach

> *physical chemistry* in its entirety, demonstrating the requisite experiments in the laboratory;
> *economic chemistry,* which rests on the former, explaining the theory and practice of its various parts, namely, dyeing, salt and saltpeter works, glass making, lime and brick firing, ceramics, porcelain making, economic metallurgy (steel making, gold and silver work, wire making);
> *mining science,* namely mining metallurgy, smelting, assaying, etc., by exhibiting the necessary models of ovens, etc. in the model collection.

In addition to his teaching, this professor would manage any of the school's factories which depended upon chemical operations

---

42. J. C. Lehmann, foreword (dated 25 March 1761) to A. S. Marggraf, *Chymische Schriften,* 2nd edition (Berlin, 1768), *1.* "Fingunt se medicos omnes" was apparently an old saying; see J. B. Mencke, *op. cit.,* pp. 37-38, 168. According to the *Deutsches Theatrum Chemicum* ..., ed. F. Roth-Scholtz (Nuremberg, 1730), *2,* 290, the following verse appeared in a work by Pantaleon of 1676:
>> Es will fast jedermann ein Alchimiste heissen,
>> Ein grober Idiot, der Junge mit dem Greissen,
>> Ein Scherer, altes Weib, ein kurzweiliger Rath,
>> Der kahl-geschorne Münch, der Priester und Soldat.

and would investigate the uses of local earths, stones, and minerals.[43]

Again, in 1776 J. G. Krünitz defined *"Chemie"* in his highly successful *Oeconomische Encyclopädie* as "that science which has as its subject the study of the nature and properties of all substances by decomposing and combining them. . . . Chemistry not only acquaints us with the nature and properties of substances, but also teaches us the correct handling of substances in order to make them useful in the world." He then went on to discuss its wide range of applications in agriculture and the crafts.[44]

Two years later the *Bergmännisches Wörterbuch* defined *"Chymie"* as "a science which investigates, purifies, transforms, decomposes, combines, and determines the natural character and effect of natural substances and makes these useful to various sciences and crafts." Significantly, it was only after giving this general definition that the article mentioned assaying and smelting, the two branches of chemistry most relevant to the intended reader.[45]

Even men whose interests were quite distant from chemistry were beginning to have favorable things to say about the subject. In 1768 the orientalist J. D. Michaelis described natural history, chemistry, and physics as the three main branches of natural science. He believed that the well-born student should give special attention to each of these subjects. Without chemistry, Michaelis claimed, one could not go far in the study of nature, especially the mineral realm. Moreover, this science was essential for mining and smelting, dyeing, minting, and other things which a man of quality should understand. Finally, nothing prevented infection from the alchemical disease better than a sound knowledge of chemistry.[46]

In 1774 the poet Klopstock also mentioned chemistry. At the beginning of his *Die deutsche Gelehrtenrepublik,* in which he

---

43. D. G. Schreber, "Entwurf von einer zum Nutzen eines Staats zu errichtenden Akademie der ökonomischen Wissenschaften," in his *Sammlung verschiedener Schriften, welche in die ökonomischen, policey- und cameral-auch andere Wissenschaften einschlagen* (Halle, 1763), *10,* 417-436. For evidence of Schreber's influence, see W. Stieda, "Das Projekt zur Errichtung einer 'Kameral-Hohenschule' in München im Jahre 1777," *Forschungen zur Geschichte Bayerns, 16* (1908), 91-93.
44. J. G. Krünitz, *Oeconomische Encyclopädie* . . . (Berlin, 1776), *8,* 53.
45. *Bergmännisches Wörterbuch* . . . (Chemnitz, 1778), p. 123.
46. J. D. Michaelis, *Raisonnement über die protestantischen Universitäten in Deutschland* (Frankfurt and Leipzig, 1768), *1,* 243,

sought to dramatize the need for a genuine national culture, Klopstock divided his Republic's inhabitants into various orders. Creative thinkers and scholars constituted the ruling order, which was made up of eleven guilds. The "great and venerable guild" of the natural scientists included the chemists. Klopstock commented, however, that the chemists could just as well have had their own guild, for in contrast to the natural scientists who merely described nature, they acted upon it through synthesis and analysis.[47]

Four years later the poet Wieland drafted a reform proposal for ailing Erfurt University. From this proposal, it is clear that Wieland regarded chemistry as a natural science which could only be taught in a laboratory by a man with special expertise. It, along with the other natural sciences, was indispensable to the physician. However, because it and natural history were also of interest and importance to "philosophers, cameralists, and many other scholars," they should be taught every year.[48]

That Michaelis, Klopstock, and Wieland, men whose central interests were so far from chemistry, should have held such favorable attitudes about the science testifies to its attainment of respectability by the 1770's. Since the 1740's the subject had changed in the eyes of educated and influential Germans from a menial and possibly dubious source of drugs to a useful and fundamental science. During this same period, such men had come to prize those activities which promoted material and intellectual progress. The combined result of these two developments was a tremendous improvement in chemistry's prestige by the 1770's. Indeed, during this and the next decade, the prevailing image of chemistry harmonized so closely with prevailing values and aspirations that the discipline came to enjoy the status of a *Lieblingswissenschaft*.[49] It appealed to "the friends of natural science, medicine, economy, and manufacturing,"[50]

47. F. G. Klopstock, *Die deutsche Gelehrtenrepublik* (Hamburg, 1774), p. 15.
48. W. Stieda, "Erfurter Universitätsreformpläne im 18. Jahrhundert," Akademie gemeinnütziger Wissenschaften zu Erfurt: *Sonderschriften*, 5 (1934), 178, 186, 211-212.
49. So far as I know, chemistry was not called a "favorite science" before the 1790's. In that decade, however, there were many in the younger generation who believed it did not merit such esteem. See below.
50. This phrase is from the title of Crell's chemical journal—*Chemisches Journal für die Freunde der Naturlehre, Arzneygelahrtheit, Haushaltungskunst und Manufakturen* (1778-1781). After publishing a quarterly with a quite different title from

to not only doctors but also "scientists, metallurgists, cameralists, and economics and finance officials,"[51] and in Berlin to "persons of all orders [including] distinguished persons of the fair sex."[52]

The developments that enabled chemistry to become a *Lieblingswissenschaft* between 1750 and 1790 underlay the growth and diversification of patronage for the science during the period. As the chemists' image of their subject spread, justifications for patronizing chemistry broadened beyond its relevance to pharmacy. And as the deepening commitment to material progress led to the founding, expansion, and reorganization of medical schools and, after the Seven Years' War, mining and administrative schools, institutional opportunities for patronizing chemistry became more numerous and diverse. Other developments, to be sure, also played a role in the growth of patronage. For instance, the sustained prosperity after the Seven Years' War and, in Catholic lands, the confiscation of Jesuit properties provided ruling elites with new resources. Still, it was their increasingly favorable estimation of the science which led them to devote some of these resources to the support of chemistry teachers and laboratories. Similarly, both the rising competition between and mounting respect for specialized scholarship in German universities redounded to the advantage of chemistry precisely because the science was esteemed for its potential contributions to learning, health, and the economy.

The favorable transformation of chemistry's public image and standing between 1750 and 1790 also underlay the growth in the number of chemists during this period. As awareness of the science's relevance to desired goals spread, public spirited men became more desirous of contributing directly to its development. And as the prestige of chemistry and its devotees climbed, men anxious for social recognition must have found participation in the science more attractive. Indeed, pharmacists, who as a group had ready access to laboratories and tended to be both public spirited and, on account

---

1781 to 1784, he returned to nearly the same form when shifting to monthly publication in 1784—*Chemische Annalen für die Freunde*. . . . This journal was usually referred to simply as the *Chemische Annalen*.

51. P. L. Wittwer, "Lebensgeschichte Dr. Jac. Reinbold Spielmann, der Arzneygelahrtheit Prof. in Strassburg," *Chemische Annalen*, *1* (1784), 563.

52. H*** in Berlin to Crell, *ibid.*, p. 342.

of doctors' snobbism, status conscious, entered chemistry at an unprecedented rate.[53] Similarly, eagerness to contribute to the commonweal, desire for recognition, plus a touch of nationalism led L. Crell to found a periodical for chemistry in the late 1770's and induced numerous others to help him make it one of the first successful discipline-oriented journals in science.

3. THE 1790's

During the last decade of the eighteenth century, the patrons of chemistry—generally men who came to maturity along with the triumph of the Enlightenment—continued to value the science as a useful and enlightening subject. J. C. Fabricius represented their viewpoint when he wrote in 1796 that chemistry "has become a favorite science (*Lieblingswissenschaft*) and also exerts considerable influence on the various productive endeavors of the inhabitants. Therefore, it merits greater support at universities."[54] J. F. Gmelin represented the same viewpoint when he wrote in 1797 that chemistry was

> the favorite science (*Lieblingswissenschaft*) of the great, to whom it promises golden mountains, the rapid restoration of ruined finances as well as ruined health, and who reward its devotees with royal generosity; the mainstay of all medicine which underlies all that happens in living man, healthy as well as sick, including the effects of drugs; the refuge of the wise who seek light and instruction; the most important auxiliary science of the naturalist, which gives him information where other disciplines desert him; the key to many secrets of nature; the chosen guiding star in the labyrinth of countless trades which nourish, bless, and enrich peoples and states; and the rational basis of smelting, metal working, the arts, and the crafts. . . .[55]

And so did Goethe when that same year he wrote Duke Carl August of Weimar urging him to employ the young chemist A. N. Scherer. Goethe began by informing the Duke of his conviction that "You

---

53. Of the twenty-nine men to "qualify" as chemists between 1770 and 1790, thirteen were apothecaries or assistants of apothecaries.
54. J. C. Fabricius, *Ueber Academien, insonderheit in Dännemark* (Copenhagen, 1796), p. 105.
55. Gmelin, *op. cit., 1*, 2.

will reap much that is useful and enjoyable from this acquisition [Scherer]." He concluded by telling the Duke that both he and A. v. Humboldt were convinced that "Your Highness can expect much good for Yourself and Your District and also perform another service for science with this acquisition."[56]

Valuing chemistry, the Duke and other potential patrons continued to establish salaried positions for chemistry during the 1790's. Indeed, they did so at a more rapid rate than ever before. This fact suggests that the concomitant decline in laboratory construction was not the result of ambivalence toward chemistry. Rather, this decline must have been due to a relative shortage of funds in the troubled decade of the French Revolution.

By contrast, it seems that the decline in the number of new chemists in the 1780's and 1790's did reflect a new ambivalence toward chemistry among its potential participants. The generation coming to maturity in this period was swept up by the Romantic and Idealist movements.[57] Those in step with these movements found the Enlightenment approach to the world unappealing, even repulsive. Many, for instance, rejected the prevailing utilitarianism, the idea of voluntarily engaging in "useful" activities, the desire to be "useful acquisitions." Wilhelm von Humboldt, Alexander's elder brother, put it succinctly when he wrote his fiancée in 1790, shortly before resigning his government post, that "the idea of usefulness is a conceit (*Eitelkeit*)."[58] Such hostility to utilitarianism must have reduced this generation's respect for chemistry since, as we have seen, material utility was the most prominent feature of the science's popular image.

Chemistry's reputation as an independent science which, because

---

56. *Briefwechsel des Herzogs-Grossherzogs Carl August mit Goethe,* ed. H. Wahl (Berlin, 1915-1918), *1,* 215-218. Scherer received a stipend for travel to Britain. Upon his return, he was made a Mining Councilor and charged with giving public lectures on chemistry in Weimar.

57. For an illuminating discussion of the high Enlightenment and the subsequent rebellion against this worldview, see H. Brunschwig, *La crise de l'état prussien à la fin du xviiie siècle et la genèse de la mentalité romantique* (Paris, 1947).

58. Quoted in R. E. Goldsmith, "The Early Development of Wilhelm von Humboldt," *The Germanic Review, 42* (1967), 45. A. N. Scherer displayed a similar hostility to utilitarianism in his praise for Gren's text as a work which properly subordinated "all mercantilistic aims." See his "Friedrich Albrecht Carl Gren," *Allgemeines Journal der Chemie, 2* (1799), 375.

of its analytical approach, yielded fundamental insights about nature also detracted from its appeal to many in the younger generation. They revolted against the notion that any one specialty or approach held the key to nature, embracing instead a holistic approach to the world. At least one of the older generation, the physicist G. C. Lichtenberg, understood and sympathized with this new feeling. He captured it in a fictitious "dream" which he published in 1794. The dream opened with Lichtenberg high above the earth, facing an old man who inspired his devotion and trust. He gave Lichtenberg a small mineral sphere, asked him to determine its nature in a nearby laboratory, and disappeared. Lichtenberg inspected the mineral, wiped it off, tested its electrical and magnetic properties, and determined its hardness and specific weight. Then he analyzed it, finding argillaceous earth, calcareous earth, silicious earth, iron, salt, and one unknown substance. He knew he had been very accurate in his analysis, for when he tallied up the components they came out exactly to a hundred. Just as he finished, the old man reappeared, studied the results with a smile, then informed Lichtenberg that the mineral sphere was really the Earth. Astounded, Lichtenberg asked what had happened to the ocean and learned that he had wiped it off. The man went on to tell Lichtenberg what his various tests had done to the Earth, then gave him a bag, told him to test its contents chemically, and again vanished. Thinking that the bag might contain the sun or a planet, Lichtenberg resolved to proceed more carefully this time. But when he opened the bag, he found only an old book in an incomprehensible script. All he could read was the title page: "Test this chemically my son and tell me what you find." Lichtenberg wondered how he could test the meaning of a book by chemical means. A chemical analysis would only reveal rags and printers ink. Suddenly, the light dawned on him. He shouted out, "I understand, I understand, Immortal Being; oh forgive, forgive me; I accept your good reproof!" Then he awoke.[59]

Such dissatisfaction with the Enlightenment's approach to nature

---

59. G. C. Lichtenberg, *Vermischte Schriften*, 2nd ed. (Göttingen, 1844-1853), *6*, 48-60. A. Schneider has shown that elements of the "dream" can be traced back to 1774 in Lichtenberg's notebooks; see his *G. C. Lichtenberg, Precurseur du Romantisme* (Nancy, 1954), *2*, 233-234.

greatly reduced the appeal of devoting oneself to a single discipline such as chemistry. Many of those who did get involved in chemistry did so only in order to be able to develop an integrated picture of nature. Some, such as A. v. Humboldt, proceeded cautiously with this enterprise, trying to weave the results of eighteenth-century specialization into a coherent whole. Others, F. Schelling for example, plunged impetuously into the task, creating a whole new *Naturphilosophie*. Hence, as the Enlightenment waned in Germany during the 1790's so did the enthusiasm of many young Germans for chemistry.

To recapitulate, I have suggested that social support for chemistry was low in early eighteenth-century Germany because its popular image as a menial, alchemical, and pharmaceutical subject was at odds with prevailing values. Only the physicians, who could distinguish it from alchemy and appreciate its empirical approach and pharmaceutical applications, esteemed chemistry. From the 1720's, they managed to secure the establishment of a growing number of salaried chemical positions in medical schools. Meanwhile, led by G. E. Stahl, the chemists were coming to see their discipline as a fundamental and broadly useful natural science. Soon they were disseminating this image of chemistry in classrooms and textbooks from their bases in the medical schools. From the 1750's, due to the spread of this new image *and* to the increasing importance attached to material and intellectual progress, social support for chemistry expanded on all fronts. During the 1770's and 1780's, when the Enlightenment and "enlightened" despotism were at their peak, patronage of and participation in this useful and fundamental science improved dramatically. In the last decade of the century, chemistry's patrons—men of the Enlightenment—continued to value and support the science much as before. By contrast, some potential participants—young men who rejected the Enlightenment—found chemistry unappealingly utilitarian, narrow, and analytical. Nevertheless, because there were others who did not go so far in rejecting the values and aspirations of the past, the turn of the century found the level of social support—participation as well as patronage—for chemistry in Germany far above its low starting point in the early eighteenth century.

Newtonian Forces and Lockean Powers: Concepts of Matter in Eighteenth-Century Thought

BY P. M. HEIMANN[*] AND J. E. MC GUIRE[**]

INTRODUCTION

The interpretation of eighteenth-century natural philosophy presents special difficulties. Though Newton's work was frequently lauded as providing the key to unlock the secrets of nature, the form of this key is not immediately apparent. Indeed, the extension, modification, and—on occasion—the rejection of Newtonian principles often appear obscure. Furthermore, the influence of other seventeenth-century systems of thought on eighteenth-century natural philosophy is still a crucial question awaiting analysis. In this study we are concerned with speculative British natural philosophy in the post-Newtonian period; but even within these limits the problems of interpretation are enormous. In what follows we attempt to explore theories of matter and force which in part arose out of the Newtonian tradition. This study is not intended as a delineation of the meaning of "Newtonianism" in eighteenth-century natural philosophy. We are concerned to examine a number of related themes in the thought of the period, themes surrounding discussions of the nature of matter. We will contend that many interpretations of nature arose from an intellectual revolution which was philosophical in character, as well as from the impact of scientific events. We will be concerned to analyze an important debate between philosophers

[*] Whipple Science Museum, University of Cambridge, Cambridge CB2 3RH, England.
[**] Program in History and Philosophy of Science, University of Pittsburgh, Pittsburgh, Pennsylvania 15213.

such as David Hume and Thomas Reid and natural philosophers such as Joseph Priestley and James Hutton. We will therefore discuss a number of connected problems that are well represented within the scientific thought of this period, but we do not intend to imply that the important change in epistemology which we delineate necessarily applies to eighteenth-century scientific thought as a whole. There is as yet no well-defined historiography of eighteenth-century science, and the conclusions of this study, which have important implications for the thought of the period, must remain suggestive.

It is a commonplace in eighteenth-century historical studies that there was a divorce between the general intellectual thought of the period and science.[1] Indeed, this view supposes that the metaphysical reorientation fundamental to seventeenth-century science provided the basic philosophical framework of eighteenth-century science.[2] The results of this study will suggest some limitations to this consensus of historical opinion. In our view the intellectual revolution in epistemology and ontology as manifested in certain features of the science of this period is as significantly "revolutionary" as that of the seventeenth century. As in the seventeenth century, new theories of science were developed in close connection with new systems of philosophy. Seen in this way, the "scientific revolution" will appear less uniquely "revolutionary" than has been supposed hitherto. The group of eighteenth-century thinkers we discuss were concerned with philosophical problems in connection with science no less than were the *virtuosi* of the seventeenth century.

In a recent study Schofield has argued for a strict dichotomy between the tradition of imponderable fluids and that of inter-

---

1. A clear statement of this opinion can be found in Peter Gay's stimulating *The Enlightenment: An Interpretation*, Vol. 2: *The Science of Freedom* (London, 1970), pp. 159 f: "The irresistible propulsion of modern scientific inquiry was towards positivism, toward the elimination of metaphysics, and the clean separation of facts and values, foreshadowed by Bacon, implied by Newton, triumphantly announced by Hume, taken for granted by the leading scientists of the late eighteenth century. Scientific thinking exacted the stripping away of theological, metaphysical, aesthetic, and ethical admixtures that had been a constituent part of science since the Greeks."
2. The extensive literature on the "scientific revolution"—for example, E. J. Dijksterhuis, *The Mechanization of the World Picture*, trans. C. Dikshoorn (Oxford, 1961), and A. Koyré, *Newtonian Studies* (London, 1965)—almost presupposes this view.

particulate forces in this period;[3] in the present paper we will question the overall validity of this interpretation. The main purpose of this paper is to show the different ways in which certain thinkers conceived the essence of matter as being constituted by attractive and repulsive "powers," a conception of nature which underlies both the tradition of imponderable fluids and that of interparticulate forces. Given this claim for the implications of the notion of "power" in the period, some indication of its significance is in order. Though eighteenth-century thinkers are not entirely unambiguous in their use of this term, it is likely that their mode of employing it arises in part from John Locke's *Essay Concerning Human Understanding* (1690). Contextually, what eighteenth-century thinkers generally mean when they use the term "power" is this: to ascribe a power to a material object is to assert what it can or cannot do in virtue of its intrinsic nature in relation to specifiable extrinsic circumstances, leaving open a complete characterization of the object's constitution in virtue of which it is held to be endowed with powers. This conception of matter involved the notion of activity in nature which contrasts with the general seventeenth-century emphasis on the passivity of material entities. This is borne out by the fact that thinkers like Descartes, Hobbes, and Boyle viewed powers as being noninherent in matter; that is, "powers" are not ascribable to bodies in and of themselves. Rather, powers are manifested only when bodies are in specifiable relations with one another; i.e., the sun has the power actively to melt wax if wax is present, and the wax to melt only when heat is present. This conception of power is very different from that widely held in the eighteenth century where powers were conceived as being substantively present in entities, thus defining the entities' essence in

---

3. Robert E. Schofield, *Mechanism and Materialism: British Natural Philosophy in An Age of Reason* (Princeton, 1970). Schofield asserts a rigid dichotomy throughout the natural philosophy of the period between "mechanists" who sought the "causation for all the phenomena of nature . . . [in] the primary particles of an undifferentiable matter . . . [their] combinations . . . their motions, and the forces of attraction and repulsion between them which determine those motions" and "materialists" who believed "that the causes of phenomena inhere in unique substances, each possessing as an essential property the power to convey . . . some characteristic quality (*ibid.*, pp. 15 f.). This dichotomy may illuminate some aspects of the period, but the ideas of many thinkers cannot be categorized in this way. For example, see Notes 43 and 193 below.

terms of inherent activity. The eighteenth-century view must also be distinguished from the doctrines of *vis insita* found in the writings of Newton and Leibniz and, in general, in theories of *conatus*.

The notion of the passivity of matter, matter as an entity embodying the lowest degree of reality and perfection, is fundamental to Newton's philosophy, since he superadded to infinitesimally small particles immaterial forces as the agents of change in nature. Newton argued that compound bodies were extremely porous. For him, forces operating in void space came to have more significance as the properties of interstitial particles became less important for explaining phenomena. Fundamental to the operation of forces was his concept of "active principle," the manifestation of God's agency in the world. The notion of "active principle" became transformed during the eighteenth century into that of "active substances," which, like the active principles in Newton's natural philosophy, were employed to explain the activity and operations of nature. However, a concept that Newton had employed to establish God's causal connection with nature became transformed into a concept used to support a theory of the balance of nature, a view which conceived nature as a self-contained system independent of divine intervention. These ideas appear very clearly in the work of James Hutton, whose system of nature falls within the tradition of theories of imponderable fluids. We will argue that Newton's aether of the 1717/18 *Opticks* was to have a significant influence on such theories. This aether was not a "mechanical"aether designed to explain phenomena by contact action, but in the limit its interstitial particles were conceived as vanishingly small, the centers of forces operating throughout void space. Thus, James Hutton associated attractive and repulsive forces with different phenomenal entities—identifying the attractive force with ordinary matter and the repulsive force with the imponderable fluids of heat, light, electricity, and phlogiston— and he went on to conceive the essence of such phenomenal entities as constituted by powers.

The theories of reality developed by men like Priestley and Hutton were consciously based on a rigorously empirical epistemology, with close attention to the foundations of scientific knowledge. In arguing that sensory experience only provided evidence of resisting powers, these thinkers denied that the essence of matter was solidity. In so doing they were to reject the traditional categories of seventeenth-century science, even one as fundamental as the primary and secondary quality distinction.

The increasingly important role which forces came to have in Newton's philosophy of nature was to have important consequences. His emphasis on the significance of forces, in his mind closely associated with the paucity of solid matter in the universe, was to lead Priestley to reject the theory that solid particles constituted the essence of matter. In addition, Priestley argued that in conjunction with extension, repulsive and attractive powers were the essence of matter. This view was to lead to the seminal idea that matter was no more than forces diffused through space, and we will indicate that this in part was to shape Faraday's concept of the physical field.

I

Without going into the well-known details of changes in the Queries of the *Opticks* early in the eighteenth century,[4] Newton's general view (both there and in related documents) of the nature of force is as follows. Appealing to the "analogy of Nature," he held that as a long-range force like gravity operates between gross bodies and their particles, so numerous short-range forces operate between various hierarchies of homogeneous particles which comprise visible matter and its *differentiae*.[5] Though he was never entirely unambiguous as to their ontological status, Newton did not consider forces

4. See A. Koyré, "Etudes Newtoniennes. II—Les queries de l'Optique," *Archs. Int. Hist. Sci., 14* (1960), 15-29; H. Guerlac, "Francis Hauksbee, expérimentateur au profit de Newton," *ibid., 16* (1963), 113-128; *id.*, "Sir Isaac and the ingenious Mr. Hauksbee," *Mélanges Alexandre Koyré*, ed. I. B. Cohen and R. Taton, 2 vols. (Paris, 1964), *1*, 228-253; *id.*, "Newton's Optical Ether," *Notes and Records of the Royal Society, 22* (1967), 45-57.

5. In the third Rule of Philosophizing, which he prefixed to the third Book of the *Principia* in 1713, Newton stated that "The qualities of bodies which cannot be intended and remitted, and which apply to all bodies on which it is possible to set up experiments, are qualities of all bodies universally," appealing to "the analogy of Nature, which uses to be simple, and always consonant to itself." (Sir Isaac Newton, *The Mathematical Principles of Natural Philosophy*, trans. B. Motte, 2 vols. [London, 1729], *2*, 203.) In the earliest draft of the Rule he wrote that "The laws (and properties) of all bodies on which it is possible to institute experiments, are laws (and properties) of all bodies whatsoever," a clear statement of a general analogy between macro- and micro-phenomena. (Trinity College Library, N.Q. 16.200.) For the full text see J. E. McGuire, "The Origin of Newton's Doctrine of Essential Qualities," *Centaurus, 12* (1968), 236.

as either secondary qualities or as material substances. In and of themselves they were closer in nature to the immaterial, though of course their effects could be ascertained in relation to material entities. Moreover, since force was not a secondary quality it did not arise from matter, nor was it in any way reducible to the interactions of bodies or particles. Like his view of space—neither a quality nor a substance—Newton tended to regard force as in a category of its own. Viewed within the Neoplatonic framework of his thought, force was closer to space than to matter, as a dynamic principle of divine operation, since force and space were respectively an expression, in the realm of nature, of God's spiritual power and existence. Moreover, as force and space were not in the category of substance they did not satisfy criteria for material existence such as impenetrability and three-dimensionality. Newton's view that force and space exist in a realm categorically different from matter will be seen as significant for later eighteenth-century thought.[6]

In terms of this ontology, Newton tended to hold that interparticulate forces were dispositional in nature, being able to transform from one "state" to another. For example, in Query 31 of the *Opticks* he stated that the "attractive Force [of particles] can reach but to a small distance from them . . . where Attraction ceases, there a repulsive Virtue ought to succeed."[7] All particles of matter were thus conceived as being surrounded by envelopes of various forces which, by entering into combinations, changed the configuration of the internal layers of particles of any given body so as to alter its manifest properties. With the primacy of force foremost in his mind, Newton was more concerned, in explaining the interactions and visible properties of matter, with the way in which particles were rearranged than with their geometrical properties. Though he was never able to quantify these interstitial forces

6. For a discussion of Newton's distinction between body and space see J. E. McGuire, "Body and Void and Newton's *De Mundi Systemate:* Some New Sources," *Arch. Hist. Exact Sci.*, 3 (1966), 206-248. For Newton's concept of force see McGuire, "Force, Active Principles and Newton's Invisible Realm," *Ambix*, 15 (1968), 154-208. For a discussion of Newton's concept of space see also F. E. L. Priestley, "The Clarke-Leibniz Controversy," *The Methodological Heritage of Newton*, ed. Robert E. Butts and John W. Davis (Oxford, 1970), pp. 34-56.

7. Isaac Newton, *Opticks or a Treatise of the Reflections, Refractions, Inflections & Colours of Light* ([Dover edition, based on the 4th edition of 1730], New York, 1952), p. 395.

by establishing measurable parameters, they were to be treated more exactly by early disciples such as the Keills and Freind. Newton's theory of the paucity of matter in the universe in relation to the void was fundamental to his theory of forces, for the forces were held to operate within the internal vacuities of matter. The paucity of matter was clearly connected in Newton's mind with the way in which forces arrange the constituents of matter. His commitment to the void was thus closely associated with his belief in the primacy of force.[8]

Important to Newton's conception of force was his notion of active principles, which were the general mode of causation of divine agency in the natural world. Though active principles were sometimes used as a cognate for force, more often they were used as a general term to denote the cause or causes of any particular force. Though at one level forces were natural agents in the world, at another they, like matter, were dependent on and manifestations of the agency of God. Though Newton tended to interchange the terms "force" and "active principle," it is important to notice that he used a contextually meaningful distinction between the two. Though forces are causes they are not, of course, a fundamental cause like God. Unlike many later thinkers, Newton held it meaningful to ask: what is the cause or causes of any given force, itself a cause at another level of being? Therefore one of his central problems was the ontological problem of the causation of force, and "active principle" was the term he used, in various contexts, to designate this category of problem. Though God, in virtue of divine concurrence, was the ultimate cause and sustainer of all natural entities, active principles, though at one stage identified with God, by the time of publication of the 1717/18 edition of the *Opticks* referred to processes in the physical world which were linked with physical and chemical phenomena.[9] This chain of thought carried the implication that many of the propensities of these principles were yet to be discovered. As there was a hierarchy of forces in nature, so presumably, for Newton, there was a hierarchy of natural causes related to them which could be known in principle. This contention is less surprising when it is realized that Newton was concerned to

8. See McGuire, "Body and Void," *op. cit.* (Note 6).
9. See McGuire, "Force, Active Principles," *op. cit.* (Note 6).

maintain the primacy of the spiritual and spirituous in nature, as is evident by his designating matter as ontologically imperfect, a conception which he associated with the paucity of matter and with the primacy of force in nature. Newton's conception of the paucity of matter is significant for understanding the eighteenth-century view of the relationship between matter conceived as "power" and the vacuity of the universe. If resistance is not related to solidity, as was the view of many eighteenth-century thinkers, and given the Newtonian assumption of the paucity of solid particles in relation to void space, natural philosophers like Joseph Priestley could consider it but a small step to reduce solid matter to powers which could affect the senses. This presupposes a non-Newtonian assumption, namely, that the nature of matter is active resistance rather than passive solidity.

It is clear that Newton uses the term "active principles" in a more embracing sense than "force." The term "aether" of the 1717 *Opticks* stands for a type of "active principle," though these terms as well are not strictly interchangeable. This aether is another attempt at solving the ontological problem of the causation of force. Thus, far from being synonymous with "force," the term "active principle" refers to yet another putative cause of forces like gravity.

A proper understanding of this aether hypothesis is imperative for comprehending much of later scientific thought in Britain. Without going into the vexed question of why this aether was introduced into the later Queries, its main characteristics are these. Being particulate in structure, it is described in the following terms: "rare," "subtile," "elastick," that which "dilates and contracts," "condenses," has graduated "density," and "vibrates."[10] By its subtlety it appears to diffuse through space, by its rarity to produce no resistance to material motions. Moreover, in terms of the graduated elastic density of the aether, in Query 21 Newton suggested that it was rarer in the pores of bodies than in surrounding space. Therefore the differential density between any two places would *cause* any two bodies to move toward one another, as from a denser to a rarer medium.

10. Newton, *Opticks,* pp. 348 ff.

There are, however, more fundamental aspects of this aether, the implications of which have yet to be appreciated, which had important historical consequences. The 1717 hypothesis is a "force-aether," and though this has been recognized by Newtonian scholars,[11] its true nature as a putative cause of gravitation has not. The key to this lies in the following passage from Query 21:

> As Attraction is stronger in small Magnets than in great ones in proportion to their Bulk, and in Gravity is greater in the Surfaces of small Planets than in those of great ones in proportion to their bulk, and small Bodies are agitated much more by electric attraction than great ones; so the smallness of the Rays of Light may contribute very much to the power of the Agent by which they are refracted. And so if any one should suppose that *Aether* (like our *Air*) may contain Particles which endeavour to recede from one another (for I do not know what this *Aether* is) and that its Particles are exceedingly smaller than those of air, or even than those of Light: The exceeding smallness of its Particles may contribute to the greatness of the force by which those Particles may recede from one another, and thereby make that Medium exceedingly more rare and elastick than Air, and by consequence exceedingly less able to resist the motions of Projectiles, and exceedingly more able to press upon gross Bodies, by endeavouring to expand it self.[12]

Without doubt Newton distinguishes between the gross aether itself, a "medium" having properties such as "rarity" and "elasticity," and its composition. The microstructure of this aether is crucially important. Arguing analogically Newton concludes that its particles are probably smaller than those of the rays of light, so that just as a small magnet has more force in proportion to its bulk than a large magnet, the tiny aethereal particles give rise to forces having great intensity in relation to their size. This relation between smallness of size and greatness of force is partly a consequence of Newton's doctrine regarding the porosity of matter. As the pores of any given physical entity decrease in size and number, so its solid parts concentrate more and more to a central point, ending theoretically in the total solidity of the remaining infinitesimal particles. Accord-

---

11. See Arnold Thackray, *Atoms and Powers: An Essay on Newtonian Matter-Theory and the Development of Chemistry* (Cambridge, Mass., 1970), p. 28.
12. *Opticks*, pp. 351 f.

ingly, since the quantity of matter is concentrated, it will give rise to a force having great intensity in relation to size; as John Keill puts it, "the attraction will not be so strong, when a particle of a given magnitude has several pores, as when it is entirely solid."[13] Size thus becomes an important factor when particles of absolute solidity are considered. In view of Newton's argument by analogy, the possibility is open that in the limit the dimensions of particles would be vanishingly small, leaving space containing interparticulate forces clustering around changing networks of foci.[14] Thus, with an extension of the argument, reduction of matter through the paucity of particles and their smallness embodies the existence of force throughout space; that is, an aethereal "medium" is dynamic in nature, not a mechanical fluid.[15] This line of reasoning is made explicit by Priestley.

From the point of view of particle size and porosity, these are the implications of the "force-aether." Moreover, Newton is not adumbrating a "mechanist" aether, but attempting to reduce, through the gross features of the aether, action-at-a-distance forces such as gravitation to a single monolithic repulsive force. In the light of this analysis it is clear that action at a distance is a doctrine which never seriously troubled Newton. For him it was an intelligible mode of action, something that actually occurs in *rerum natura*. While "brute inanimate" matter could not act across space, interstitial forces could, since their mode of existence was that of filling and operating in space. These are the basic entities of the aether, constituting its distinctive ontological character, and they are privileged in status since

---

13. See John Keill, "In qua Leges Attractionis aliaque Physices Principia traduntur," *Philosophical Transactions,* 26 (1708/9), 104 (Theorem 14).

14. See McGuire, "Force, Active Principles," *op. cit.* (Note 6).

15. This distinction between the aethereal medium which embodies the existence of force throughout space and a mechanical fluid underlies much eighteenth-century natural philosophy. Bryan Robinson's *Dissertation on the Aether of Sir Isaac Newton* (Dublin, 1743) is in many ways a unique work. It develops a mechanical treatment, in terms of the sizes of aether particles, the density of the aether, and the force between the particles, of how the aether is supposed to cause gravitation, elasticity, heat, cohesion and fermentation, and the phenomena of optics. Robinson's aether must be sharply distinguished from theories involving the imponderable fluid "aethers" of light, electricity, and heat. As we will show, such theories have relation to Newton's "force aether" rather than to a "mechanical fluid" of the kind supposed by Robinson.

they cannot in turn be explained.[16] Thus, we are in a position to see that not only was Newton's aether nonmechanical but that it was truly dynamic in character with respect to its operation between the aether's particles. Hence the gross properties of the aethereal medium are merely phenomenal manifestations of a microstructure embodying only repulsive forces acting between vanishingly small particles. Viewing the aether in this way, Newton's theory of a nearly matterless universe is preserved. This interpretation of the aether is strengthened by remarks on a manuscript of six folio sheets in which Newton develops some "Observations" on Hauksbee's experiments on electrical phenomena, intended for the 1717/18 edition of the *Opticks*. In defining his "very subtle active substance or medium," Newton says: "To distinguish this Medium from the bodies which flote in it, & from their effluvia & emanations & from the air, I will hence forward call it Aether & by the word bodies I will understand the bodies which flote in it, taking this name not in the sense of the modern metaphysicians, but in the sense of the common people."[17] Newton's firm reference to what he understands by the term "body" is almost certainly related to definitions of body and void which were intended to preface Book III of the third edition of the *Principia*. These were probably written late in 1716 at the same time

---

16. With important differences of emphasis, we are in basic agreement with Laurens Laudan regarding the nature of Newton's aether. See Roger H. Stuewer, ed., *Historical and Philosophical Perspectives of Science* (Minneapolis, 1970), pp. 230-238. Newton's thought certainly functioned on many levels—the descriptive, the architectonic, the regulative, the teleological—and appealed to the transcendental harmonies of nature in wrestling with the problem of the intelligibility of action-at-a-distance. See Gerd Buchdahl, "History of Science and Criteria of Choice," *ibid.*, pp. 204-230. Nevertheless, the 1717/18 aether theory taken on its own merits seems to claim that the mechanists are maintaining that action at a distance, to be made intelligible and acceptable, must be reduced to, or explained in terms of, collision, impacts, or some other sort of contact action. But as is clear from the above analysis, Newton's aether is basically structured in terms of intense repulsive forces *between* the aether particles. Thus, these sorts of forces explain the gross features of the macro-aether. And it is crucial to mention that Newton never offered aetherial or mechanical explanations for short-range distance forces with which the *Opticks* abounds. It is difficult to believe, then, that Newton was attempting to meet his critics by supplying them with a "mechanical" explanation of gravity. The basic ontology of his aether is nonmechanist, and it is in no way necessarily related to the theological model which embodies God operating on matter with void space as his sensorium. For Newton this sort of explanation is in the category of teleology.

17. University Library, Cambridge, Add. 3970.9, fol. 623r.

as the "Observations." Like the "Observations," these definitions oppose the metaphysical views of the Cartesians and Leibnizians regarding body. Body is defined as "everything that can be moved and touched, in which there is resistance to tangible things . . . it is indeed in this sense that the common people always accept the word."[18] Thus Newton makes a clear and unambiguous distinction between matter and the aether. Moreover, since the basic action of this aether is by means of repulsive forces between infinitely small particles, the analogy with the proportionality of any given quantity of matter to gravitation does not apply to its microstructure. This is clear since the smaller the particle the more intense is the force. Moreover, the analogy applies only to the density of the aether in a descriptive sense, since in placing the aether in an entirely different category from matter the proportionalities between mass, inertia, and gravitation are not ascribable to the former. Indeed, there is evidence to show that Newton may have thought the 1717 aether to be noninertial in character.[19] Moreover, since impulsive action is a necessary condition of a mechanical hypothesis, the 1717 aether is clearly not mechanical. It might be thought, however, that Newton is merely outlining a possible explanation of force in terms of force, and thus opening a regress. This is not the case. The repulsive force of the aether is not only different in its mode of action from large-scale attractive forces, but it is related to matter in a fundamentally dif-

18. University Library, Cambridge, Add. 3965.13, fol. 422ʳ. For full Latin text and translation of the definition see McGuire, "Body and Void," *op. cit.* (Note 6), pp. 246 f, 220 f.

19. Any mechanical aether would give rise to conceptual difficulties. The differential density of the aether was supposed to cause gravity, but the exact application of the inverse-square laws indicated a nonresisting void (see J. Lohne, "Newton's 'Proof' of the Sine Law and Mathematical Principles of Colours," *Arch. Hist. Exact Sci., 1* [1961], 402). Again, the mode of transmission of such impulses is obscure (see McGuire, "Body and Void," *op. cit.* [Note 6], 231). A "mechanical" fluid aether of the kind supposed by the Cartesians is, for Newton, a physical impossibility, and for this reason he states the difference between "aether" and "body"; indeed, he even speculates in a draft letter to Leibniz that this aether is "a substance in which bodies move and flote without resistance & which therefore has no *vis inertiae*, but acts by other laws than those that are mechanical" (U.L.C. Add. 3965.17, fol. 257ʳ, quoted in McGuire, "Force, Active Principles," *op. cit.* [Note 6], 203). To escape from some of the difficulties arising from the supposition of an impulsive aether, William Jones argued that by impulse Newton meant an "incorporeal" cause (*An Essay on the first Principles of Natural Philosophy* [London, 1762], p. 24), supposing "*immaterial impulses* in a vacuum" for the explanation of gravity (*ibid.*, p. 75).

ferent way. It is not proportional to the total quantity of matter, but is more intense the smaller the particle. Hence, the increase of force with the decrease of particle size is not merely a consequence of the inverse proportionality with distance. Thus the impulsive action of the aether as a function of its elasticity and density is an apparent, not real, mode of action. Therefore with respect to its microcharacteristics the aether offered in terms of its repulsive mode of action an economy of explanation. Moreover, such action is clearly an active principle, a cause of the gross properties of the aether to which it gives rise, properties such as elasticity and differential density. That it could be considered as an active substance with respect to its own properties by eighteenth-century thinkers is also clear from Newton's claim that the "elastick force" of this aethereal medium is higher in proportion to its density than is air and other elastic substances. It was Newton's theory of the microstructure of the aether as the embodiment of repulsive forces that was to enable a number of eighteenth-century thinkers to transform his concept of "active principle" into that of "active substance."

It is clear that Newton made an ontological distinction between the ultimate primordial particles and perceptible bodies. Indeed, his doctrine of the vacuity of matter was intimately associated with the view that light would be "stifled and lost"[20] within vacuities, so that the ultimate primordials could not interact with light, and so would be beyond experience. Nevertheless, by his third Rule of Philosophizing he was concerned to stress that certain characteristics of macrobodies—the essential qualities—could be transferred to the primordial particles. Without going into the difficulties surrounding Newton's doctrine of essential qualities or the justificatory principles he adumbrated to sanction inferences from what is observable to what is in principle unobservable, it must be emphasized that for Newton it was the essential qualities of hardness, impenetrability, mobility, and inertia which were common to all matter, whether

20. Newton, *Opticks*, p. 340. Newton's belief that the smallest particles of matter are in principle physically impossible to see—"it seems impossible to see the more secret and noble Works of Nature within the Corpuscles by reason of their transparency" (*ibid.*, p. 262)—hinges on the conception that matter is almost vacuous; because of the smallness of the corpuscles and this vacuity, the primordials will not reflect light. The primordials are beyond observation, and contextually in Newton's writings they have a different physical status from other levels of matter.

perceptible or imperceptible. Indeed, for Newton the essential qualities of matter are a reflection of God's immutability. In rejecting Newton's doctrine of essential qualities, eighteenth-century thinkers such as James Hutton were to reject Newton's view that the insensible primordials had properties in common with those of sensible bodies. For these thinkers such an explanation of the nature of matter was inappropriate: to explain the properties of gross bodies by ascribing similar properties to the constituent primordials was merely to explain one unknown by another. For Newton, however, such an argument was sanctioned by the "analogy of Nature"; the gap between bodies and their constituents was bridged by the postulation of a chain of bodies between the macro- and micro-worlds. Newton transformed the doctrine of the chain of being into that of the *scala naturae* so as to unify matter by the analogy of nature, and to bridge the ontological gap between bodies and their constituent particles.[21]

II

A central feature of seventeenth-century discussions of matter was the doctrine of primary and secondary qualities. Without going into the well-known epistemological difficulties surrounding this doctrine,[22] it can be stated that natural philosophers like Gassendi, Descartes, Charleton, and Boyle were concerned to establish the essential or primary characteristics of matter, which they held to exist independently of human perception. Qualities such as sensation of color, on the other hand, were held to arise as a result of a relation between the perceiving mind and the primary qualities of matter which provided their causal nexus. For ontological and epistemological reasons they accepted the distinction between absolute qualities (i.e., primary qualities) and relational qualities. Many leading eighteenth-century thinkers, we shall see, were to reject the validity of this distinction with respect to theories of matter, and with it the validity of the theory of primary and secondary qualities and related

21. See McGuire, "Atoms and the 'Analogy of Nature': Newton's Third Rule of Philosophizing," *Studies in History and Philosophy of Science, 1* (1970), 3-58.
22. For an interesting discussion of this question, see J. Bennett, "Substance, Reality and Primary Qualities," *Am. Phil. Quart., 2* (1965), 1-17.

doctrines. For them the essential characteristics of matter were entirely to be found in the seventeenth-century domain of secondary qualities, or as they saw it, in the realm of mind-dependent and relational properties.

The seeds of this fundamental shift in the presuppositions associated with theories of matter are to be found in the writings of many seventeenth-century *virtuosi* both in England and on the Continent. For example, though it was held that conceptions of matter are to be based on sensory experience, it was nevertheless contended that the essential qualities of microparticles, which were beyond such experience, could be known in principle.[23] As has been indicated, such arguments were fundamental to Newton's theory of matter; and we will argue that such epistemological ambivalence was to be mitigated in certain respects during the course of the eighteenth century. Nevertheless, even seventeenth-century thinkers were led to an implicit awareness of the eighteenth-century view that the characteristics of matter could only be established by means of the *ways* in which it aroused our sensations.

In his *Origine of Formes and Qualities* (1666) Robert Boyle explicitly states the view that ordinary material agents, ontologically speaking, are no more than the totality of their constituent particles which are known through their disposition to produce certain effects:

> I do not deny, but that Bodies may be said, in a very favourable sense, to have those Qualities we call Sensible, though there were no Animals in the World: for a Body in that case may differ from those Bodies, which now are quite devoid of Quality, in its having such a disposition of its constituent Corpuscles, that in case it were duely apply'd to the Sensory of an Animal, it would produce such a sensible Quality . . . so if there were no Sensitive Beings those Bodies that are now the Objects of our Senses, would be but *dispositively,* if I may so speak, endow'd with Colours, Tasts and the like, and *actually* but onely with those more Catholick Affections of Bodies, Figure, Motion, Texture, etc.[24]

23. See McGuire, *op. cit.* (Note 21).
24. Robert Boyle, *The Origine of Formes and Qualities, according to the Corpuscular Philosophy* (Oxford, 1666), pp. 47 ff. Boyle also states that "if there were no animals in the world, there would be no pain, yet a pin may upon account of its figure be fitted to cause pain" (Boyle, *ibid.*, p. 47). This last sentence is revealing of Boyle's view of dispositions. In considering the "power" that

Though Boyle's discussion is within the framework of primary and secondary qualities, he connects the latter with the primary qualities as associated dispositional properties. Since bodies merely have a disposition to produce sensations of colors and tastes under certain conditions of perception, there is nothing actually in them which corresponds to our ideas of these secondary qualities. Moreover, his view of dispositional properties is relational in character: they are not substantive, essential entities inherent in bodies.

In his *Essay Concerning Human Understanding* (1690), Locke developed this line of reasoning. Chapters 8 and 21 of Book II are concerned with an analysis of the idea, implicit in the "New Science," that objects have the disposition to produce sensations in the mind which do not correspond to anything in the object itself. Thus, besides having "original" or primary properties defining their essence, Locke held that material objects and particles possessed "imputed" and relational properties called "powers": "For the power in fire to produce a new colour, or consistency, in *wax* or *clay*—by its primary qualities, is as much a quality in fire, as the power it has to produce in *me* a new idea or sensation of warmth or burning, which I felt not before—by the same primary qualities, viz. the bulk, texture, and motion of its insensible parts."[25] Since secondary qualities "are powers barely, and nothing but powers, relating to several other bodies, and resulting from the different modifications of the original qualities,"[26] Locke not only designates secondary qualities as relational properties, but also tends to construe them as exemplifying an undefined notion of causation. In the *Essay* the discussion of causation is separated from that of power, perhaps because Locke wished more closely to associate the latter

---

a key has to open a door Boyle points out that there is nothing in the key over and above its shape and size and the fact that it fits a particular lock. This shows three things about Boyle's conception of power: (1) that it is *relational,* (2) that powers are *not* entities distinct from the primary qualities nor are they inherent properties in objects, and (3) that they are distinct from the effects objects have on one another by means of their inherent qualities.

25. John Locke, *An Essay Concerning Human Understanding,* ed. A. C. Fraser, 2 vols. ([Dover edition], New York, 1959), *1,* 171.

26. *Ibid.,* p. 179. Though there are passages to the contrary, Locke's basic view of powers is relational. Like Boyle he holds that dispositional powers like malleability, solubility, and fragibility must be analyzed in relation to other objects. Both thinkers were also aware that the primary/secondary-qualities distinction does not do full justice to the nature of such predicates.

with change in any given substance rather than with interaction between substances as in his analysis of causation. And though he relates power to the persistence of, and consistent changes in, our ideas either of sensation or reflection, the term nevertheless is employed contextually in an "active" causal sense. For Locke, the mind looks for the agent of change: with secondary qualities this is the efficacy of powers inherent in physical objects. According to this conception the primary or original qualities give rise to causal powers either actually efficacious or able so to act. To say of matter that it possesses the property of being red is the same as to say it is capable of producing under suitable conditions an idea of redness in our awareness. Thus, a chair does not have the color red as well as the power to produce the idea in the mind. To have a causal power is not the same as to have a property like red. With respect to primary qualities, however, Locke holds that they exist absolutely and categorically in the object and also have the "power" to produce ideas of primary qualities which in the mind "resemble" those qualities themselves. From the epistemological point of view it seems that Locke must deny the coherence of the ordinary idea of material objects being colored, for neither they nor their interstitial parts are colored, having only the disposition to produce such sensations. Since, however, we see bodies as colored surfaces, experience still allows the formulation of the everyday conception of objects, and Locke's account analyzes how the corpuscular hypothesis explains our impression that objects are colored in terms of "theoretical" matter possessed only of primary qualities.[27]

The notion of "power" is of crucial importance in eighteenth-century natural philosophy, and it seems clear that Locke's discussion of this problem was influential. Though Locke tends to use the term "natural powers" contextually in close association with the notion of causal efficacy, he does deny, however, that powers inherent in physical objects are active. Drawing on the traditional distinction between "active" and "passive," Locke states:

> A body at rest affords us no idea of any active power to move; and when it is set in motion itself, that motion is rather a passion than an action in it. For, when the ball obeys the motion of a billiard-stick,

---

27. See also John W. Yolton, *Locke and the Compass of Human Understanding: A Selective Commentary on the 'Essay'* (Cambridge, 1970), pp. 21 ff.

it is not any action of the ball, but bare passion. Also when by impulse it sets another ball in motion that lay in its way, it only communicates the motion it had received from another, and loses in itself so much as the other received: which gives us but a very obscure idea of an *active* power of moving in body, whilst we observe it only to *transfer*, but not *produce* any motion.

Thus, strictly speaking, bodies are not ontologically causative in nature, and it seems to Locke that "we have, from the observation of the operation of bodies by our senses, but a very imperfect obscure idea of *active* power; since they afford us not any idea in themselves of the power to begin any action, either motion or thought."[28] Nevertheless, if anyone claims that he has a clear idea of active power from the observation of bodies, Locke is not willing to reject this on epistemological grounds.[29] In the final analysis, however, the clearest idea of this notion comes from two related sources: a "consideration of God and spirits,"[30] and the mind "from reflection on its own operations."[31] Though Locke is somewhat uneasy about such concepts as "action," "change," and "power" and about their relationships, especially as they apply to physical objects, he never doubted that it was possible to have, though

---

28. Locke, *op. cit.* (Note 25), *1*, 312. There are numerous passages in his *Principles of Philosophy* where Descartes seemed to recognize the difference between powers and qualities. Of the secondary qualities he states that they are "nihil aliud esse . . . quam dispositiones quasdam in magnitudine, figura & motu consistentes," *Oeuvres de Descartes*, ed. Charles Adam and Paul Tannery, 12 vols. (Paris, 1897-1905), *8*, cxcix, 323. Also: "non etiam a nobis animadverti, ea, quae in objectis externis, luminis, coloris, odoris, saporis, soni, caloris, frigoris & aliarum tactilium qualitatum, vel etiam formarum substantialium, nominibus indigitamus, quicquam aliud esse quam istorum objectorum varias dispositiones, quae efficiunt ut nervos nostros variis modis movere possint" (*ibid.*, cxviii, p. 322). This is similar to Boyle's position.

It is interesting also that in his *De Corpore* Hobbes associates the notion of "power" with that of causal efficacy, arguing that the *power of the agent* and the *efficient cause* are the same thing, and goes on to argue that "these powers . . . are but conditional, namely, *the agent has power, if it be applied to a patient; and the patient has power if it be applied to an agent;* otherwise neither of them have power, nor can the accidents, which are in them severally, be properly called powers." (*The Metaphysical System of Hobbes in Twelve Chapters From Elements of Philosophy Concerning the Body*, selected by Mary Whiton Calkins, 2nd ed. [LaSalle, Illinois, 1963], pp. 76 ff.) Hobbes, like Descartes, thus denies that bodies are ontologically causative in nature, that the powers are inherently active in bodies. They exist only insofar as bodies are related in certain specifiable ways.

29. Locke, *op. cit.* (Note 25), *1*, 313.
30. *Ibid.*, p. 310.
31. *Ibid.*, p. 313.

imperfectly, a "simple" idea of natural power. And on the whole he tends to conceive powers in the relational sense, though passages such as those cited could be construed, and were by eighteenth-century thinkers, as supporting a substantive conception of powers.

It remained for Berkeley, Hume, and Reid to contest the Lockean analysis. Working from different points of view, they isolated a central difficulty which has a twofold aspect. Can there be any clear and distinct idea of power from sensory experience? Even granted an affirmative answer to this, there is seemingly no warrant in terms of Locke's doctrine of primary and secondary qualities for ascribing the notion of powers to insensible particles of matter. Both these problems were fundamental to natural philosophers like Hutton and Priestley.

III

Newton's theory of forces and Locke's doctrine of powers were to have a profound but diverse influence on eighteenth-century thought. Two thinkers early in the eighteenth century who developed these ideas were the New England divine Jonathan Edwards and the Cambridge theologian and natural philosopher Robert Greene. Edwards' natural philosophy was primarily an extension of ideas implicit in the Queries to Newton's *Opticks*, and a reformulation of the theory of primary and secondary qualities (probably influenced by Locke and Berkeley).[32] Thinking within a generalized Newtonian framework about matter and force, Edwards concluded that the essence of matter was resisting power. This theory, though not always derived from the same presuppositions, was to be advanced many times in the course of the eighteenth century.

Theological in orientation and following in the tradition of the Cambridge Platonists, Edwards closely identifies space with God, arguing that "Space is this necessary, eternal, infinite, and omnipresent being."[33] Since Edwards identifies being in general with

32. See George Rupp, "The 'Idealism' of Jonathan Edwards," *Harvard Theological Review*, 62 (1969), 209-226.
33. Harvey G. Townsend, ed., *The Philosophy of Jonathan Edwards from His Private Notebooks* (Eugene, Oregon, 1955), p. 2.
Although Edwards' ideas as expressed in his notebooks were very likely not widely known, similar views regarding matter and space are apparent in his influ-

God, all things are from God, and through Him subsist their characteristics and mutual relationships. Thus it is not so much *that* the reality of all things is *in* God which interests Edwards, but *how* that reality is *comprehended* in God. Accordingly his identification of being in general with divine presence does not result in pantheism.[34] Crucial in the thought of Edwards, as in that of Newton, is the relationship of space with God. Though he identifies being in general with consciousness, believing it inconceivable that anything exist and nothing be conscious of it,[35] Edwards does not wish to affirm Berkeley's conclusion that for a finite "material" entity *esse* is *percipi*. Thus the real and necessary existence of absolute space shows *how* all things are comprehended in God. But Edwards repudiates Berkeley's conclusion that space is relative, since for Berkeley it is the case "either that real space is God, or else that there is something besides God which is eternal, uncreated, infinite, indivisible, immutable."[36] For this reason alone, Edwards is not a "subjective idealist" but more an empiricist holding that entities and their relations dynamically projected in time through space manifest an objective order which Newton's natural philosophy describes and explains.[37]

Edwards' view of matter follows directly from his conception of the total dependence of all existence on God. In a non-Lockean move Edwards argues that all primary qualities are merely manifestations of one and the same thing; i.e., solidity, impenetrability, and indi-

---

ential *Freedom of the Will* ed. Paul Ramsey (New Haven, 1957), pp. 384-396. In this work Edwards developed doctrines of causation and determinism which not only shaped his views on moral responsibility, but also set the framework for his philosophy of nature. Edwards' views on free will and causation were widely known in Scotland, as he commented on the determinism advocated by Lord Kames. (*Ibid.,* pp. 453-479.) Priestley held a high opinion of the New England divine, whose ideas on divine necessity he considered in association with Kames, Hartley, Collins, and Hume. (*Works, 3,* 449-540 [Note 84, below].)

34. Though Edwards holds that nature is an expression of God's attributes he does not contend that it is identical with God. On the contrary, he is at pains to stress that there is a reality separate from human perception which God created as the object of knowledge.

35. Edwards' view that it is absurd to suppose anything exists and nothing be aware of it basically characterizes the ubiquity of God. In Edwards' view the creation of anything presupposes a consciousness of it.

36. *The Works of George Berkeley, Bishop of Cloyne,* ed. A. A. Luce and T. E. Jessop, 9 vols. (London, 1948-1957), 2, 94.

37. Rupp, *op. cit.* (Note 32), and Priestley, *op. cit.* (Note 6).

visibility are similar phenomena, since "bodies resist division and penetration only as they obstinately persever to be."[38] They remain in being, however, since

> resistance or solidity are by the immediate exercise of divine power, it follows that the certain unknown substance which philosophers used to think subsisted by itself and stood underneath and kept up solidity and all other properties (which they used to say it was impossible for a man to have an idea of) is nothing at all distinct from solidity itself. Or, if they must needs apply that word to something else that does really and properly subsist by itself and support all properties, they must apply it to the divine being or power itself. And here, I believe, all those philosophers would apply it if they knew what they meant themselves. So that this substance of bodies at last becomes either "nothing" or nothing but the Deity acting in that particular manner in those parts of space where he thinks fit. So that speaking most strictly there is no proper substance but God himself.[39]

There are a number of interesting points arising from this passage. Here and elsewhere Edwards rejects by implication the need for intermediary entities like an aether or fluids of various sorts; his close connection between God, space, and power obviates the need for such mediating principles. Thus, solidity, the essence of matter at the level of sensory experience, "results from the immediate exercise of God's power causing there to be indefinite resistance in that place where it is."[40] With matter thus reduced to a resisting power—a resistance to being annihilated—there is no epistemological and ontological need for the existence of an unknown substratum: atoms and visible bodies are fundamentally characterized as focal points of divine energy.[41]

This line of reasoning can be illuminated by further considering Edwards' view of the distinction between primary and secondary

38. Townsend, *op. cit.* (Note 33), p. 13.
39. *Ibid.*, p. 17.
40. *Ibid.*, p. 16.
41. It is fascinating to observe that Edwards' conception of parts of empty space becoming by divine fiat centers of resisting power is similar to Newton's doctrine, as expressed in an unpublished manuscript *De Gravitatione* (see A. Rupert Hall and Marie Boas Hall, eds., *Unpublished Papers of Isaac Newton* [Cambridge, 1962], pp. 90-156), of spatial loci manifesting through divine power "imperviousness" to penetration. It may be that Edwards used hints scattered throughout Newton's published writings in developing his own position.

qualities. The only true substance is God, according to Edwards, and since all reality is comprehended in God, space is the "Deity acting," and the nature of this action is the law of space and gravity. From the epistemological point of view, Edwards, in denying the existence of a substratum, can hold that ideas of primary qualities do not relate to qualities inherent in such an entity. Thus, they and secondary qualities in principle can be known, since the primary qualities are not in a privileged ontological position in the sense of inhering in an unknowable substratum. Theologically speaking, given that primary qualities reduce to solidity or resistance and that the latter is a direct expression of divine activity, no qualities attributable to such a phenomenon will be in a privileged epistemological position. Reasoning in this manner, Edwards reached an interesting conclusion which unites Newton's theory of forces with the idea of matter as a collection of resisting powers. Space being the most general condition of existence, the manifestation of the dynamics of God's action, and all things by its mediation being comprehended in divine nature, Edwards' general view of reality is fundamentally relational. Gravitation is merely a means of characterizing centers of resisting power with respect to general relations like motion, just as sensory qualities are a means of characterizing divine power as it shows itself in parts of space by making these impenetrable. Thus solidity, in Edwards' sense, and gravitation are two faces of the same reality, for the "laws of nature" were "the stated methods of God's acting with respect to bodies and the stated conditions of the alteration of the manner of His acting."[42] These laws of the manifestation of divine power, Newton's laws, could only be known through sensory experience as solidity or resistance.

While Edwards reasoned within a Newtonian framework and developed a theological argument to demonstrate that the essence of matter was resisting power, Greene rejected Newton's corpuscularian philosophy of homogeneous matter and the void and developed a philosophical argument to show that matter was resisting power. Thus, Greene, along with Berkeley, was one of the first English natural philosophers to attempt to overthrow the fundamental principles of the new science, though he accepted in principle the

42. Townsend, *op. cit.* (Note 33), p. 19.

theory of universal gravitation. We have already noted the close association of force and void space in Newton's thought, and it is significant that while Priestley was to develop a concept of matter as powers by extending Newton's theory of the paucity of matter and the primacy of force, Greene was led to a similar conclusion by rejecting Newton's doctrine of the void, but nevertheless basing his argument on a theory of active forces. In his *Principles of the Philosophy of the Expansive and Contractive Forces* (1727) he argues that the doctrine of the void was not confirmed by sensory experience. After examining Newton's ideas regarding the porosity of matter, he concludes:

> And now I ask could any Authority, besides that of the present Philosophy, ever support so absurd an assertion as this, that Gold and consequently all other dense Substances here mention'd have more Pores than solid Parts? Why, we must bid adieu to our Senses, and to all our Notices communicated from thence, if we must acknowledge this for a Truth, and affirm that Solidity, Plain, Visible, and palpable Solidity is nothing but mere Space.[43]

Greene thus holds the concept of void space to be gratuitous. A similar epistemological view was to be developed by Kant in a more sophisticated way, for Kant was to maintain that empty space is not a possible object of direct experience.[44] With Greene it is clear that

---

43. Robert Greene, *The Principles of the Philosophy of the Expansive and Contractive Forces or An Inquiry into the Principles of Modern Philosophy: that is, into the Several Chief Rational Sciences, which are Extant* (Cambridge, 1727), p. 5. For somewhat different treatments of Greene's thought see Schofield, *op. cit.* (Note 3), pp. 117-121, and Thackray, *op. cit.*, (Note 11), pp. 126-134, who fail to delineate the philosophical argument that was fundamental to Greene's conception of nature. Schofield's failure to do this leads him to regard Greene as a "materialist" (supposedly believing causes to inhere in substances), whereas Greene's notion that the essence of matter is action or force would seem to indicate that Greene should be regarded as a "mechanist" (in the emphasis on forces), to use Schofield's terms. In our view Schofield's categories of "mechanism" and "materialism" are simply inadequate here. See Note 3 above.

44. In itself there is nothing remarkable about this claim: probably no one has ever claimed that empty space was an object of experience. Kant, however, was the first to take this seriously in his consideration of the foundations of Newtonian science. When we explain densities in terms of comparative ratios of void intensities to solid particles composing bodies, as do the Newtonians, the explanation turns on something that is not directly experienced. This is essentially Greene's point. Of course Greene had nothing like the Kantian critical philosophy at his disposal, but both in respect to being a plenist and in his view of force he is close to Kant in spirit. In the *Metaphysical Foundations of*

his critique of the new science was founded on a more empirical view of knowledge which itself, ironically enough, more nearly satisfied the ideology of seventeenth-century science.

In place of the theory of void space, Greene held that there was a plenum in nature. He stated that there are four possible positions which can be held regarding the conception of matter: matter could be "Similar or Dissimilar, and there must be a Plenum or a Vacuum; a Plenum and a Vacuum it is Evident cannot Subsist together, nor a Similar and Dissimilar Matter; consequently there are only Four Cases Remaining, and therefore there are only Four Possible Hypotheses, which Philosophers can Espouse, or Maintain."[45] After characterizing the possible combinations of similar and dissimilar matter, a plenum, and a vacuum, Greene asserts that he will maintain and defend the conception of "Dissimilar Matter and a Plenum"

Natural Science, trans. James Ellington (New York, 1970), Kant tells us: "The familiar question as to the admissibility of empty spaces in the world—cannot be dispatched. For space is required for all forces of matter; and since space also contains the conditions of the laws of the diffusion of these forces, it is necessarily presupposed before all matter," and he adds: "For all experience gives us only comparatively empty spaces to cognize; these can be perfectly explicated from matter's property of filling its space by an expansive force greater or progressively smaller to infinity, in all possible degrees without requiring empty spaces" (p. 94). Thus, space is necessarily posited before matter. In the *Critique of Pure Reason* [A:24] we are told that we can never represent to ourselves the absence of space though we can think it empty of objects. Space for Kant is a form of pure intuition. As Gerd Buchdahl has pointed out in a private communication, space as pure formal intuition should not be confused with empty intuition. However, in the *Critique* [A:166-170; B:208-212] Kant tells us that in an epistemological context "Every reality therefore in a phenomenon has intensive quantity, that is, a degree." And [A:170-174; B:212-216] "if therefore all reality in perception has a certain degree, between which and negation there is an infinite succession of ever smaller degrees, and if every sense must have a definite degree of receptivity of sensations, it follows that no perception, and therefore no experience is possible, that could prove—the complete absence of all reality in a phenomenon. We see therefore that experience can never supply a proof of empty space—because the total absence of reality in a sensuous intuition can itself never be perceived." Again, in the third analogy of experience Kant puts his position on void space: "Empty space may exist where perception cannot reach, and where therefore no empirical knowledge of coexistence takes place, but, in that case, *it is no object for any possible experience*," *Critique* [A:212-216; B:259-263]. Italics supplied. In the *Metaphysical Foundations*, Kant argues against the mechanical mode of explaining phenomena. He conceives reality as filled with the intensive forces of repulsion and attraction. Though both are ontologically the same, repulsion with respect to experience is epistemologically prior. (*Ibid.*, pp. 57-58.)

45. Greene, *op. cit.* (Note 43), p. 934.

which hitherto has had no advocates. Greene thus rejected both Newton's doctrine of void space and his theory of the homogeneity of matter, and the crux of his acceptance of a plenum was his rejection of "Similar Matter and a Vacuum."[46] In addition, he rejected Newton's doctrine of essential qualities, those qualities which did not intend and remit of degrees. Speaking of motion, Greene argues that the postulation of a vacuum is not necessary, since the plenum will not offer "infinite Resistance" to bodies:

> if all Matter was the same, there would indeed be an infinite Resistance in a Plenum; but if it is not, and there are various kinds of Matter, and of differing Forces, and Aether should be suppos'd to have an infinitely small Resistance, such an Objection will be of no Validity against a Plenum, and consequently no Argument in Favour of a Vacuum; nor can it be said, if the world was full of the minutest Resistance, that taken together such a Resistance would be Infinite, and consequently Motion impossible; since the Quantity of Resistance cannot be estimated from the Quantity of Matter which resists, but from the Degree of Resistance, which belongs to any greater or less portion of it . . . for the intrinsick Nature of the Fluid is to be consider'd (and not the Quantity) to determine its Quality and Force.[47]

Hence by means of his theory that matter manifests various gradations and intensities, the plenum—its "intrinsick Nature"—is something that can be intended and remitted. This is a very important argument. According to Newton, objects can only fill or occupy space in an absolute sense;[48] in order to explain why of any two bodies one is heavier than the other, we must suppose that it is more tightly packed with solid particles and consequently contains less pores or empty space.[49] In rejecting the Newtonian doctrine of void space, Greene offers an alternative to Newton's argument.

---

46. *Ibid.*, pp. 934 f. Greene also argued that the concept of a plenum in nature was in consonance with divine power (*ibid.*, p. 652).

47. Robert Greene, *The Principles of Natural Philosophy, In which is shown the Insufficiency of the Present Systems, to give us any Just Account of that Science: And the Necessity there is of some New Principles, In order to furnish us with a True and Real Knowledge of Nature* (Cambridge, 1712), pp. 117 f.

48. See McGuire, *op. cit.* (Note 5).

49. For interesting discussions of this conception of matter and Kant's alternative view, see Jonathan Bennett, *Kant's Analytic* (Cambridge, 1966), pp. 170-176, and Patrick Suppes, "Some Extensions of Randall's Interpretation of Kant's Philosophy of Science," *Naturalism and Historical Understanding* (New York, 1967), pp. 109-120.

He rejected the Newtonian theory of intention and remission as applied only to *qualities,* and argued that two bodies may be of the *same size,* take up exactly the same volume, while having different degrees of hardness or rigidity. Thus the way in which these two volumes could contain two different amounts of "stuff" would not depend on the ratio of particles to void space. Thus every space can be thought of both as full and yet as filled in varying degrees. That is, for Greene, "bodies" can intend and remit of degrees, and the distinction between material bodies and the plenum is merely one of degree of intensity. Material bodies present more "intense" degrees of resistance in nature than does the plenum, and can be conceived to pass through a celestial plenum having "minutest Resistance." Thus, by viewing forces as manifestations of a plenum with degrees of intensity in its action on the senses, Greene, as Kant later, could bypass the question of the nature of the centers to and from which these forces act. It was not necessary to ask to what these forces were to be ascribed. The novelty of Greene's thought is partially obscured by the fact that he is still using the traditional distinction between substance and attribute. Moreover, it is not that Greene saw himself as providing a *Principia* containing a set of scientific techniques, but as outlining the possibility of a physics based on different principles from those of the corpuscularian philosophy.

This interpretation of a universal plenum is borne out by his conception of "dissimilar matter"; for in the context of rejecting atoms and the void he says:

> I have also shewn that such a Similar and Homogeneous Matter is a mere Hypothesis of the Mind, and wholly Incapable of solving any one Appearance we can name, or any one Quality of that Matter, which is assum'd and taken to account for; on the contrary, we propose from these Intrinsick forces in Matter, and from an Universal Action in all Nature . . . to Explain, the Various Phenomenons which occur to us and which we are fully satisfied may be better and more Rationally done than from fanciful and humorous Abstractions of a void Space, of a mere Length, Breadth, and Thickness, and of an Hypothetick, Extended, Solid, Divisible and Moveable Matter, without one Real Force belonging to it.[50]

50. Greene, *op. cit.* (Note 43), p. 62.

We could find no more uncompromising repudiation of Newtonian principles than this. The traditional primary qualities are for Greene merely "Abstracted" ideas,[51] and thus matter as generally conceived is merely a hypothetical construct having no existence in reality.

According to Greene's theory of matter, however, dissimilarity in physical objects and the differential intensities of the universal plenum are to be explained by the ratio of two different kinds of force. Matter is an active substance and is thus known through its action which "we have also said is distinguished into the Expansive and Contractive Forces, which, and the Different Combinations of them, are the occasion of those Diversities of Matter we Feel and See to Exist in Like and Equal Portions of Space."[52] For Greene, "Nature is active, and . . . matter itself is so,"[53] and he argues that "Action or Force in General is the Essence or Substratum of Matter."[54] It is merely a prejudice of the "Corpuscular Way" that philosophers have argued that "Action is Inconceivable without some Solid Substance to support such Action, and in which it should Inhere."[55] Thus, the active substratum is expansive and contractive forces, and Greene is quite explicitly stating two related theses which are found in later writers. First, the essence of matter is resistance or action, and second, matter is merely a set of active powers, not adhering in a substratum and capable of resisting by impulse. Moreover, Greene holds that the abstracted ideas of primary qualities themselves arise from resistance, since we could not see or feel extension, impenetrability, or solidity unless bodies were "Endued with some Force or Action."[56] Greene is therefore groping his way toward the notion that resistance *qua* active force is epistemologically *prior* to the idea of solidity as a mode of contact between material bodies conceived as filling regions of space.

Greene did not develop his novel ideas with clarity or rigor, and they lay buried in his rambling and diffuse writings. It is likely that Greene generalized the line of reasoning of Boyle and Locke analyzed above, for though Greene could not accept Locke's ontology

51. Greene, *op. cit.* (Note 47), p. 120.
52. Greene, *op. cit.* (Note 43), p. 409.
53. Greene, *op. cit.* (Note 47), p. 391.
54. Greene, *op. cit.* (Note 43), p. 286.
55. *Ibid.*, p. 409.
56. *Ibid.*, p. 287.

or epistemology—and he devotes an entire book of his major treatise to the philosophy of Locke's *Essay*—it is probable that his own theory of matter arose from Locke's analysis of secondary qualities arising from causal powers. For in his examination of Locke's account of the simple ideas of color, taste, and smell, Greene concludes:

> Lastly as to Peculiar, Extensions of Bodies by which they Appear in such a Certain Manner to Fill the Eye with a Greater or Less Constipation of their Parts, and to our Feeling to have a certain Degree of Roughness or Smoothness, of Solidity or Fluidness, of Rarer or Denser, and of all the Tangible Qualities of Bodies, by which they are Infinitely Diversify'd to us; They seem to arise from the Filling of their Dimensional Spaces with Infinitely Different kinds of Matter, that is, with Different Sorts of Actions, Communicated to our Senses. ... All which theory of Matter only proceeds upon this One Plain Axiom, that it is Impossible for us to have any Sensations from Matter, but by some Kind of Action, or Other, Impressed upon our Minds from it; And that, it is Impossible, we should have Different Sensations but by such Different Impressions or Actions.[57]

Thus, for Greene, a material object is no more than a set of active powers able to produce visual and tactile experience, and Locke's powers are assimilated into the realm of mind-related experience from the invisible realm of particles. Thus such Lockean occult qualities as substance and substratum are avoided, as indeed is the Newtonian category of forces as causes, and Greene's conception of matter is squarely based on sensory evidence. As we shall see, Robert Young in his *Essay on the Powers and Mechanism of Nature* (1788) was to indicate that his conception of matter as active powers was derived from Locke, linking these causal powers to an active substance, an immaterial entity possessed of inherent activity. It seems clear that Greene viewed the content of sensations as varying not only in kind but in intensity, and he conceived matter as a counterpart comprising varying combinations of moving force impinging on the senses as action or resistance. Thus, apart from the intrinsic difficulties which Greene associates with the atomic hypothesis, he could well have been led to the

57. *Ibid.*, p. 659.

theory that matter and the plenum manifest differential intensities arising from active powers by a theory of knowledge based directly on the character of the contents of sensation. Though this conception of matter would not have satisfied Hume, it does have striking parallels with the more sophisticated views of Kant. Nevertheless, it is significant that Greene developed his conception of matter in terms of the categories of the Newtonian theories of force, space, and essential qualities (which he rejected) and of the Lockean doctrine of powers (which he implicitly extended to the realm of mind-related experience). Both Edwards and Greene developed theories of matter which are best understood in terms of their relation to Newtonian forces and Lockean powers.

IV

The problems of Locke's analysis of matter and of his discussions of causal powers and the relation between primary and secondary qualities were to receive a sustained analysis in the work of Berkeley and Hume. Both were concerned to reduce primary qualities to the perceptible domain of the secondary, and both held that the idea of power was not established either from direct sensory experience or from reflection. These arguments were to be reflected in the epistemology of natural philosophers like Priestley and Hutton, who were to make the realm of seventeenth-century secondary qualities the basis for doctrines regarding the essential characteristics of matter. At the same time these natural philosophers did not accept the views of Berkeley and Hume on the ontological status of power, regarding the essence of matter as being constituted of "powers." However, the legitimacy of the concept of power was defended by Thomas Reid, and the interpretation of Berkeley's and Hume's arguments is essential to an understanding of the thought of Priestley and Hutton.

On the basis of his critique of abstract ideas and a nominalistic ontology of unrelated particulars,[58] Berkeley rejects the view that

58. In discussing the possibility that ideas represent objects, Berkeley emphasizes that "an idea can be like nothing but an idea" (*op. cit.* [Note 36], 2, 44). Thus, ideas can only represent other ideas, because they only resemble one another in being ideas. For Locke, however, ideas resemble objects, and he appeals to God for certainty that our ideas resemble things.

power or activity is inherent in physical objects. In the *Principles of Human Knowledge* (1710) he opposed Locke's doctrine of abstract general terms (held to correspond to abstract ideas), rejecting the Lockean distinction between things and ideas. Analogous to this was his rejection of a substratum of invisible particles, which the new philosophy held was responsible both for the properties of physical objects and for the mind's sensations and ideas. For Berkeley, all that is known of external reality is what is perceivable, and *qua* ideas these are concrete and unrelated particulars, not a spurious realm of invisible particles. Hence the corpuscles of the new science exist only as reified concepts.[59] In the same way he rejects the distinction between primary and secondary qualities: "For my own part, I see evidently that it is not in my power to frame an idea of a body extended and moved, but I must withal give it some colour or other sensible quality which is acknowledged to exist only in the mind. In short, extension, figure, and motion, abstracted from all other qualities, are inconceivable. Where therefore the other sensible qualities are, there must these be also, to wit, in the mind and no where else."[60] Thus both sorts of quality are on the same ontological footing; both are mind-dependent, and he applies the same argument to Lockean powers: "All our ideas, sensations, or the things which we perceive, by whatsoever names they may be distinguished, are visibly inactive, there is nothing of power or agency included in them . . . since they [ideas] and every part of them exist only in the mind, it follows that there is nothing in them but what is perceived. But whoever shall attend to his ideas, whether of sense or reflexion, will not perceive in them any power or activity; there is therefore no such thing contained in them."[61] Thus, Berkeley concludes that an idea of power cannot be obtained from experience; ideas are passive and can reveal neither activity nor power. It follows that the primary qualities can neither be the cause of sensations nor "the effects of powers" resulting from the configuration of corpuscles. For Berkeley, agency

59. For an interesting and rather different view of Berkeley's treatment of abstract general ideas and the substratum of invisible particles see Gerd Buchdahl, *Metaphysics and the Philosophy of Science; The Classical Origins: Descartes to Kant* (Oxford, 1970), pp. 279 ff, 309.
60. Berkeley, *op. cit.* (Note 36), 2, 45.
61. *Ibid.*, p. 51.

and change are to be located "in an incorporeal active substance or spirit"[62] to accord with his conception of divine causation. Not only did Hume criticize doctrines of causation apparent in seventeenth-century thinkers, but he also rejected the theological view of agency found in the writings of Berkeley, and with that the view of Malebranche and the Cartesians.[63]

In the work of Hume there is a sustained *philosophical* analysis of the concept of power, with the consequence that theological and many natural philosophical presuppositions are rejected as being without meaning for such concepts as change and cause. In the *Treatise of Human Nature* (1739) Hume presents a devastating attack on the principles of the new science, especially those relating to the theory of primary qualities.[64] Having accepted that secondary qualities have no real independent existence, Hume argues further that we have no idea of solidity and therefore none of matter, so the primary qualities do not afford us a distinct idea of body:

> The idea of solidity is that of two objects, which being impell'd by the utmost force, cannot penetrate each other; but still maintain a separate and distinct existence. Solidity, therefore, is perfectly incomprehensible alone, and without the conception of some bodies, which are solid, and maintain this separate and distinct existence. Now what idea have we of these bodies? The ideas of colours, sounds, and other secondary qualities are excluded. The idea of motion depends on that of extension, and the idea of extension on that of solidity. 'Tis impossible, therefore, that the idea of solidity can depend on either of them. For that wou'd be to run in a circle, and make one idea depend on another, while at the same time the latter depends on the former. Our modern philosophy, therefore, leaves us no just nor satisfactory idea of solidity; nor consequently of matter.[65]

---

62. *Ibid*, p. 52.
63. See Richard H. Popkin, "Berkeley and Pyrrhonism," *Review of Metaphysics*, 5 (1951), 223-246; *id.*, "David Hume and the Pyrrhonian Controversy," *ibid.*, 6 (1952), 65-81; A. A. Luce, *The Dialectic of Immaterialism: An Account of the Making of Berkeley's Principles* (London, 1963); and Richard A. Watson, *The Downfall of Cartesianism 1673-1712: A Study of Epistemological Issues in Late 17th Century Cartesianism* (The Hague, 1966).
64. Hume argues not only that we only know "impressions" and "ideas," but that only impressions and ideas exist; thus "The idea of a substance as well as that of a mode, is nothing but a collection of simple ideas, that are united by the imagination, and have a particular name assigned to them." (David Hume, *A Treatise of Human Nature*, ed. L. A. Selby-Bigge [Oxford, 1888], p. 16.)
65. Hume, *op. cit.* (Note 64), pp. 228 f.

Therefore there "remains nothing, which can afford us a just and consistent idea of body."[66]

Drawing on his argument that activity is not a necessary characteristic of a physical thing's continuing existence, Hume rejects the view that we have an idea of power, force, or efficacy. Instances of power in bodies are not observed under any specifiable conditions: "All ideas are deriv'd from, and represent impressions. We never have any impression, that contains any power or efficacy. We never therefore have any idea of power."[67] He also criticizes natural philosophers who attribute power to the activity of God: "For if every idea be deriv'd from an impression, the idea of a deity proceeds from the same origin; and if no impression, either of sensation or reflection, implies any force or efficacy, 'tis equally impossible to discover or even imagine any such active principle in the deity."[68] Hence, even if force or power did exist as an independent entity, it could not be a characteristic of, or a substitute for, matter, the existence of which, as described by the new science, Hume also doubted. With this went the rejection of Lockean powers as attributable to invisible particles. Though the natural philosophers of the eighteenth century could accept the reduction of primary qualities to the perceptible domain of the secondary, they could not accept Hume's uncompromisingly skeptical conclusion:

> When we reason from cause and effect, we conclude, that neither colour, sound, taste, nor smell have a continu's and independent existence. When we exclude these sensible qualities there remains nothing in the universe, which has such an existence.[69]

In denying the reality of the external world, Berkeley and Hume clearly went too far for natural philosophers such as Priestley and Hutton, for though the natural philosophers wished to interpret nature in terms of principles squarely related to experience and though they rejected the invisible realm of seventeenth-century thought, the principles of these philosophers threatened the very possibility of a realist view of scientific knowledge. In addition, the natural philosophers related their work, in one sense or another, to

66. *Ibid.*, p. 229.
67. *Ibid.*, p. 161.
68. *Ibid.*, p. 160.
69. *Ibid.*, p. 231.

that of Newton, whereas Berkeley and Hume showed little or no interest in discussing Newton's methodology or philosophy of science.[70] The Scottish Common-Sense school of philosophy was in marked contrast to this; not only did Thomas Reid attempt a repudiation of what he characterized as Hume's uncompromising skepticism, but he also undertook the first systematic analysis of Newton's methodology and philosophical principles, having interests both in natural philosophy and metaphysics.[71] For these reasons, Reid's philosophy was an important element of the general intellectual climate to which many natural philosophers were indebted.[72]

Reid's appeal to common sense appears very clearly in his examination of the concept "power" in the thought of Locke, Hume, and, by implication, Berkeley. In the *Essay on the Active Powers of the Human Mind* (1788), Reid considers terms like "agency," "efficacy," "action," "cause," and "change," which in everyday language are closely associated with power.[73] He argues that the idea of power cannot be derived from experience, and though he agrees with Hume that we have no idea of power either from sense or reflection, he holds, nevertheless, that we have such a concept and moreover that it is meaningful, clear, and distinct.[74] For Reid, the idea of power arises mainly from the operations of the mind:

> The only distinct conception I can form of active power is, that it is an attribute in a being by which he can do certain things if he wills. This, after all, is only a relative conception. It is relative to the effect,

---

70. The British empiricists did not, of course, ignore Newton: Berkeley's *De Motu*, for example, continued a critique of Newton's concepts of absolute space and time; but these philosophers paid little attention to Newton's methodological writings.

71. See L. L. Laudan, "Thomas Reid and the Newtonian Turn of British Methodological Thought," *The Methodological Heritage of Newton*, ed. Butts and Davis (Oxford, 1970), pp. 103-131; G. E. Davie, "Hume and the Origins of the Common Sense School," *Revue Internationale de Philosophie*, 6 (1952), 213-221; and Ernest C. Mossner, *Life of David Hume* (Austin, 1954).

72. The natural philosophy of the Scottish Enlightenment has received scant attention by historians. For common-sense philosophy see Andrew Seth, *Scottish Philosophy* (Edinburgh and London, 1885); S. A. Grave, *The Scottish Philosophy of Common Sense* (Oxford, 1960); and also George Elder Davie, *The Democratic Intellect: Scotland and Her Universities in the Nineteenth Century* (Edinburgh, 1961).

73. Thomas Reid, *Essays on the Active Powers of the Human Mind*, introduction by Baruch Brody (Cambridge, Mass., 1969), p. 28.

74. *Ibid.*, p. 39.

and to the will of producing it. Take away these, and the conception vanishes.[75]

Active power is thus known only relative to its effects. Reid denies that we can know power in the Lockean sense of arising from collocations of primary qualities, but argues that we derive the idea of power from an instinctive disposition to see nature as uniform with respect to change and from attention to the operations of the mind, though "we neither perceive the agent nor the power, but the change only."[76] Thus though the concept of power is not found in sensory experience, it, like causation and the uniformity of nature, anticipates and structures our experience. Accordingly, Reid provided a justification of the concept of power, since the genesis of the idea is located in an area of human nature not treated explicitly by Hume's philosophical principles. This is clearly stated in the following passage from Reid's most mature writing:

> It is not easy to say where we first get the notion or idea of power. It is neither an object of sense nor of consciousness. We all see events one succeeding another; but we see not the power by which they are produced. We are conscious of the operations of our minds; but power is not an operation of mind. If we had no notions but such as are furnished by the extreme senses and by consciousness, it seems to be impossible that we should ever have any conception of power. Accordingly Mr. Hume, who has reasoned the most accurately upon this hypothesis denies that we have any idea of power, and clearly refutes the account given by Mr. Locke of the origin of this idea.
>
> But it is in vain to reason from an hypothesis against a fact, the truth of which every man may see by attending to his own thoughts. It is evident that all men, very early in life, not only have an idea of power, but a conviction that they have some degree in themselves, . . . without which no man can act the part of a reasonable being.[77]

Nevertheless, Reid's analysis failed to meet Hume's doubts regarding the origin of the notion, if it were claimed to arise from general experience.[78]

75. *Ibid.*, pp. 38 f.
76. *Ibid.*, p. 33.
77. Thomas Reid, *Essays on the Intellectual Powers of Man* [1785], ed. A. D. Woozley (London, 1941), pp. 382 f.
78. For a discussion of the differences between Hume and Reid on the notion of causality, see Laudan, *op. cit.* (Note 71), pp. 127 ff.

NEWTONIAN FORCES AND LOCKEAN POWERS

Reid went on to discuss "those active powers which Philosophers teach us to ascribe to matter,"[79] and he argued that for Newton, science was restricted to establishing laws connecting antecedent and consequent conditions, so that a law becomes a cause.[80] For this reason, it is not the concern of natural philosophy to discover causes having dispositional powers, and so natural philosophers can avoid the ambiguities inherent in terms like *"cause, agency,"* and *"active power."*[81] The notion of active power is meaningful, and efficient causes actually exist in reality even though the mind can never know "what their nature, their number, and their different offices may be."[82] Nevertheless Reid held that bodies could be conceived as sets of active powers: "I conclude, then, that colour is not a sensation, but a secondary quality of bodies, in the sense we have already explained; that it is a certain power or virtue in bodies, that in fair daylight exhibits to the eye an appearance, which is very familiar to us, although it hath no name."[83]

The similarity to Locke's analysis of causal powers giving rise to secondary qualities is apparent, and it is possible that Reid is indebted to the former for his dispositional theory of material objects. In this way, Reid can be seen to have provided a justification for the ascription of powers to matter; and it was this doctrine that was fundamental to the thought of Priestley and Hutton. Both, however, were critical of Reid. Priestley, though he tends to misrepresent Reid's position, criticizes his doctrine of instinctive principles of the mind by arguing that they can be reduced to Hartley's principle of association,[84] itself derived from Locke. In general, Priestley defends Locke's doctrines of ideas and sensation and his view of existence of the external world against Reid. Hutton, on the other hand, simply rejects Reid's claim that knowledge is to

79. Reid, *op. cit.* (Note 73), p. 41.
80. *Ibid.*, pp. 45 ff.
81. *Ibid.*, p. 41.
82. *Ibid.*, p. 47.
83. Thomas Reid, *An Inquiry into the Human Mind on the Principles of Common Sense* [1764], ed. T. Duggan (London, 1970), p. 101.
84. Priestley's "Introductory Essays" to his *Hartley's Theory of the Human Mind, on the Principle of the Association of Ideas; with Introductory Essays Relating to the Subject of It* (London, 1775), in *The Theological and Miscellaneous Works of Joseph Priestley*, ed. J. T. Rutt, 25 vols. (London, 1817-1831), *3*, 183 ff.

be found directly in sensation. Such crude empiricism, as Hutton saw it, went against a rationalistic approach to natural knowledge. Neither thinker, however, rejected Reid's notion that it was meaningful to speak of powers existing in matter.

V

A full understanding of Priestley's thought regarding matter demands an appreciation of a number of intellectual movements, which can only be indicated here. Of importance is the early reaction to Berkeley's *Principles,* for Berkeley was seen as a skeptic and immaterialist and was considered in many quarters, both in Britain and the Continent, as an atheist in consequence of his denial of the existence of matter.[85] As Berkeley denied the existence of matter according to the common-sense conception of a brute, irreducible entity, so his thought endangered the traditional theological dichotomy between the soul and matter. In his *Enquiry into the Nature of the Human Soul* (1733), Andrew Baxter developed the first sustained and serious critique of Berkeley's *Principles,*[86] with the avowed aim of defending the metaphysics of Christianity. Using the Newtonian idea of forces as the manifestation of God's power—gravity being "the virtue and power of an immaterial cause, or being, constantly impressed"[87] upon matter—Baxter sought to reassert the importance of the distinction between mind and matter, and thus to combat the alleged atheistic consequences of Berkeley's ideas. Baxter's work was well known to the Scottish Common-Sense school; and Priestley was aware both of Baxter's writings and of the reaction of philosophers like Reid to the skepticism of Berkeley and Hume. Priestley's response was an attempt to combat the immaterialism and skepticism of Berkeley, but—ironically—his materialism gave rise to the same fears as had Berkeley's immaterialism; for Priestley's philosophy also denied the traditional mind and matter dichotomy, though in a very different way than Berkeley's did. Priestley argued that "matter" was the only ontological reality, whereas for Berkeley it

85. See Harry M. Bracken, *The Early Reception of Berkeley's Immaterialism 1710-1733,* rev. ed. (The Hague, 1965).
86. *Ibid.,* pp. 59-81.
87. Andrew Baxter, *An Enquiry into the Nature of the Human Soul* (London, 1733), p. 15.

was "spirit." In the wider context both thinkers were against the dualist doctrine of man in traditional Christianity. Their monist viewpoint was not so much concerned with concepts of matter as with differing views regarding the nature of spirit as distinct from matter, and of the relationship between the two.

Finally, the extent to which Priestley's thought was indebted to his examination of academic philosophers such as Reid, Beattie, Oswald, and Hume has not been sufficiently appreciated. Priestley accepted the general philosophical framework of Locke and especially his doctrine of ideas. Moreover he upheld Hartley's theory of the mind, itself derived from Locke, in which all mental operations were held to be subject to the law of the association of ideas.[88] Thus, in terms of the empirical philosophy of Locke and Hartley, Priestley rejected many of the doctrines of the Common-Sense school of philosophy: their criticisms of Lockean ideas, their avowal of instinctive principles of the mind, and their analysis of sensation. In rejecting Hume's doctrine of causation and his general philosophical position, Priestley defended Locke's doctrines of power and causation.[89] Priestley's *Examination of Dr. Reid's Inquiry into the Human Mind* was written in 1774, before Priestley's mature statement of his philosophy of nature and theology in the *Disquisitions Relating to Matter and Spirit* (1777). His critical reactions to the Scottish philosophers—there are many references to Hume in the *Examination*—almost certainly helped to form his views on the nature of matter. Priestley's thought admirably illustrates the widespread interaction between academic philosophy and natural philosophy, evident during the course of the eighteenth century.

The system of thought developed by Priestley in the *Disquisitions Relating to Matter and Spirit* can be characterized as a philosophical monism. Basing his thought on a close examination of sensory experience he propounded a philosophy which purported to reinterpret the relationship between nature and man, and which he held to be in accordance with the pristine sense of Scripture. His principal object was:

---

88. See Joseph Priestley, *Examination of Dr. Reid's Inquiry into the Human Mind on the Principles of Common Sense* (London, 1774).
89. See Priestley's *Letters to a Philosophical Unbeliever* (London, 1780), in *Works* (Note 84), 4, 398.

> to prove the uniform composition of man, or that what we call *mind,* or the principle of perception and thought, is not a substance distinct from the body, but the result of corporeal organization, . . . for whatever matter be, I think I have sufficiently proved that the human mind is nothing more than a modification of it.[90]

His view of mind as a modification of matter, his denial of the preexistence of souls, his rejection of free will by his conception of philosophical necessity are the central doctrines of what Priestley called materialism. These and "that which is commonly called *Socinianism*" formed one system of thought based on nature and Scripture. He asserts that "whoever shall duly consider their *connection* and *dependence on one another,* will find no sufficient consistency in any general scheme of principles, that does not comprehend them all."[91] Though in general conditioned by his theological and religious outlook, Priestley's conception of matter is nevertheless important for his general philosophy of nature and man.[92] While he maintains that there are close links between the doctrines of materialism, Socinianism, and necessity, each was capable of separate demonstration; as scriptural exegesis led to Socinianism, the systematic and rational use of Newton's first two rules of philosophizing supported materialism. With a strict adherence to a theory of knowledge related to the contents of sensory experience, Priestley rejects the traditional theories of matter with their insistence on the absolute existence of primary qualities like solidity, arguing that "*resistance,* on which alone our opinion concerning the solidity or impenetrability of matter is founded, is never occasioned by *solid matter,* but by something of a very different nature, viz. a *power of repulsion* always acting at a real, and in general, an assignable distance from what we call the body itself."[93] Thus all that can be truly gathered about external reality is that "all resistance can differ only in degree, this circumstance . . . [leading] to the supposition of a greater or less repulsive power, but never to the supposition of a cause of resistance entirely different from such a power."[94]

90. Joseph Priestley, *Disquisitions Relating to Matter and Spirit,* 2nd edition, 2 vols. (Birmingham, 1782), *1,* iv, in *Works* (Note 84), *3,* 220.
91. Priestley, *Works, 3,* 221.
92. For a different view of Priestley see Schofield, *op. cit.* (Note 3), pp. 261 ff.
93. Priestley, *Disquisitions,* in *Works* (Note 84), *3,* 223.
94. *Ibid.,* p. 227.

Newton's third Rule of Philosophizing is implicitly denied, since an invisible realm beyond the experience of senses, though it may exist, is not a possible object of knowledge.

This conception of matter, Priestley is keen to assert, is not incompatible with the characteristics of the mind. On his view, matter, being "destitute of what has hitherto been called *solidity*," is "no more incompatible with sensation and thought, than that substance, which, without knowing any thing farther about it, we have been used to call *immaterial*."[95] Employing this type of reasoning, Priestley concludes: "Man, according to this system, is no more than what we now see of him. His being commences at the time of his conception, or perhaps at an earlier period. The corporeal and mental faculties, inhering in the same substance, grow, ripen, and decay together; and whenever the system is dissolved, it continues in a state of dissolution, till it shall please that Almighty Being who called it into existence to restore it to life again."[96] Hence in Priestley's theory of matter, nature and man formed a coherent unity, and the skeptical challenge is thus abated: there is no need to posit the existence of a soul beyond experience, no need to entertain such theological doctrines as the preexistence of souls, and no need to ponder how the immaterial can act on the material. In knowing matter, manifested by varying degrees of intensity in relation to the content of immediate experience, we know the wisdom of God in creating a reality that can be utilized for human ends. And in knowing reality we comprehend all that is necessary for an understanding of human nature. As he rejected the absolute nature of primary qualities traditionally conceived as inhering in an invisible substratum, he rejected an immaterial and invisible soul. In this way not only does the mind avoid the extravagances of vain imagining, but it propounds an interpretation of nature that is consistent with the primitive sense of Scripture.

We must now give a more complete analysis of Priestley's arguments in relation to his natural philosophy. As has been noted, in his *Disquisitions Relating to Matter and Spirit* (1777) he wished to abolish the separate categories of matter and spirit, denying that "there are *two distinct kinds of substance . . . matter* and *spirit*."

95. *Ibid.*, p. 230.
96. *Ibid.*, pp. 256 f.

In Priestley's view matter had "been said to be possessed of the property of *extension* . . . and also of *solidity* or *impenetrability*, but it is said to be naturally destitute of all powers whatever," while spirit had been defined as "a substance entirely *destitute of all extension*, or *relation to space*, so as to have no property in common with matter; and therefore to be properly *immaterial*, but to be possessed of the powers of *perception*, *intelligence*, and *self-motion*." Moreover, he denies that two substances could be "capable of *intimate connection* and *mutual action*" unless they had common properties, and argues that "matter is not that *inert* substance that it has been supposed to be; that powers of *attraction* or *repulsion* are necessary to its very being, and that no part of it appears to be impenetrable to other parts."[97] In abolishing the dichotomy between matter and spirit Priestley argues that matter could not be defined as separate from its powers of attraction and repulsion:

> I therefore, define it [matter] to be a substance possessed of the property of *extension*, and of *powers of attraction or repulsion*. And since it has never yet been asserted, that the powers of *sensation* and *thought* are incompatible with these (*solidity*, or *impenetrability* only, having been thought to be repugnant to them) I therefore maintain, that we have no reason to suppose that there are in man two substances so distinct from each other, as have been represented.[98]

The essence of matter was therefore extension together with inherent powers of attraction and repulsion conceived to exist in a substantive sense and not merely relationally or dispositionally as in seventeenth-century thought. Appealing to Newton's first two Rules of Philosophizing, he went on to argue that the apparent impenetrability and solidity of matter were not essential properties of matter. Resistance was not due to the impenetrability and solidity of matter but to a power of repulsion; it was the powers which were "essential to the *actual existence* of all matter."[99] By denying that solidity and impenetrability were essential properties of matter, he concludes that solidity and impenetrability were due to the powers: "The reason why *solid extent* has been thought to be a complete definition of matter, is because it was imagined that we could separate from

97. *Ibid.*, pp. 218 ff.
98. *Ibid.*, p. 219.
99. *Ibid.*, p. 223.

our idea of it every thing else belonging to it, and leave these two properties independent of the rest, and subsisting by themselves. But it was not considered, that, in consequence of taking away *attraction,* which is a *power, solidity* itself vanishes."[100] Rejecting Locke's argument that solidity constituted the "essence of matter" he argues that the powers are essential to the existence of matter:

> I by no means suppose that these powers, which I make to be essential to the being of matter, and without which it cannot exist as a material substance at all, are *self-existent* in it. All that my argument amounts to, is, that from whatever source these powers are derived, or by whatever being they are communicated, matter cannot exist without them. . . . Whatever *solidity* any body has, it is possessed of it only in consequence of being endued with certain *powers.*[101]

Thus, solidity and substance were the mere effects of the powers, and all that was known of matter was powers and extension. He argues that "we know nothing more of the nature of substance than it is something which supports properties."[102] He supposes that the powers are essential to matter: "take away attraction and repulsion, and matter vanishes."[103] Priestley made a number of moves here. Not only did he argue that matter was known only through its powers, but he advanced an ontological statement that matter was known this way because it possessed extension and powers of attraction and repulsion. Without these powers it would be nothing except vacuous extension. Thus powers, rather than impenetrability or solidity, made matter what it was.

It is important to realize that Priestley did not explain the way in which powers could *"inhere in"* or *"belong to"*[104] matter and that he regarded the powers of attraction and repulsion as dispositional properties which by their action gave the appearance of impenetrability and solidity to matter. Thus, in effect Priestley was arguing that matter was a set of powers with respect to extension, not merely collapsing the traditional primary and secondary qualities distinction into the world of appearances, but abolishing it all together. It is

100. *Ibid.,* p. 224.
101. *Ibid.,* pp. 224 f.
102. *Ibid.,* p. 233.
103. *Ibid.,* p. 238.
104. *Ibid.*

little wonder that his critics failed to understand him, since like Berkeley's his thought lay essentially outside the traditional logic of substance and attribute.

This notion of matter as a set of powers and the denial of solidity as constituting the essence of matter can be found in Robert Young's *Essay on the Powers and Mechanism of Nature* (1788), a work in which Priestley's influence was acknowledged explicitly. Young argues that "Body . . . may be said to contain, or consist of, all its primary qualities, extension, solidity, figure, inactivity, and mobility, together with a power to produce certain effects."[105] He continues by emphasizing that "it appears to me as little justifiable to say solidity is in body, as to say heat is in the fire,"[106] arguing that "We can only conceive of solidity as being a resistance of the parts of any body, to a power which endeavours to separate them, or to bring them nearer together," denying that "bodies are in any sense solid, [other] than as having a power to resist."[107] The solidity of bodies is therefore held to be the result of the resisting powers of bodies, and Young goes on to maintain that "all which is real, positive, and peculiar to body, are certain active powers."[108] His position with respect to the content of sensations in relation to matter is similar to that of Greene. Young argues that "Our ideas of the differences of densities in bodies, is that of different fulnesses," and he goes on to emphasize that "Fulness is an idea capable of intention and remission; the same extension may be filled with different quantities of the filling substance; it may be more or less full, in all possible degrees."[109] Young, like his contemporary Kant and his predecessor Greene, is denying the corpuscularian hypothesis that the differences in densities of bodies is to be explained by varying ratios of pores to solid parts; rather, the hypothesis of varying intensity of "filling substance" satisfies the evidence of the senses and does not turn on the gratuitous assumption of empty space.

---

105. Robert Young, *An Essay on the Powers and Mechanism of Nature, intended By a Deeper Analysis of Physical Principles, To extend, improve, and more firmly establish, The Grand Superstructure of the Newtonian System* (London, 1788), pp. 11 f.
106. *Ibid.*, p. 15.
107. *Ibid.*, p. 17.
108. *Ibid.*, p. 20.
109. *Ibid.*, p. 34. Young explicitly applies the notion of intension and remission to bodies in a very similar way to Greene: see Greene, *loc. cit.* (Note 47).

In regarding matter as a complex of intensive powers capable of producing effects—a set of dispositional properties—Young also advanced ideas similar to Priestley's, to which he referred with approval, accepting Priestley's argument "inasmuch as it denies solidity."[110] Young indicated that his conception of matter as a set of active powers was derived from Locke's discussion of powers, explicitly referring to Locke's discussion of "powers in the bodies,"[111] and he explicitly rejected Berkeley's argument that "there was no intermediate agency between our minds and the supreme mind."[112]

Priestley's theory of matter as powers in conjunction with extension, his denial of solidity, and his supposition of the *"mutual penetrability of matter"*[113] were extremely influential. Priestley's arguments were taken up almost immediately by William Nicholson in his *Introduction to Natural Philosophy* (1782), where Nicholson states that "Matter is known to us only by its properties . . . we are totally ignorant of the substance in which these properties are united."[114] For Nicholson these properties are powers rather than qualities such as impenetrability. He goes on to argue that all the effects of impenetrability could be ascribed to the action of a repulsive force, noting that all bodies "exert a repulsive force on each other, and that the common effects which are attributed to contact and collision are produced by this repulsion: and, if so, why not attribute all the effects of the same nature to this cause, which we know exists, instead of supposing an impenetrability that can never be proved? If the force of repulsion be sufficiently great, it may not be in the power of any natural agent to overcome it; and, consequently, all the effects of a real impenetrability will take place, though the substance of matter itself may not be impenetrable, or even extended,"[115] Nicholson goes farther than Priestley in admitting the possibility of nonextended matter. In interpreting contact and collision in terms of repulsive action, Nicholson was advancing a

110. Robert Young, *op. cit.* (Note 105), p. 65.
111. *Ibid.*, p. 11n.
112. *Ibid.*, p. 64.
113. Priestley, *Works* (Note 84), *3*, 232.
114. William Nicholson, *An Introduction to Natural Philosophy*, 4th ed., 2 vols. (London, 1796), *1*, 7.
115. *Ibid.*, p. 15.

mode of analysis which was, as we shall see, treated more fully by John Leslie, John Robison, and Dugald Stewart in their rejection of contact action. With respect to micromatter, Nicholson's position was equally radical: "If by the first rule of philosophizing we are to admit no more causes of natural things than are sufficient to explain the phenomenon, then we know that a sphere of repulsion exists as the proximate cause of our ideas of impenetrability and extension, why should we add to this an extended atom existing in the centre of the sphere of repulsion?"[116] With his rejection of the necessity of extended atoms or centers of force, Nicholson goes beyond Boscovich and hints at a possibility not realized until the nineteenth century of a truly dynamical model of force entailing the denial of microparticles conceived as centers of force.

It is significant that Priestley develops his theory of matter as a set of powers in terms of the Newtonian theory of the primacy of force and the paucity of matter in the world. In arguing that the powers and forces in nature were all that constituted matter he refers to the Newtonian doctrine that space contained very little solid matter:

> The principles of the Newtonian philosophy were no sooner known, than it was seen how few, in comparison, of the phenomena of nature were owing to *solid matter,* and how much to *powers,* which were only supposed to accompany and surround the solid parts of matter. It has been asserted . . . that all the solid matter in the solar system might be contained within a nut-shell, there is so great a proportion of *void space* within the substance of the most solid bodies. Now, when solidity had apparently so very little to do in the system, it is really a wonder that it did not occur to philosophers sooner, that perhaps there might be nothing for it to do at all, and that there might be no such thing in nature.[117]

Thus, Priestley explicitly connects his denial of solidity and his theory that matter was a set of powers with the Newtonian doctrine of the paucity of matter and vacuity of the universe; powers and forces thus became the primary agents in nature. Priestley extended the Newtonian theory of the paucity of matter and propounded a

---

116. *Ibid.,* pp. 16 f.
117. Priestley, *Works* (Note 84), *3,* 230.

theory of Newtonian forces in the context of his interpretation of the Lockean doctrine of causal powers.

The implications of Priestley's denial of solidity and his emphasis on the forces in nature have been noted for chemical theories in the eighteenth century.[118] It has not, however, been sufficiently realized that Priestley developed his ideas in the context of optical theory. This is extremely significant, in that one of Newton's most important statements of the porosity of matter was propounded—as we have noted—in relation to the penetration of bodies by light, light being "stifled and lost" within the vacuities. Priestley denied the solidity and impenetrability of matter, and he argued that if one supposed the *"penetrability of matter"*[119] then one could explain the penetration of bodies by light, for "the particles of light are never found to impinge upon . . . or to be obstructed by"[120] dense bodies. Priestley stated that his theory of the penetrability of matter was also that of John Michell. Michell regarded light particles as ponderable entities, and his theory of the *"mutual penetration of matter"*[121] enabled him to explain the penetration of bodies by these particles.[122] Priestley also claimed that this theory of the penetrability of matter was similar to that of Boscovich.[123] Though Boscovich did not, in fact, replace matter by powers, since he maintained nonextended centers from which repulsion and attraction operated, nor did he argue for the penetrability of matter, nevertheless Priestley's connection of his own views with those of Boscovich shows the relation of his

118. Arnold Thackray, "'Matter in a Nut-shell': Newton's *Opticks* and Eighteenth-century Chemistry," *Ambix, 15* (1968), 29-53; see also Thackray, *op. cit.* (Note 11).
119. Priestley, *Works* (Note 84), *3,* 231.
120. *Ibid.,* p. 228.
121. *Ibid.,* p. 232.
122. Joseph Priestley, *The History and Present State of Discoveries relating to Vision, Light, and Colours* (London, 1772), p. 391. Regarding light particles as ponderable entities he attempted to "ascertain the momentum of light" (*ibid.,* p. 387); he also calculated the short-range force between light and matter (*ibid.,* pp. 790 f.), and the gravitational retardation of the sun's light (*ibid.,* pp. 787-790). See also Russell McCormmach, "John Michell and Henry Cavendish: Weighing the Stars," *Brit. J. Hist. Sci., 4* (1968), 126-155. Though generally regarding the attraction of light by bodies as analogous to gravitational attraction, he also speculated that "it is just also possible, that light (and perhaps too the electric fluid, which seems to be in some degree allied to it, etc.) may not be so much affected by gravity, in proportion to their vis inertia, as other bodies" (Michell to Cavendish, 20 April 1784, Cavendish MSS., Chatsworth).
123. Priestley, *op. cit.* (Note 122), pp. 391 f.

theory of powers to the Newtonian tradition of forces operating in an almost matterless universe. The basis of Boscovich's system was a force law operating between mathematical points.[124] This law involved alternating zones of attractive and repulsive forces, undoubtedly an extension of Newton's notion of atoms as surrounded by envelopes of attractive and repulsive forces.[125] In arguing that light particles did not impinge on dense bodies, Priestley was using Newton's idea that a ray of light is reflected at a distance by an envelope of repulsive power surrounding a body. These arguments, as we will see, were important for a number of writers in the tradition of Newtonian forces.

Priestley's account of this theory of the penetration of bodies by light was taken up by William Herschel, whose unpublished speculations on the nature of matter date from about 1780 and were prompted by Priestley's *Disquisitions*. Herschel rejected the theory of the mutual penetrability of matter and the system of alternating spheres of attracting and repelling forces postulated by Boscovich, arguing that each particle of matter was endowed with a system of central forces and that phenomena were produced by the joint effect of the different forces. The innermost sphere (attraction of cohesion) "would effectively stop every particle that comes within its compass," and this would explain the absorption of light by bodies.[126]

Thomas Young referred to the theory "that matter itself is penetrable, that is, immaterial"[127] in his *Lectures on Natural Philosophy*

124. For a discussion of Boscovich's system see J. Brookes Spencer, "Boscovich's Theory and its Relation to Faraday's Researches: An Analytic Approach," *Arch. Hist. Exact Sciences, 4* (1967), 187-194.
125. Boscovich's system is also based on the Leibnizian "Law of Continuity" which he "considered as existing in Nature," *A Theory of Natural Philosophy*, trans. J. M. Child (London, 1922), p. 45.
126. William Herschel, "Observations on Dr. Priestley's Optical Desideratum—'What becomes of light?'" [unpublished paper, 1780], *The Scientific Papers of Sir William Herschel*, ed. J. L. E. Dreyer, 2 vols. (London, 1912), *1*, lxx. Herschel relates his argument quite explicitly to Newton, Boscovich, and Michell:

> Sir Isaac Newton says that reflexion and refraction may be caused by the powers of *repulsion and attraction* belonging to bodies and extending to certain distances beyond their surfaces.
>
> Mr. Boscovich goes a little farther and maintains that matter consists of physical points only, endowed with powers of attraction and repulsion taking place at different distances; that is, surrounded with various spheres of attraction and repulsion, in the same manner as solid matter is supposed to be. And that it acts upon light by these powers. Mr. Michell, also, is of this latter opinion (*ibid.*, lxix).

127. Thomas Young, *A Course of Lectures on Natural Philosophy and the Mechanical Arts*, 2 vols. (London, 1807), *1*, 458.

(1807). He argued that this theory supposed that the particles of light could "penetrate the ultimate atoms of other matter,"[128] but he did not regard this notion with favor. He pointed out that the wave theory of light avoided the necessity of explaining the penetration of bodies by light by a theory which required such an "astonishing degree of porosity"[129] of matter, for this was what was implied by the emission theory of light. Thus, matter theory was again related to optical problems. While Michell and Priestley used the problem of the transmission of light through bodies to support their theory of the penetrability of matter, Young regarded such a theory of matter with disfavor. Starting from different conceptual presuppositions, Young considered it "probable that the particles of matter are absolutely impenetrable to each other,"[130] and for this reason he could well regard the emission theory of light as decidedly inferior to the wave theory.

Priestley's theory that the essence of matter was not to be found in its solidity and impenetrability, but in extension and its powers, in its disposition to give the effects of solidity and impenetrability, was developed within the framework of a Newtonian theory of forces and of the porosity of matter. With respect to his theory of powers he could extend the Newtonian theory of the material vacuity of nature and argue that there was no such thing as solidity in nature, only powers. Though Priestley in general is critical of Reid's philosophy, his view of the origin of the concept "power" was similar. In his *Introductory Essays* to *Hartley's Theory of the Human Mind* (1775), Priestley states:

> the idea of *power* seems at first sight to be a very simple one; but it is in fact exceedingly complex. A child pushes at an obstacle, it gives away . . . in like manner he practises a variety of other bodily and mental exercises, in which he finds that it only *depends upon himself* whether he performs them or not; and at length he calls that general feeling, which is the result of a thousand different impressions, by the name of *power*. . . . Even inanimate things have certain invariable *effects,* when applied in a particular manner. Thus a rope sustains a

---

128. *Ibid.*, p. 607.
129. *Ibid.*, p. 458.
130. *Ibid.*, p. 611. For Young the essential properties of matter include extension, impenetrability, and inertia (*ibid.*, 607).

weight, a magnet attracts iron, a charged electrical jar gives a shock, &c. From these and other similar observations, we get the idea of *power, universally and abstractedly considered;* so that in fact, the idea of power is acquired by the very same mental process by which we acquire the idea of any other property belonging to a number of bodies, viz. by leaving out what is peculiar to each, and appropriating the term to that particular circumstance or appearance, in which they all agree.[131]

Priestley goes on to claim that "the idea of *solidity*, or *impenetrability* [is] what could not be deduced from *sense*, but must have its origin in the understanding."[132] He concludes in support of his position that "we see in the case of Father Boscovich, and Mr. Michell that the very idea of the proper impenetrability of matter may be disputed."[133] In his *Letters to a Philosophical Unbeliever* (1780) he observes in a critical discussion of Hume's doctrine of causation that "I think I have sufficiently shown in the third of the *Essays* prefixed to my edition of *Hartley's Theory of the Mind* that there is nothing in the idea of *power* or *causation* (which is only the same idea differently modified) that is not derived from the impressions to which we are subject, this being to be ranked in the class of *abstract ideas*, where it does not appear that Mr. Hume ever thought of looking for it."[134] The influence of Locke is evident. Moreover, Priestley is not willing to grant that Hume's rejection of the term power has any validity for scientific thought. In 1778 Price raised the following query: "Since experiments, do not furnish us with the ideas of *causation*, and *productive power*, how come we by these ideas, and how does Dr. Priestley know they have any existence? How, in particular, does he avoid the sceptical system which Mr. Hume has advanced?"[135] Priestley answered: "my idea of *causation*, and of its *origin* in the mind is . . . the very same with that of other persons; but we all distinguish between *primary* and *secondary* causes, though speaking strictly and philosophically, we call secondary causes mere effects, and confine the term *cause* to the primary

---

131. Priestley, *Works* (Note 84), *3*, 191.
132. *Ibid.*, p. 191.
133. *Ibid.*, p. 192.
134. *Ibid.*, *4*, 398.
135. *Ibid.*, p. 106, in Joseph Priestley, *A Free Discussion of the Doctrines of Materialism and Philosophical Necessity, in a Correspondence between Dr. Price and Dr. Priestley* (London, 1778).

cause."[136] For Priestley this is the deity, "the same first cause from which the powers of the magnet, and all the powers of nature are derived."[137] Thus action in nature is in reality divine power. In this respect Priestley differs from deists like Hutton by denying that nature is a self-regulating system. In any event, the two traditions of Newtonian forces and Lockean epistemology of powers merge in Priestley's theory of matter, as they did—in different ways—in the thought of Edwards and Greene. Priestley did more than formulate a philosophical foundation for the description of the essence of matter: he also developed a theory of matter such as to account for a major problem in the theory of optics.

VI

With James Hutton the Lockean doctrine of power was developed in a somewhat similar way—from the philosophical point of view— as in Priestley's *Disquisitions*. The striking differences between their arguments result from the very different theories of nature and matter in the work of these two natural philosophers. With Hutton the basis for interpreting nature must begin from metaphysical first principles which provide the general framework within which specific disciplines can be developed. Thus for him the business of natural philosophy is "to investigate the powers or laws of action."[138] These, however, can only be investigated when we have a theory of matter based on established principles of knowledge. Hutton, more explicitly than Priestley, recognized that the basic properties of material things cannot be established by observation alone. Moreover, he also realized that any particular scientific theory of matter presupposed the problem at issue, which concerns the conceptual status of the terms employed. Thus it is necessary that a theory of knowledge illuminate the relationships between concepts like impenetrability, solidity, and resistance. Both the terms *body* and *matter* have been abstracted from "our compound natural perceptions." Body is "the thing conceived as subsisting independent of our

136. Priestley, *Works* (Note 84), *4*, 106.
137. *Ibid.*, p. 107.
138. James Hutton, *Dissertations on Different Subjects in Natural Philosophy* (Edinburgh, 1792), p. ix.

thinking principle"; "*body* is made of matter," and matter as opposed to body does not "signify anything that may be immediately examined, but denotes something inferred, or judged of, from things which appear."[139] In his *Dissertations on Different Subjects in Natural Philosophy* (1792), Hutton begins his analysis of matter and body by establishing what he takes to be the principles necessary for knowing their existence and properties. Thus he makes a consistent distinction between sensible and perceptible qualities, the first arising from the immediate effect of sensation, and the second carrying an existential import by the action of the mind; he goes on to distinguish these qualities from judged or inferred qualities which "proceed in reason from those sensible and perceptible qualities."[140] It is from these judged qualities that our knowledge of matter must arise, for sensible and perceptible qualities only characterize gross bodies which are themselves made up of "matter." Thus Hutton makes a clear distinction, in terms of his division of qualities, between matter and body; as we shall see, this is a point of crucial significance. Again, sensible and perceptible qualities cover those distinguished by the traditional primary and secondary qualities dichotomy, an ontology which Hutton rejects, since he cannot accept the doctrine of absolute and independent qualities that it entails.[141] Qualities in bodies such as extension and impenetrability are only "conditional," depending on the state of contraction or dilation of the resisting powers from which they arise:

> instead then of saying that matter, of which natural bodies are composed, is perfectly hard and impenetrable, which is the received opinion of philosophers, we would affirm, that there were no permanent properties of this kind in a material thing: but that there were certain resisting powers in bodies, by which their volume and figures are presented to us in the actual information, which powers, however, might be overcome. . . . [Thus] the extension of the most solid body, would be considered only as a conditional thing.[142]

Hence properties like extension and resistance do not "arise from the absolute nature of the thing"[143] since intrinsic powers of bodies

139. *Ibid.,* pp. 278 f.
140. *Ibid.,* p. 281.
141. *Ibid.,* pp. 290-292.
142. *Ibid.,* p. 290.
143. *Ibid.,* p. 292.

can be modified or even annihilated. Thus knowledge of bodies and matter is relational, that is, relative to conditions obtaining in the powers external to the mind. Intrinsically powers and the sensible properties to which they give rise are conditional. The absolute and relational quality distinction has no place in the thought of Hutton. All properties ontologically speaking are relational. Because he holds this view, he is in clear disagreement with the metaphysics of seventeenth-century science.

Hutton holds that the idea of motion is derived from attending to changes in the content of sensation. We can have no concept of motion without that of moving things which denote external action; this necessarily involves ideas of magnitude, figure, space, situation, and time. The mind is "informed by means of observed motion, of design; for when a regular order is observed in those changing things, whereby a certain end is always attained, there is necessarily inferred an operation somewhere, an operation similar to that of our mind, which often premediates the exertion of a power and is conscious of design."[144] The teleological cast of Hutton's thought is apparent, and it is fundamental to his philosophy of nature in general. For Hutton "the proper purpose of philosophy is to see the general order that is established among the different species of events, by which the whole of nature, and the wisdom of the system, is to be perceived."[145] Hutton denies that final causes cannot be discovered, asserting that they are "the proper object of our knowledge." It is only when final causes are discovered that "we may be said to understand those things, when we see the end for which they are intended in the system of this world, and perceive the means by which, in the wisdom of nature, the end is certainly effected."[146] With respect to Hutton's theory of knowledge and of matter, this teleological approach is of special importance. Hutton draws an analogy between the intentions of the mind and the ends of external change through the motion of things. Thus there is a *de facto* harmony between things and powers without us "actuated by design"[147] and the anticipations of the mind which is intensely

144. *Ibid.*, p. 285.
145. *Ibid.*, p. 262.
146. *Ibid.*, p. 624.
147. *Ibid.*, p. 286.

conscious of them. Thus the order and design constitutive of the mind is reflected in the structure of nature teleologically oriented, such that the operations of powers producing the sensible qualities of bodies can be premediated in judgment. In the *Investigation of the Principles of Knowledge* (1794) Hutton agrees with Locke and Berkeley that ideas are not innate,[148] and with Berkeley that primary and secondary qualities are on the same ontological level.[149] He does not agree, however, with Berkeley that we can only know our ideas,[150] the reliability of which depends on God, or with Hume's skeptical view that since causation is no more than belief associated with present impressions, the knowledge of external reality is custom based on constant conjunction. Nor does he agree with Reid that the truth lies in things and that sensory knowledge is certain as a matter of fact. Rather, scientific knowledge is gained through a slow process of mental operations, which organize, order, and structure sensory impressions and subsequent conceptualizing into a systematic and consistent representation of reality; for "Truth is not a thing and truth is not a fact."[151] He goes on to argue that "truth and falsehood are things which cannot properly be said to exist in nature, being only distinctions which take place with regard to the mind of man."[152] Thus Hutton's view of truth and knowledge is one of internal consistency, the organizing power of an active mind structuring reality. Accordingly, he believes that the mind can premediate nature's operation and that the principle of economy is reflected by the economy of forces in nature. There is no doubt that Hutton is an *a priorist* rather than an inductive thinker: "Order is in thought not thing."[153] His position closely approaches the view that the *esse* of material things is *concipi*.

148. James Hutton, *An Investigation of the Principles of Knowledge, and of the Progress of Reason, from Sense to Science and Philosophy*, 3 vols. (Edinburgh, 1794), *1*, 86 ff.
149. *Ibid., 1,* 132.
150. *Ibid., 1,* 359.
151. *Ibid., 2,* 258.
152. *Ibid., 2,* 279.
153. *Ibid., 2,* 106. The annotations in his own copy of Volume I of Hutton's *Investigation*, now in the British Museum, show that Samuel Coleridge linked Hutton with the philosophical approach of Kant: "There's a great metaphysical talent displayed in it; and the writer had made an important step beyond Locke, Berkeley and Hartley, and was clearly on the precincts of the Critical Philosophy with which and the previous treatises of Kant he appears to have had no acquaintance."

Hutton's theory of knowledge is more sophisticated than that of Greene, who also based his theory on the content of experience to formulate his concept of matter. But where Greene holds that primary and other qualities, being "abstracted" ideas and hence having no external reality, are irrelevant to the conception of matter, Hutton uses them as characteristics of body to infer the nature of invisible matter manifested to the senses as power. This, for him, is the proper object of our knowledge, since "material things exist not in the form which we ourselves imagine; but . . . they exist in power and energy; and . . . the effect of that external power and action is passion and knowledge in our mind."[154]

The notion of "power" is fundamental to Hutton's theory of matter. In the *Investigation* he emphasizes that "when it relates to effects which are known not to have been in consequence of our action, then, power properly denotes a thing, a real existence; for, this is all that is known of the external cause";[155] indeed, with respect to experience, "power, the cause of our sensation, is to be considered as a first cause."[156] He goes on to point out that when power is exerted there must be "a substance existing, in which that power should reside."[157] Arguing that the term "substance" requires explanation, he examines the doctrine of essential qualities, that is, the Newtonian view that certain qualities cannot be intended and remitted of degree. Thus, he argues that

> according to the philosophy which is now considered, or the principles generally adopted by philosophers, the substance of a body is that thing which, however it may be divided or changed in its figure, always preserves its volume; and therefore, in this philosophy, it is believed, that every material substance has necessarily a certain volume, which fundamentally is unalterable. Therefore, solidity, as opposed to vacuity, must appear to be the proper idea of substance in that philosophy.[158]

This is Newton's view that bodies fill or occupy space in an absolute sense, and it is this doctrine which Hutton rejects as untenable. He states that "magnitude and figure have no other existence than in the conceiving faculty of our mind, and that these qualities are truly

---

154. *Ibid.*, *3*, 49.
155. *Ibid.*, *2*, 385 f.
156. *Ibid.*, *2*, 387.
157. *Ibid.*, p. 389.
158. *Ibid.*, p. 392.

ideas formed upon certain occasions and according to established rules, then, this philosophical idea of substance falls to the ground, and, together with it, all the material system built thereon." Thus, like Priestley, Hutton denied that solidity was the essence of substance, and it was power which was "the cause of our knowledge ... subsisting externally in relation to our mind or thought." Nevertheless, power was "to be considered as a term implying an unknown thing in action."[159]

It is here that Hutton's distinction between matter and body is important, for it was *body* which possessed extension and figure, while *matter*—which constituted body—"is considered as the substance, essence, or principles of external things,"[160] and as having neither magnitude, figure, nor inactivity. Emphasizing that "the term substance comes to be, in some respects, equivalent to that of matter,"[161] he concludes that "Matter, in this view, will appear to be a thing absolutely different from that external thing which is perceived by our mind; and the proper attribute of matter will be, the having power to affect our mind in making us to know. This is all that matter has in relation to our mind or knowledge; and this is the proper metaphysical idea of matter."[162] On the level of experience, "power and matter are found to mean the same thing, matter being properly the thing, and power the attribute thereof."[163] Thus, with respect to experience, power is to be considered as the first cause and to denote existing things.

It is here that Hutton makes an important distinction between the "physical and metaphysical ideas of matter."[164] The physical idea of matter was derived from the idea of power, while the metaphysical idea of matter was to be distinguished from our perceptions of external things. This distinction relates to his distinction between matter and body, and to his emphasis on the fact that to assume that matter was composed of extended solid particles—inert, hard atoms— was to assume as the principles of bodies "nothing but bodies

159. *Ibid.*, p. 393.
160. *Ibid.*, p. 399.
161. *Ibid.*, p. 394. He goes on to say that "the term matter is general to the whole of external things; the term substance is the matter of a particular body or class of bodies."
162. *Ibid.*, p. 407.
163. *Ibid.*, p. 403.
164. *Ibid.*, p. 407.

themselves under the pedantic designation of atoms or corpuscles."[165] This argument was based on an implicit denial of Newton's third Rule of Philosophizing, for Hutton argued that the principles of bodies must be different from the bodies themselves. With Hutton, as with other contemporary thinkers, Newton's third Rule was an ill-conceived attempt at disguising the fallacy of composition. Nevertheless, a physical conception of matter must be founded on properties "by which perceived things are made known to us."[166] He argues that though we cannot literally ascribe sensible qualities to invisible matter, as the third Rule prescribes, there being no absolutes in reality, it is from these sensible qualities that we are to infer the characteristics of powers and the unobservable substratum: "nothing is to be allowed, as belonging to matter, that is not authorized in the strictest examination of actual things. For science, at least that of physics, consists not of imagining what may be; but in the investigation of what is actually found in nature."[167]

Hutton's theory of material existence would appear to have three distinct levels: body, matter in its physical aspect (that is, matter as manifested by powers), and matter in its metaphysical aspect (matter as a nonspatial substance, powers being its attribute). It is at this third level, which is only indirectly manifested through the action of powers (which, with respect to experience, are held to constitute the first cause), that the qualities of matter as a substratum can only be characterized negatively: "But whatever matter is of itself, it must be considered as the cause of motion and resistance in natural bodies; and this is all that we are permitted to judge of in the science of physics. We never shall learn to know what matter is in itself; nor have we any occasion for that knowledge. . . . But though we know not what matter truly is, we certainly may know what it is not."[168] Thus *qua* substratum, matter is destitute of bodily form, cannot change place in space, is nonsolid, nonspatial, and incessantly active. Its mode of action, therefore, can only be likened to a metamorphosis in time, not a geometrical translation through space. Though its ontological status is somewhat unclear, it seems in the final analysis

---

165. Hutton, *op. cit.* (Note 138), p. 669.
166. Hutton, *op. cit.* (Note 148), 2, 406.
167. Hutton, *op. cit.* (Note 138), p. 300.
168. *Ibid.*, p. 315.

that for Hutton the substratum is nonmaterial in nature. That there is a harmony between the constitution of the mind and the processes of nature assures us that we may have some idea of matter in itself through the design of its operations manifested by powers.

With reference to the third Rule, Robert Young made a distinction similar to Hutton's between compounds and simples. In his *Essay on the Powers and Mechanism of Nature* (1788), he argues that "the separate elements of matter . . . cannot answer to the definition of matter,"[169] for "the elements of matter are not of the nature of matter";[170] his term "matter" corresponds to Hutton's term "body" and his phrase "elements of matter" to Hutton's term "matter." More clearly than Hutton, Young identified the substratum with an immaterial entity, an "active substance" which was "possessed of active power."[171] Also like Hutton, with respect to the substratum Young attempted to adumbrate a level of existence which is neither mental nor physical. This he could only do negatively. For Young, "Matter is a being, as a whole quiescent, and inactive, but constituted of active parts"; on the other hand, "Mind is a substance which thinks." Young concludes: "A being which should answer to neither of these definitions would be neither matter nor mind, but an immaterial, and, if I may so say, an *immental* substance. Such is the active substance."[172] This active substance was the substratum, the elements of matter, and so was of a different nature from matter. In constituting matter it "puts on . . . the form of matter, and becomes material, solid, and inert."[173] For Hutton, too, the contrast between inactive, passive body (to use his term) and active matter was fundamental; he emphasized that the law of inertia applied only to bodies, for "there is not any evidence of *inertia* being proper to the matter,"[174] again implicitly rejecting Newton's third Rule as applying to the totally different category of Huttonian "matter."

169. Robert Young, *op. cit.* (Note 105), p. 154.
170. *Ibid.,* p. 156.
171. *Ibid.,* p. 1.
172. *Ibid.,* pp. 84 f.
173. *Ibid.,* p. 150.
174. Hutton, *op. cit.* (Note 138), p. 297. He emphasizes that "it is not in the *matter,* which constitutes natural bodies, that the law of *inertia* has been investigated, but in the *bodies* themselves." Hutton thus again emphasizes the distinction between the macro- and microworlds. In his *Dissertation upon the Philosophy of Light, Heat and Fire* (Edinburgh, 1794), p. 262, Hutton argues

NEWTONIAN FORCES AND LOCKEAN POWERS

There are a number of traditions merging in Hutton's natural philosophy. The first is the notion of power, which probably derives from Locke, since Hutton devotes considerable space to his writings in the *Investigation*. Hutton's use of the concept of power is strikingly original, but his argument that power was to be construed as a first cause, as regards experience, was characteristic of the tradition which we have been discussing. Hutton also develops the Newtonian theory of forces in an interesting and complex way, and we will now turn to this problem in his thought. Here his argument that "When a power is found to be exerted, it is commonly thought that there must be a substance existing, in which that power should reside"[175] is crucial, for his concept of force was intimately connected with his view of the nature of this substance, that is, of matter. Hutton supposed two kinds of matter: gravitational matter which acted by the principle of attraction, and matter emanating from the sun—the "solar substance"—which acted by the principle of repulsion. The hardness and cohesion of a body was due to the gravitating matter, which acted on the principle of attraction, while heat, for example, was "a power in bodies by which the uniting principle perceived in cohesion, gravitation, and hardness, is opposed and resisted" by a principle of repulsion. Gravitational matter could not be separated from a body, while the solar substance could be considered as having existence apart from a body in one or other of its modifications. For Hutton, light, heat, and electricity were different modifications of the solar substance:

> Thus light, heat, and electricity, appear to be three different modifications of the same matter. Light is considered as being perfectly disconnected with the body from whence it moves; and, the moment it is again connected with a body, so as to lose its proper motion, it ceases to be light. Heat, on the contrary, is perfectly connected with a body, and forms part of its substance; heat being removed from this internal connection of bodies, ceases to be heat. Electricity is a modification that will appear to be a medium between light and heat, considered as

---

that bodies "acquire inertia in the balance of those opposite powers" of attraction and repulsion. See Note 19 above for Newton's discussion of a noninertial aether.

175. Hutton, *op. cit.* (Note 148), 2, 389.

extremes; for neither is it unconnected with bodies, nor internally connected with their substance.[176]

Phlogiston was explained in terms of a union of the matter of light and heat with some of the chemical substances of bodies; the destruction of phlogistic matter when bodies burn was balanced by its regeneration in plants, for it was in plants that "the combination of the solar substance is made for the production of phlogistic matter."[177] This, for Hutton, was an example of "a system of things which seem all to be connected together by design," illustrating the "admirable contrivance of the system in which we are placed."[178] Thus he stressed the notion of natural cycles, especially the great cycle of the organic and the inorganic. This contained heat, light, water, earth, and fixed air, supporting the plants which in their turn would produce vital air and phlogiston to maintain animal life. When in turn animal life decayed, it sufficiently renewed the stocks of fixed air, water, heat, earth, and light for the great cycle to begin anew. The great cycles of nature were themselves manifestations of the "attractive and repulsive powers" of the gravitational and solar substances opposing and balancing one another, thus securing a dynamic tension in nature.[179] When the gravitational power is supposed to prevail, bodies would form an inert mass, and when the power of heat prevails bodies are dispersed throughout the universe; but "by a just combination of those two different powers, we find moveable and moving bodies properly disposed in a great and connected system of things."[180]

Thus for Hutton, the processes and balance of nature were due to the operation of two kinds of material entity, one associated with attractive and the other with repulsive power. As we have shown, matter for Hutton was an active substratum, and his theory of "acting powers"[181] as attributes of a substratum can be seen to have relation to Newton's conception of the aether in the 1717 *Opticks*. As we pointed out above, Newton regarded the microstructure of the aether as the embodiment of repulsive forces acting between van-

176. Hutton, *op. cit.* (Note 138), pp. 505 f.
177. *Ibid.*, p. 229.
178. *Ibid.*, p. 233.
179. *Ibid.*, p. 265.
180. *Ibid.*, p. 263.
181. *Ibid.*, p. 501.

ishingly small particles, and for Hutton, too, his nonspatial, noninertial, effectively immaterial substratum could be considered only in terms of attractive and repulsive powers. While Newton's concept of active principle was used to explain the ontological problem of the causation of force in terms of the causation of divine agency in the natural world, for Hutton the balance of nature was secured by the operations of the attractive and repulsive powers alone. Though God is "the author of nature," God must be distinguished from nature, for the system of nature comprises all powers and relations in the universe (though "nature" itself is a figurative term):

> We do not necessarily ascribe this power of the universe to God, as the immediate cause; for, we consider every person as possessed of a certain power and influence in this system of moving things, being able at will to move and resist to a certain degree, however limited. Reasoning, therefore, by analogy, we consider a certain being interposed between the superintending mind, or first cause, and those effects of power which we perceive; and, this imaginary Being is called Nature. Hence, every thing that is observed either to act or suffer, is said to be a natural thing; and nature is always employed in every change that happens to those things. Now, as a thing cannot be known except by either acting or resisting, in both of which cases there is power; so, everything that is known, is considered as belonging to nature; and, every action that takes place, is considered as being performed by this power, which is, in relation to human art, supreme.[182]

182. Hutton, *op. cit.* (Note 148), 2, 415. Hutton's deistic emphasis on nature as a self-contained system must be sharply distinguished from the Hutchinsonians' rejection of Newton's conception of nature. For the Hutchinsonians, Newton had employed divine action in explaining the processes and operations of nature as an "occult cause." Regarding nature as a mechanism, they rejected all "occult causes" for physical phenomena. Nevertheless, nature was ultimately totally subordinate to God: "If God, instead of framing the universe into so curious a piece of perpetual motion, had designed to transact all by his own immediate presence, there had been no use for . . . such regular laws"; the "perspicuity and perfection" of nature manifests God's power, so there was no need for God "to work by occult qualities." (*An Abstract from the Works of John Hutchinson, Esq. being a Summary of his Discovery in Philosophy and Divinity*, 2nd ed. [London, 1775], p. 142.) Hutchinson argues that occult qualities detract from "the essential attributes of JEHAVAH ALEIM" (*ibid.*, p. 155).

As William Jones puts it, the Hutchinsonians maintained "an agency of material and secondary causes, under the direction of God, the moral governor of the world, and the Supreme cause of all things" (*op. cit.* [Note 19], p. 55). For Hutton, the order of nature manifests the wisdom of its creator; for the Hutchinsonians, nature was subject to the direct agency of spiritual causes. For

Nature was thus a self-contained system comprising the totality of effects in the universe, these effects being produced by attractive and repulsive powers, that is, by the active substratum. Unlike Newton, who conceived forces as acting dynamically *between* particles, and Priestley, who saw the essence of matter as power *and* extension, in Hutton's view matter is fully dynamic in the sense that it is neither predicated on point centers conceived atomically nor on nonextended Boscovichean entities. Nor is extension a necessary condition of its existence. Newton's conception of active principle as a cause of the forces operating in nature has here been conflated into the concept of powers which are themselves held to manifest activity. Hutton, therefore, transformed Newton's theory of forces, with respect to his theory of powers, into a conception of nature as a self-contained, active system, which in its structure manifested the wisdom and design of God. But the system of nature was to be explained purely in terms of the operations and balance of the attractive and repulsive powers, and its operations were not subject to the providential intervention of God. On the contrary, because he was a rational deist believing profoundly in a religion of nature, its self-sufficiency operated by chains of unbroken laws, which were not, for Hutton, to be abrogated by any form of divine intervention. Since nature as created is sufficient for divine ends, the powers implanted in it are adequate for the production of design. And if nature represents the perfection of God, its present state indicates that the forces now at work do so with the same intensity as they did during the earlier history of physical processes. Hutton's uniformitarianism is thus conditioned by his deism and the belief that should nature have ever been different from its present state it would be an imperfect

---

a discussion of the Hutchinsonians see Albert J. Kuhn, "Glory or Gravity: Hutchinson vs. Newton," *J.H.I.*, 22 (1961), 303-322.

It is interesting to observe that the Scot John Robison in his *Proofs of a Conspiracy against all the Religions and Governments of Europe* (Philadelphia, 1798), contends that Priestley's materialist system goes against the Newtonian philosophy which is the best support for traditional religion. For Robison, Newtonianism affords the best basis of arguments for God's existence in that it posits immaterial forces acting in nature on passive and contingent matter. In the supplement to the third edition of the *Encyclopædia Britannica* he links also Priestley with the atheistic and subversive thought which led to the excesses of the French Revolution. By implication Hutton is associated with the subversive trends undermining Christian mankind in a nonutopian society.

system. Thus nature was a system of organic processes, manifesting cycles among cycles arising from the resultant tension of repulsive and attractive power producing a state of equilibrium. These are the fundamental principles constitutive of Hutton's system. And his view of uniformitarianism and the doctrine of economy of geological forces are, for Hutton, closely linked to these principles.

This conception of nature provided the framework for Hutton's analysis of geological processes as developed in his *Theory of the Earth* (1795). The subterranean heat of the earth is the central agent of Hutton's theory.[183] It is the compensating mechanism which regenerates land for the preservation of organic life and which balances the gradual degenerative process of denudation. Not only does heat fuse rocks in the bowels of the earth, but it is the means whereby the new strata are elevated out of the sea. This view was supported, Hutton argued, by signs of elevation on existing rocks—folding, fractures, contortions, faults—which indicated that heat was responsible for their elevation. On the principle of the economy of forces, earthquakes, volcanoes, and mineral and metallic veins also were caused by the action of heat. This agent was identical with the expansive, active operations of repulsive power, and was balanced by the contractive power of gravitation.[184]

Hutton's conception of matter as powers was widely known in Scotland in the late eighteenth century. Indeed, in his *Philosophical Essays* (1794) Dugald Stewart supports Hutton's theory by arguing that Locke had implicitly conceived the essence of matter in terms of repulsive powers. Stewart gives a clear statement in support of the conception of matter advanced by Priestley and Hutton: "The effects . . . which are vulgarly ascribed to actual contact, are all produced by repulsive forces, occupying those parts of space where *bodies* are perceived by our senses; and, therefore, the correct idea that we ought to annex to *matter,* considered as an object of percep-

---

183. See John Playfair, *Illustrations of the Huttonian Theory of the Earth* (Edinburgh, 1802), pp. 181 f, 187, for the association of Newton's Query 11 of the *Opticks* with Hutton's theory of heat.

184. James Hutton, *Theory of the Earth with Proofs and Illustrations* (Edinburgh, 1795), pp. 3-32; and Hutton, "Theory of the Earth; or an Investigation of the Laws Observable in the Composition, Dissolution, and Restoration of Land Upon the Globe," *Trans. Roy. Soc. Edinb.,* 1 (1788), 269 ff. For Hutton's deism see R. Hooykaas, *The Principle of Uniformity in Geology, Biology, and Theology* (Leiden, 1963).

tion, is merely that of *a power of resistance,* sufficient to counteract the compressing power which our physical strength enables us to exert."[185] Referring to Book II, Chapter IV of Locke's *Essay* ("Of Solidity"), Stewart maintains that Locke, in analyzing cohesion and compressibility, implicitly held that the essence of matter could be conceived in terms of repulsive powers. Thus Stewart is consciously connecting Locke's theory of powers with the conceptions of matter of Priestley and Hutton. Stewart concludes that the views of Boscovich, Priestley, and Hutton "with respect to *matter,* so far as hardness or relative incompressibility is concerned, offer no violence to the common judgments of mankind, but aim only at a more correct and scientific statement of *the fact* that is apt to occur to our first hasty apprehensions."[186]

The view that the essence of matter is power led, in Scotland, to a critique of contact action. The argument is clearly put by John Playfair in his "Biographical Account of James Hutton":

> But if this be granted, and if it be true that in the material world every phenomenon can be explained by the existence of power, the supposition of extended particles as a *substratum* or residence for such power, is a mere hypothesis, without any countenance from the matter of fact. For if these solid particles are never in contact with one another, what part can they have in the production of natural appearances, or in what sense can they be called the residence of a force which never acts at the point where they are present? Such particles, therefore, ought to be entirely discarded from any theory that proposes to explain the phenomena of the material world.
>
> Thus, it appears, that power is the essence of matter, and that none of our perceptions warrant us in considering even body as involving anything more than force, subjected to various laws and combinations.[187]

---

185. Dugald Stewart, *Philosophical Essays,* 3rd ed. (Edinburgh, 1818), p. 123.
186. *Ibid.,* p. 133.
187. Playfair, "Biographical Account of James Hutton, M.D.," *The Works of John Playfair,* 4 vols. (Edinburgh, 1822), 4, 85. It is interesting to observe that the Scottish writers on contact action, unlike Maupertius and Boscovich, do not, with the exception of John Leslie, argue explicitly that such action violates continuity in nature. Nevertheless, in general the view that matter is essentially powers manifesting degrees was related to the continuity of action and reaction. The Scottish concern with power, cause, and continuity in nature can be traced from the time of Rankenian Club in the early eighteenth century. These problems were discussed by Lord Kames and David Hume. See Mossner, *op. cit.* (Note 71).

Playfair goes on to state that matter conceived in this way was "indefinitely extended" through all space as "is proved by the universality of gravitation." Playfair's views on contact action were also held by Dugald Stewart, John Robison, and John Leslie. While affirming that action at a distance was unintelligible, these Scots also affirmed their belief in the nonintelligibility of contact action.[188] As the passage from Playfair shows, if interstitial particles can be conceived as being unable to come into contact, they serve no important role in explaining the properties of things, since these can be shown to arise from intrinsic powers. He made the point clearly with respect to gravitation. It is not "according to this system . . . the action of two distant bodies upon one another, but it is the action of certain powers, diffused through all space, which may be transmitted to any distance."[189] Thus conceiving the essence of matter as powers was seen to be not only incompatible with contact action but also with the theory of invisible particles.

Though Stewart, in his conception of matter, held that powers were more fundamental than solidity, like Hume he nevertheless denied that the former concept was derived from experience. In arguing that ideas of color can only "reside in a mind" he holds that "In the same way we are led to associate with inanimate matter, the ideas of *power, energy,* and *causation* which are all attributes of mind, and can exist in a mind only."[190] Also influenced by the Humean analysis of power was the Edinburgh physician James Gregory. In analyzing the work of Locke, Reid, and Priestley, Gregory constantly affirms that the term power can only be applied to inanimate things metaphorically,[191] for only agents which can deliberate have the volition to act toward a preconceived end. Stewart and Gregory both attest to the widespread use of the philosophical connotations of the term power, reflect also the Humean critique of the notion, and provide clear evidence that the notion was derived from Locke's *Essay.*

188. See Richard Olson, "The Reception of Boscovich's Ideas in Scotland," *Isis, 60* (1969), 91-103.
189. Playfair, *op. cit.* (Note 187), *4,* 86.
190. *The Collected Works of Dugald Stewart,* ed. Sir William Hamilton, 10 vols. (Edinburgh, 1854-1858), *2,* 98.
191. James Gregory, *Philosophical and Literary Essays,* 2 vols. (Edinburgh, 1792), *1,* x-cclxvi.

VII

Hutton's deistic philosophy of nature raises issues that extend beyond the limits circumscribed by the themes of this study, but his interpretation of nature in terms of the operation and balance of attractive and repulsive powers is of great significance for the traditions we have been discussing. We will now attempt to put Hutton's system of nature clearly in the tradition of Newtonian forces, by examining more fully the development of that tradition and Hutton's relation to it. Thus far we have been mainly concerned with the connections between the traditions of Newtonian forces and Lockean powers, but theories of force were not always explicitly developed in the context of a doctrine of powers. The significance of Hutton's system as a development of a Newtonian theory of forces can be seen by comparing his theory with an earlier work, Gowin Knight's *Attempt to Demonstrate, that all the Phaenomena in Nature may be Explained by Two Simple Active Principles, Attraction and Repulsion* (1748). Knight argues that the essential or primary qualities of matter are immutable and are universal properties of all bodies, their immutability being secured by the immediate will of God: "All immediate Causes, being the Effects of God's Will, must necessarily be constant, immutable and irresistible by any finite force. From hence appears the Truth of Sir *Isaac Newton's* third Rule." Thus the "Existence, Extension, Impenetrability, mobility, and *Vis Inertiae* of Matter, are apparently the immediate Effects of God's Will." The very existence of matter, as well as these general properties, was therefore held to be the direct effect of God's will, and so Knight followed Newton in regarding the essential qualities as an expression in nature of the perfection of God; the essential qualities were a reflection of God's immutability. Knight goes on to argue that the essential qualities are inactive and do not admit of intension and remission of degrees, and that motion, not being immutable, is not an essential property of matter. From this he concludes that there is in nature "some Active Principle, or Principles capable of producing and continuing Motion in the Universe."[192]

---

192. Gowin Knight, *An Attempt to Demonstrate, that all the Phaenomena in Nature may be Explained by Two Simple Active Principles, Attraction and Repulsion: wherein The Attraction of Cohesion, Gravity and Magnetism, are shown to be one and the same; and the Phaenomena of the latter are more particularly explained* (London, 1754), pp. 4 f.

These active principles are those of attraction and repulsion, and are themselves the manifestations of divine activity in nature.

Thus far Knight's conception of nature would seem to have little relation to that of Hutton. But despite their different theological ideas, they had many ideas in common; this appears clearly with Knight's association of the active principles of attraction and repulsion with material entities. In addition he goes on to conclude that "attraction and repulsion cannot both, at the same time, belong to the same individual substance, being contraries. . . . Therefore we must conclude, that there are in Nature two kinds of Matter, one attracting, the other repelling."[193] Though Knight's view of the nature of the attractive and repelling substrata is determined by his interpretation of the third Rule—and so his primary particles possess the essential qualities of matter—his theory of attractive and repelling substances has obvious affinities with Hutton's system.

Knight therefore uses the third Rule to argue for material substrata, one possessed of attractive and the other of repulsive force. The essential qualities of bodies—the immutable qualities to which the third Rule was held to apply—were insufficient to explain the activity of nature. The active principles of attraction and repulsion were required, and these principles were associated with two different substrata. His primary particles, which he holds to be "originally of the same Size, and all round,"[194] are of two types: those which attracted one another, and those which repelled each other but were attracted by the attracting particles. Light, heat, electricity, and magnetism are all explained in terms of the operations of this repellent matter. Thus he supposed light to be the propagation of a vibrational tremor through a series of repellent particles; and magnetism is explained in terms of a "perpetual Motion" of the repellent matter from one side of a body to the other, a "circulation of the repellent fluid"[195] which had no connection with the vibratory motion which was the cause of light. Fundamental to his system is

---

193. *Ibid.*, p. 10. Schofield, *op. cit.* (Note 3), pp. 176 ff, fails to emphasize sufficiently the relation of Knight's ideas to the tradition of Newtonian forces. As in his treatment of Greene (see Note 43) his dichotomy of "mechanism/materialism" has lead him to emphasize only one aspect of Knight's thought. On Knight see also Mary Hesse, *Forces and Fields* (London, 1961), p. 182, and Thackray, *op. cit.* (Note 11), pp. 141-147.
194. Knight, *op. cit.* (Note 192), p. 12.
195. *Ibid.*, pp. 66 f.

the argument that every particle of attracting matter was surrounded by "as many repellent particles as will just ballance its attracting force,"[196] and gravitational inverse square law forces were explained as arising from the superposition of attracting force (decreasing inversely with distance) and repelling force (decreasing with distance though no precise force law is given). The operations and balance of nature can be explained in terms of the attractive and repulsive forces. Here again there are obvious similarities with Hutton's system.

In developing theories in which nature was to be explained in terms of the action of active substrata of attractive and repulsive forces, Hutton and Knight were employing the imponderable fluid theories of eighteenth-century natural philosophy. It has long been established that these theories, in which electricity, for example, was explained in terms of the action of subtle, mutually repulsive electrical particles diffused through common matter, show the influence of Newton's theory of the aether as developed in the Queries to the *Opticks*.[197] The striking feature of the systems proposed by Hutton and Knight is the association of the principles of attraction and repulsion with substrata, these substrata being considered as "active" and regarded as entities forming the "matter" which underlies gross bodies. Significantly for the interpretation of the aether of the 1717 *Opticks* proposed in this study, both Hutton and Knight regarded the microstructure of their attractive and repulsive principles as merely the embodiment of attractive and repulsive forces. It was these attractive and repulsive forces which explained the operations and balance of nature, and accounted for the manifestation of various fluids. Just as the aether of the 1717 *Opticks* was, with respect to its microstructure, force embodied in space, the attractive and repulsive substrata supposed by Hutton and Knight were forces opposing and balancing one another.

It is therefore plain that Hutton and Knight were developing theories of the primacy of force. Hutton's "matter" and Knight's attractive and repellent particles were very different entities. Hut-

---

196. *Ibid.*, p. 19.
197. See I. Bernard Cohen, *Franklin and Newton: An Inquiry into Speculative Newtonian Science and Franklin's Work in Electricity as an Example Thereof* (Philadelphia, 1956); and also Schofield, *op. cit.* (Note 3), pp. 157-190.

ton's "matter" was nonspatial, noninertial, and effectively immaterial, while Knight's primary particles possessed the same essential qualities as gross matter. Nevertheless, they were both conceived as the embodiment of attractive and repulsive forces, being analogous to the aether of the 1717 *Opticks*. Again, though Hutton envisaged nature as a self-contained system (his "matter" being an active substance) and Knight supposed his active principles of attraction and repulsion to be the immediate effects of divine action in nature, in the works of both these theorists we see the conflation of the Newtonian categories of "force" and "active principles." Though Knight's active principles are the immediate effects of God's will, they are themselves the attractive and repulsive forces; and Hutton's attractive and repulsive powers could manifest activity in and of themselves. Despite their very different theological conceptions of nature, both Hutton and Knight rejected the hierarchical, Neoplatonic framework of Newton's thought. While Newton asserted the primacy of force and regarded matter as embodying the lowest degree of reality and perfection, for Knight and Hutton the concepts of "force" and "matter" were intimately associated with one another.

The association of principles of attraction and repulsion with different substances as a basis for a system of nature was not uncommon in eighteenth-century natural philosophy. Bryan Higgins proposed a similar system in his *Philosophical Essay Concerning Light* (1776). For Higgins, earth and water were forms of attractive matter, while electricity, light, and phlogiston were forms of repellent matter. Using Newton's third Rule, he argues for the immutability of atoms, and supposes the "powers [of attraction and repulsion] implanted in the atoms" as being "incessant and immutable," these powers being endowed by God.[198] The repellent matter is here considered nongravitational, and this notion is maintained by Patrick Dugud Leslie in his *Philosophical Inquiry into the Cause of Animal Heat* (1778).[199] The association of the two principles of attraction

198. Bryan Higgins, *A Philosophical Essay Concerning Light* (London, 1776), p. 23.
199. Patrick Dugud Leslie, *A Philosophical Inquiry into the Cause of Animal Heat: with Incidental Observations on Several Physiological and Chymical Questions, connected with the Subject* (London, 1778). For Leslie, "phlogiston is fire and light, or a certain subtile elastic fluid, upon the modifications of which the phenomena of heat and light depend." The sun communicates phlogiston to

and repulsion with different substances also appears in Robert Harrington's *New System on Fire and Planetary Life* (1796), the two principles being those of light and fire (the principle of motion) and of earth (the principle of inactivity), the former consisting of mutually repelling and the latter of mutually attracting particles. Again, the balance of nature was secured by the action of attractive and repulsive principles. Harrington is quite clear as to the implications of this idea, pointing out that whereas Newton had been "obliged to bring in the immediate hand of the Deity" to account for gravitation, his own theory did away with the necessity of supposing divine action.[200] He explained the gravitational action of the earth and the sun in terms of the circulation of fire (the repellent principle) between the earth and the sun and its attraction by the matter of the earth and sun. Similar ideas can be found in Adam Walker's *System of Familiar Philosophy* (1799), for Walker supposes that *"attraction and repulsion are the great acting principles of the universe,"*[201] light, fire, electricity, and phlogiston being "modifications of one and the same principle,"[202] the principle of repulsion. He emphasises that the operations of nature were "determined by a balance of those two powers"[203] of attraction and repulsion: "But of all opposing or antagonistic principles, none exhibit so general an enmity as fire and attraction. These two enemies are in a state of unceasing warfare: attraction drawing the particles of matter into a closer and closer union; while fire (or

---

bodies, "the same matter, which in a separate state constitutes fire and light, when modified in bodies is the cause of the inflammability" (*ibid.*, p. 104 f.). This matter is "the chief cause and principle of activity" in the universe (*ibid.*, p. 9), and is "exempted from the common laws of gravitation" (*ibid.*, p. 119). For Higgins on nongravitational matter, see his *A Syllabus of Chemical and Philosophical Enquiries* (London, [1775]), p. xlviii. The relation to Hutton's ideas is clear. Higgins, Leslie, and Hutton were all associated with Joseph Black: see J. R. Partington and D. McKie, "Historical Studies on the Phlogiston Theory. III. Light and Heat in Combustion," *Annals of Science, 3* (1938), 337-371. Cf. Michell, who regarded light particles as ponderable (Note 122, above).

200. Robert Harrington, *A New System on Fire and Planetary Life; shewing that the Sun and Planets are inhabited, and That They Enjoy the same Temperature as our Earth* (London, 1796), p. 18.

201. Adam Walker, *A System of Familiar Philosophy*, rev. ed., 2 vols. (London, 1802), *1*, 1.

202. *Ibid.*, p. 14.

203. *Ibid.*, p. 6. The idea of the balance of nature is itself a topic of investigation.

caloric, in the language of modern chemistry) is still striving to set those particles more and more at a distance."[204] The distance of the earth from the sun is determined by a balance of the power of gravity and the impulse of light from the sun, and Walker follows William Herschel in supposing the sun to be a planet surrounded by an atmosphere of fire.[205] The principles of attraction and repulsion were identified with material entities; by "matter" was meant "everything solid or fluid in nature, and we conceive this matter to be made up of particles . . . [which are] infinitely small."[206]

In the writings of these theorists we can see the influence of the Newtonian conception of the aether as the embodiment of force, and of the rejection of Newton's doctrine of active principles with its replacement by a conception of nature as a closed system, its operations secured by attractive and repulsive forces. In discussing the aether of the 1717 *Opticks* we pointed out that this aether—its microstructure embodying forces between vanishingly small particles—must be viewed in the context of Newton's theory of a nearly matterless universe regulated by forces. Indeed, his theory of force and paucity of matter and his theory of the "force-aether" were expressions of his concept of the primacy of force in the universe. Gowin Knight explicitly linked his theory of attractive and repulsive forces to the Newtonian doctrine of the paucity of matter in the universe, arguing that "All bodies whatsoever, whether solid or fluid, must contain more Pores than solid Parts. The Truth of this Proposition has been sufficiently proved by most of the Philosophers of this and the last Century, from Facts and Experience."[207] These theories, then, must be viewed in the context of the Newtonian

---

204. *Ibid.*, p. 18.
205. *Ibid.*, p. 12. See William Herschel, "On the Nature and Construction of the Sun and Fixed Stars," *Philosophical Transactions*, 85 (1795), 46-72; reprinted in Herschel, *Papers* (Note 126), *1*, 470-484. Herschel argues that the sun was a planet, sunspots being its "real solid body" which was covered by an atmosphere of "various elastic fluids" some of which are of "shining brilliancy" (*ibid.*, pp. 472 f.). See also Harrington, *op. cit.* (Note 200), p. 2.
206. Walker, *op. cit.* (Note 201), *1*, 37.
207. Knight, *op. cit.* (Note 192), p. 37. Cf. Walker: "the whole matter of the universe is supposed capable of being compressed into a walnut" (*op. cit.* [Note 201], *1*, 371); and also Higgins: "some authors suppose that the pores of [bodies] . . . may be much larger than their solid impenetrable parts; and the ingenious *Boscovich* finding this insufficient, supposes that matter is not impenetrable according to our conception of it" (*op. cit.* [Note 198], p. 246). Walker and Higgins clearly reflect Priestley's influence (*loc. cit.* [Note 117]).

tradition of the primacy of force. The influence of these ideas, developed in a different way than in Knight's theory, but nevertheless having relation to it, can be seen in John Rowning's *Compendious System of Natural Philosophy* (1738–1745). Like Knight, Rowning denies that the operations of inactive matter can be explained by a mechanical cause. They are "the Act of an *immaterial Cause,* in Virtue of which *inactive Matter* performs the offices for which it was designed,"[208] this cause being "the continued acting of God upon Matter, either mediately or immediately."[209] Like Knight, Rowning identifies the attractive and repulsive forces as "Powers or active Principles" which were not "essential to its Existence, but impressed upon it by the Author of its Being."[210] The influence of Newton is clear, though neither Rowning nor Knight maintained his relative distinction between force and active principles. Rowning attempted to explain a variety of phenomena using these principles, extending Newton's statement in Query 31 on envelopes of forces surrounding particles by supposing each particle to be surrounded by "three Spheres of Attraction and Repulsion, one within another."[211]

These ideas on spheres of force, to be extended by Boscovich and further developed by Priestley into a theory of the primacy of force, clearly have relation to the theories of attractive and repulsive powers developed by theorists such as Hutton and Knight. Despite the apparent differences, all these theories form part of a tradition of Newtonian forces. The eclecticism of these eighteenth-century theorists is paralleled by Newton's own varying views on the nature of force. Over and above their debt to Newton, these theorists also rejected much of his natural philosophy. This appears most clearly with respect to Newton's concept of active principles, for the activity of nature came to be regarded as being inherent in the natural realm. The concept of active substances provides a clear example of this; and the identification of a substratum with a principle of activity occurs in much of the natural philosophy we have discussed. A further example, which yet again emphasizes the relation of theories of this kind to the concept of power, can be

208. John Rowning, *A Compendious System of Natural Philosophy*, 4th ed. (London, 1745), Part 1, p. vi.
209. *Ibid.*, p. xxxix.
210. *Ibid.*, p. 12.
211. *Ibid.*, Part 2, p. 6n.

found in the work of Cadwallader Colden. In his *Explication of the First Causes of Action in Matter* (1748), Colden explained phenomena in terms of three kinds of powers—resisting, moving, and elastic powers—and he identified these powers as "agents or acting principles."[212] He considered that "We have no idea or conception of any thing other than of its power or force,"[213] for in the attempt to "describe matter without action, power, or force, the whole description must consist of negatives ... it must be the description of *nothing.*"[214] Colden regarded these "agents or acting principles" as "species of matter,"[215] for, as he put it in his *Principles of Action in Matter* (1751), action had to inhere in something: "motion is a property or quality, or more properly an action; we cannot then conceive it without supposing that it exists in something, which has in itself the power of moving." As he made clear, action not only existed in matter—or could be regarded as an attribute of matter—but was itself a substance, a species of matter. Indeed, matter as such could not be considered apart from action, power, or force. Colden quite explicitly associates his theory of motion as an action, existing in a material entity itself possessed of the power of moving, with a rejection of Newton's theory of bodies being maintained in motion by the continued action of the deity. For he stated that "It seems to be a very unphilosophical method of reasoning to suppose, that motion comes immediately from the Divine Being."[216] Once again, a theory of powers, forces, and active substances was conceived in terms of a conception of nature as a self-contained system.

A theory of a balance of powers can also be found in Thomas Exley's *Principles of Natural Philosophy* (1829). Arguing that matter is perceptible by means of its powers "which in themselves are in continual operation, and appear to constitute the very essence of matter,"[217] he affirms that the powers of attraction and repulsion

---

212. Cadwallader Colden, *An Explication of the First Causes of Action in Matter, and of the Cause of Gravitation* (London, 1746), p. 25.
213. *Ibid.*, p. 38.
214. *Ibid.*, p. 26.
215. *Ibid.*, p. 25.
216. Cadwallader Colden, *The Principles of Action in Matter, the Gravitation of Bodies and the Motion of the Planets explained from those Principles* (London, 1751), p. 73. For a rather different account of Colden see Schofield, *op. cit.* (Note 3), pp. 130-133.
217. Thomas Exley, *Principles of Natural Philosophy: or, a New Theory of Physics, founded on Gravitation, and applied in explaining the general proper-*

secure a balance in nature: "These powers are denominated attraction and repulsion. Their nature is not known, but the laws of their operation have been at least partially developed: that both these belong to matter is incontrovertible: did attraction exist without repulsion, matter would be conglomerated into one body, and if there were repulsion only, all bodies would be universally dispersed."[218] Exley's theory of powers is based on his view that the powers constitute the essence of matter. Arguing that "we know nothing of *matter*, but by the forces which it exerts, and which doubtless constitute its nature,"[219] he goes on to state that all that could be known of an atom was

> a balance of forces on every side of a central point, and this is all we can understand of it . . . for it is nothing but mere hypothesis, the effect of imagination, and a vulgar notion, to judge, that there is a minute solid impenetrable mass necessary to constitute an atom of matter, on which forces act . . . we know nothing of such little solids, we have never seen them, nor felt them, nor perceived them by any one of the senses; if they do exist at all, we have not been affected by them but only by the forces of attraction and repulsion . . . the forces are considered as constituting the essence of matter.[220]

These arguments attest to the widespread use of the idea of a balance of attractive and repulsive forces, particularly in relation to the philosophical implications of the notion of power.

CONCLUSION

In this study we have attempted to delineate the relationships among a number of connected themes underlying eighteenth-century British natural philosophy; these themes were centered on the epistemological and ontological problems of matter theory. We have shown that many of the fundamental principles of late seventeenth-century natural philosophy—the primary and secondary qualities distinction and Newton's notion of essential qualities, the connection of unobservables to observables—were rejected by a significant group

---

ties of matter, the phenomena of Chemistry, Electricity, Galvanism, Magnetism, and Electro-Magnetism (London, 1829), p. vi.
218. *Ibid.*, p. vii.
219. *Ibid.*, p. 470.
220. *Ibid.*, pp. 473 f. Unlike Hutton, Exley maintains that "matter exists continually by the power of its great Author" (*ibid.*, p. xxvii).

of eighteenth-century natural philosophers. This important shift in the foundations of scientific knowledge was intimately associated with the rise of new systems of science. In addition, Newton's hierarchical conception of nature and his notion of divine providence were rejected: nature came more to be seen as a self-contained system with activity ascribable to its intrinsic characteristics rather than to the operations of divine energy. In attempting to show how these ideas can be viewed in a tradition deriving from Newton and Locke we have sought to characterize a number of apparently diverse currents of thought, indicating common themes which were fundamental to the problems of scientific explanation. In tracing these ideas by no means have we sought to imply that these problems pervaded all of eighteenth-century British natural philosophy. However, in linking diverse systems of natural philosophy to the traditions of Newtonian forces and Lockean powers we have shown that these ideas appear in systems of nature which are characteristic of the natural philosophy of the period. For this reason alone, the conclusions of this study have wide implications for the history of natural philosophy in the post-Newtonian period.

The main theme of this study has been the doctrine that the essence of matter is constituted by powers. We have established the pervasive influence of this notion in eighteenth-century thought, and in conclusion we will indicate that this idea can be traced into the nineteenth century. Indeed, this notion was employed by Michael Faraday and was to underlie his development of the theory of the primacy of lines of force. Thus, one of the characteristic ideas of nineteenth-century field theory was developed from the view that matter consisted of powers and forces which extended continuously throughout space. Faraday argued that the hardness of matter was due to the *"force of repulsion . . .* in the particles," and "if we recognize matter by its *hardness,* what do we other than recognize by our sensations a force exerted by it."[221] Matter was known only by the forces it exerted, and Faraday considered that these forces constituted the essence of matter. Disinguishing between the particles of matter and the powers associated with these particles, Faraday defined "the particles of matter away from the powers a, and the system of powers or forces in and around it m," arguing that

221. Michael Faraday, MS on "Matter," quoted in T. H. Levere, "Faraday, Matter and Natural Theology," *Brit. J. Hist. Sci., 4* (1968), 105.

the properties of a substance belong to it "in consequence of the properties or forces of the *m*, not of the *a*, which, without the forces, is conceived of as having no powers. But then surely the *m* is the *matter*." Faraday concluded that "the substance consists of the powers or *m*," and he went on to suppose that this notion of matter supposed "the mutual penetrability of matter."[222] For by virtue of its powers and forces—which constituted its essence—"matter" extended continuously throughout space, and interactions between material particles were conceived in terms of the interactions between forces diffused through space. From this notion of matter as powers and forces diffused through space Faraday was led to develop his theory of the ontological primacy of lines of force.[223] It was this theory of Faraday's that Maxwell was to adopt in his first attempt to develop Faraday's ideas in "Faraday's Lines of Force" (1856), and though the problem of Maxwell's adoption of concepts from Faraday is extremely complex, his use of Faraday's theory of lines of force—which treat "the distribution of forces in space as the primary phenomenon"[224]—was fundamental to his development of Faraday's work.[225] Seen in this way, the concept of the physical field can in part be traced through the speculative tradition in British thought discussed in this paper to Newton's emphasis on the role of forces in nature.

ACKNOWLEDGMENTS

We wish to thank Gerd Buchdahl and Laurens Laudan for comments, Jack Morrell for useful suggestions, Roy Porter for discussion on Hutton, Alan Shapiro for drawing our attention to a passage in Hobbes, and other colleagues for their kindness in reading the manuscript.

222. Michael Faraday, "A Speculation touching Electric Conduction and the Nature of Matter" (1844), *Experimental Researches in Electricity*, 3 vols. (London, 1839-1855), *2*, 290 ff.
223. The results of this study clearly challenge L. Pearce Williams' interpretation of Faraday in his *Michael Faraday* (London, 1965). For a critical analysis of Williams' views see J. B. Spencer, *op. cit.* (Note 124), pp. 184-202. For a different interpretation of Faraday to that of Williams see P. M. Heimann, "Faraday's Theories of Matter and Electricity," *Brit. J. Hist. Sci., 5* (1971), 235-257.
224. Draft MS on "Faraday's Lines of Force," University Library, Cambridge, Add. 7655. The paper was published in *Trans. Camb. Phil. Soc., 10* (1856), 27-83.
225. See P. M. Heimann, "Maxwell and the Modes of Consistent Representation," *Arch. Hist. Exact Sci., 6* (1970), 171-213.

The Impact of the Neutron: Bohr and Heisenberg

BY JOAN BROMBERG*

1. INTRODUCTION

Niels Bohr and Werner Heisenberg received word of Chadwick's discovery of the neutron in the middle of March 1932. In this paper, I examine the impact of the neutron on their ideas about the atomic nucleus. The period I cover is the three years before Chadwick's announcement and the six months after it. I have attempted to bring my examination to bear on two problems. First, within three months of hearing of the neutron, Heisenberg succeeded in using it as the basis of a semiquantitative explanation of the composition and stability of nuclei. A comparison of Heisenberg and Bohr makes it possible to elucidate and, in part, to explain Heisenberg's achievement. Second, I have taken up the question of the place of nuclear theory within theoretical physics as a whole. This problem has not yet been dealt with by historians of science; I treat only a part of it here—the state of affairs as Bohr and Heisenberg saw it. This is, however, a significant part, for because of their abilities, age, and situation, Heisenberg and, above all, Bohr were in a position in 1929–1932 to have an unusually comprehensive view of the whole of their subject.

The conclusions I have been led to may be best summarized by answering the second question first. At the end of the twenties, nuclear physics was a part of relativistic quantum mechanics. To put the situation too simply for the moment, the reason for it was this. Only two fundamental particles were known: the proton and the electron. To account for nuclear masses and charges, therefore, as

---

* Department of the History and Philosophy of Science, Hebrew University, Jerusalem.

well as for the emission of beta-rays (electrons), electrons were assumed to be in nuclei in addition to protons. The uncertainty relations, however, require that if particles as light as electrons are retained in volumes as small as nuclei, they must have kinetic energies that are so large that relativistic effects become important. Therefore, nuclear electrons and, hence, nuclei must be treated by a relativistic theory.

It is thus necessary to look at the situation in relativistic quantum mechanics and in the related field of quantum electrodynamics at this time. When quantum mechanics was created in 1925 and 1926 physicists expected that it would be capable of being extended to the electromagnetic field. These early hopes proved too optimistic. Moreover, the theory was not relativistic, and attempts to make a relativistic theory of charged particles met with difficulties. By 1929 the state of this part of physics was widely felt to be unsatisfactory. This feeling was strengthened by the puzzling results of experiments on nuclei, which were seen as part of the same problem-complex. Above all, the experiments on beta-ray energies and on nuclear statistics contributed to the sense of mystery.

Physicists began to feel that another major transformation in physical thought would be necessary to deal with these problems, a transformation as radical with respect to the ordinary quantum mechanics as the latter had been with respect to the physics of the early twenties. As a consequence, among the concerns of Bohr and Heisenberg in this period were the search for a new physics and for a precise definition of the boundary between its domain of applicability and that of ordinary quantum mechanics.

The implication this has for the historian is that statements about nuclear phenomena may not be interpreted solely in terms of the ideas and experiments internal to *nuclear* physics. A case in point—one that is central to this paper—is Bohr's well-known suggestion that the principle of energy conservation fails for beta-ray disintegrations. It is usually—and incorrectly—regarded as giving a measure of the gravity with which physicists viewed the difficulties created by experiments on the nucleus for nuclear theory alone.[1]

---

1. For example, this is the interpretation of Edward M. Purcell in his pioneering study, "Nuclear Physics without the Neutron; Clues and Contradictions," *Xth International Congress of the History of Science, Proceedings* (Ithaca, 1962), *1*, 128. See also D. M. Brink, *Nuclear Forces* (Oxford, 1965), p. 8.

Instead, Bohr's hypothesis must be seen within the context of the search for a new theory. It was part of the overarching attempt he was making to judge which parts of the old physics were likely to be retained in such a theory and which were not.

This same state of affairs also produced a more complicated attitude towards nuclear electrons than I suggested above. For if the need to put electrons in the nucleus was a chief reason for connecting nuclear physics with relativistic quantum theory, the connection itself facilitated proposals for eliminating electrons from the nucleus. That such proposals were made will be demonstrated below. That they had to make use of bold hypotheses—at a time when only electrons and protons were known—is clear from the fact that the mass of nuclei, measured in units of the proton's mass, is roughly twice the nuclear charge, measured in units of the proton's charge.[2] Strange ideas were encouraged by an atmosphere in which a new physics was expected. As a result, it is not correct to say that nuclei were universally assumed to contain electrons in this period. Attitudes were more open; the hypothesis that electrons were in nuclei and the intuition that they were excluded were held simultaneously.[3]

It is now easier to see the character of Heisenberg's achievement. Broadly put, the discovery of the neutron made it possible to change the relation between nuclear physics and the domain of unsolved problems.[4] A substantial number of nuclear problems now became solvable by ordinary quantum mechanics. The achievement of Heisenberg was to see this possibility and find a way to give it formal expression.

---

2. The same thing is also clear from beta-decay. One must devise a way to get electrons out of nuclei without having them inside. This part of the problem was not affected by the discovery of the neutron and required bold ideas. It is not surprising that in the ideas of Bohr before 1932 we shall also find elements of the solution Fermi gave in his beta-decay papers of 1933 and 1934.

3. Here also I differ from Purcell, who wrote that with the exception of an article written by J. Dorfman, "every nuclear model had to be loaded with those electrons before the discussion could begin and it seemed that every other principle would be sacrificed before their presence would be questioned." ("Nuclear Physics without the Neutron," p. 131.) However, my interpretation is, above all, suggested by the unpublished sources, while Purcell based his on published papers.

4. Compare S. Devon's "Comments" to the paper of Emilio Segrè in the *Xth International Congress . . . Proceedings*, *1,* 155-158.

Sections 2 and 3 below deal with Bohr and Heisenberg, respectively, in the three years before Chadwick's discovery. They present evidence for the picture I have given of the state of nuclear theory. The documents examined also illustrate the contrast between Bohr's and Heisenberg's intellectual styles. The fourth section begins with a description of Bohr's reactions to the neutron, placing Heisenberg's work in sharper relief, and proceeds to an analysis of Heisenberg's articles. I have tried, among other things, to trace the effect of Heisenberg's style in these articles. Bohr's own contribution to nuclear structure theory—the compound nucleus model—did not come until the end of 1935. It is interesting that although Bohr's model depended essentially on the work of Heisenberg and others, work that placed nuclear structure under quantum mechanics, it was inspired by the same conviction of a profound difference between nuclear and atomic systems that is expressed in such a different form in the documents studied here.[5]

2. BEFORE THE NEUTRON: BOHR

"Fate," Bohr wrote W. Pauli on 1 July 1929, "has truly been very ungracious with respect to the completion of the various small notes I had promised to send. . . . So that you may see that the promise was not a complete falsehood, I enclose two fragments. . . . The second is a little piece on the beta-ray spectra, that has long been on my mind, and that I have had recopied in the last days, without, however, being able to persuade myself to send it off, as it gives so little positive, and is so roughly executed."[6]

The disintegration of radioactive nuclei by the emission of beta-rays showed a significant difference from the disintegration by emission of alpha-rays (helium nuclei). In the latter case, all the emitted particles have one (or a few) well-defined velocities. In the former case, the particles have all velocities from zero on. The early dispute between C. D. Ellis and L. Meitner as to whether the electrons actually emerged from the nuclei with a continuous

---

5. See Bohr's "Neutron Capture and Nuclear Constitution," *Nature, 137* (1936), 344-348, particularly paragraph 6, with its explicit reference to the Faraday Lecture, and paragraph 4.

6. Bohr to Pauli, 1 July 1929, Bohr Scientific Correspondence (hereafter abbreviated BSC), Danish original. All translations in this article are mine.

310

distribution in their energy values, as Ellis maintained, or only acquired this energy distribution as a result of secondary processes, as Meitner held, had been settled to Ellis' satisfaction in 1927 with the calorimeter experiments he conducted with W. A. Wooster. Meitner and W. Orthmann confirmed these results in an article they submitted in December 1929. At the same time, they disposed of the suggestion that part of the energy released by the decaying nuclei was carried off by gamma-rays.[7] This left the question of why seemingly identical parent nuclei could decay into seemingly identical daughter nuclei with the emission of differing amounts of energy. Bohr's letter and accompanying note on beta-rays were written at a time when the data on energy loss in beta decay had almost attained its most puzzling form.[8]

The immediate occasion for the composition of the note seems to have been provided by two articles by G. P. Thomson in 1928 and 1929.[9] As the most plausible interpretation for the results of Ellis and Wooster, Thomson advocated that "energy is not conserved in the individual act of emission." He further maintained that this was a "natural consequence of the wave theory of matter."[10] Thomson argued that the emission of a nuclear electron could be regarded as "sudden." Consequently, the emerging electron should be described by a well-localized wave packet. This wave packet would quickly spread, so that the front of the packet would be increasingly ahead of the rear. Thomson implicitly associated the front of the wave group with fast electrons and the rear with slow ones. Some electrons would therefore speed up after emission, and others would slow down.[11]

---

7. Purcell, p. 128. C. D. Ellis and W. A. Wooster, "The Average Energy of Disintegration of RaE," *Proc. Roy. Soc., 117* (A) (1927), 109-132; L. Meitner and W. Orthmann, "Über eine absolute Bestimmung der Energie der primären β-Strahlen von Radium E," *Zeits. f. Phys., 60* (1930), 143-155.

8. The important result that the upper limit of the beta-ray spectrum is equal to the difference in the energies of the parent and daughter nuclei was not established until 1933. See C. D. Ellis, "The β-Ray Type of Radioactive Disintegration," *International Conference on Physics, London 1934 . . . Papers and Discussions* (Cambridge, England, 1935), *1*, 46.

9. G. P. Thomson, "The Disintegration of Radium E from the Point of View of Wave Mechanics," *Nature, 121* (1928), 615-616; "On the Waves Associated with β-Rays and the Relation between Free Electrons and Their Waves," *Phil. Mag., 7* (1929), 405-417 (dated 7 Jan. 1929).

10. Thomson, "On the Waves Associated with β-Rays," p. 406.

11. Thomson, "The Disintegration of Radium E," p. 615.

In his second long article, Thomson made the ideas underlying his argument explicit: "The conception of a particle in motion is almost meaningless unless it can be supposed to have a definite velocity at a definite time. . . . We can, however, keep both the conception of moving particles and the whole analytical machinery of the wave mechanics (at least for free electrons) if we are prepared to allow the possibility of an electron changing speed in force-free space."[12]

It was characteristic of Bohr that an argument he thought erroneous could engage his attention and start him on a multisided consideration of the problem on which the argument touched. Bohr himself had suggested in 1924, before the emergence of quantum mechanics, that energy conservation is violated in the Compton effect. Subsequent events led him to reconsider this, however, and at least by the time of the Como Conference in 1927, he had satisfied himself that energy conservation is valid in quantum mechanics.[13] His note, entitled, "β-Decay Spectra and Energy Conservation,"[14] began therefore by arguing against Thomson's idea that a violation of the energy law can be derived from wave mechanics. From this, however, Bohr did not conclude that energy must be conserved in beta-decay, but rather that beta-decay is outside the competence of wave mechanics: It is unlikely, he wrote, that there is "any simple explanation of the continuous β-spectra based on the ordinary ideas of wave mechanics."

Bohr next took up the question of whether there were any theoretical grounds on which one could "defend a violation of the principles of conservation in radioactive processes." He suggested

12. Thomson, "On the Waves Associated with β-Rays," p. 413.
13. N. Bohr, H. A. Kramers and J. C. Slater, "The Quantum Theory of Radiation," *Phil. Mag.*, 47 (1924), 785-802. See Martin J. Klein, "The First Phase of the Bohr-Einstein Dialogue," *Historical Studies in the Physical Sciences*, 2 (1970), 1-39. I do not know of any historical study on Bohr's reconsideration of the Compton effect in the years when matrix and wave mechanics and their interpretation took shape. The Como address was published as "The Quantum Mechanics and the Recent Development of Atomic Theory," *Nature, 121* (1928), 579-590.
14. Bohr Manuscript Collection (hereafter abbreviated BMC). The manuscripts are catalogued by year and title. The manuscript is in English and appears from its contents and from the reply of Pauli's cited below to be a slightly different version of what Bohr sent Pauli. It consists of five pages: three pages of undated, consecutive narrative, a second page "3" dated 21 June 1929, and a "3a" page correcting the dated "3."

that such grounds might exist within the context of physical ideas in which the "problem of the constitution of the elementary electric particles" is treated. The latter, he pointed out "has so far escaped a proper treatment on the basis of classical electrodynamics." Implicit in the last remark is one of the reasons Bohr was willing to entertain solutions like nonconservation in this circle of problems. In the proper limit, ordinary quantum mechanics gives the results of classical physics. However, there is no satisfactory result in classical physics for the problem of the electron's structure, and where we do not know the limiting, classical behavior we are particularly liable to surprises. Bohr then mentioned one of the results of recent work in quantum electrodynamics and its possible implications for a violation of energy conservation in the nucleus.

Not only the emitted betas, Bohr wrote, but the nuclear electrons also "seem to fall entirely outside the field of consistent application of the ordinary mechanical concepts, even in their quantum theoretical modification." The reference is partly to the high kinetic energies that must be attributed to nuclear electrons, but more to the growing puzzle of the "anomalous" measurements of nuclear spins and statistics. "From this point of view," Bohr continued, "the disintegration of the nucleus should rather be regarded as the creation of the dynamical individuality of the electron expelled. If, therefore, experimental evidence should really corroborate [that the conservation principles fail] . . . we can hardly reject this suggestion on purely theoretical grounds." The violation of energy conservation, Bohr suggested, might also explain the production of energy in stars. Under the right initial conditions, a process involving both the capture of betas and their subsequent reemission might result in a net gain of energy.

Pauli's reaction to this was immediate and unfavorable. After expressing his enthusiasm for the first of the manuscripts Bohr had sent him, he continued, "It is otherwise with the note on beta-rays. I must say that I am *little* satisfied by it." Pauli went on to give specific objections, and concluded, "*In any case, therefore, lay this note aside for a good long while. And let the stars radiate in peace!*"[15] A little more than one year later, Pauli was to offer his

---

15. Pauli to Bohr, 17 July 1929 (BSC), German original.

own hypothesis of beta-decay, which was to stand in the early thirties as the principal rival to Bohr's interpretation. In Pauli's theory, energy and momentum are conserved, but the unobserved portions are carried off by one (or more) very light, neutral particles, to which Fermi later gave the name of "neutrino."[16]

Other aspects of Bohr's thoughts on beta-decay in this period can be seen from an exchange of letters with Dirac at the end of 1929.[17] "Dear Dirac," the first letter began, "From Gamow I hear that you are now back in England again, and that you have made progress with the mastering of the hitherto unsolved difficulties in your theory of the electron. As we have not yet heard about any details, Klein and I should be very thankful if you would be so kind to tell us something of your present views." The principal difficulties Bohr had in mind, as the letter later makes clear, were those associated with the so-called "Klein paradox" that Oskar Klein, then in Copenhagen, had just discovered. It may be stated as follows: According to Dirac's wave equation, if an electron is subject to an electrostatic field where the potential increases by more than the electron rest mass, mc^2, in a distance of less than the Compton wavelength, h/mc, it has a considerable probability of passing through the potential barrier into a region where it has a negative kinetic energy, that is, where it behaves like a particle of negative mass. After posing his question, Bohr went on to speculate on a connection between the Klein paradox, the beta-ray spectrum, and stellar energies. He suggested that a new kind of complementary relation might come into effect, one between the validity of the energy and momentum conservation laws on the one hand and the concept of particle permanence on the other. Just as the complementarity between position and momentum is governed by Planck's constant, the classical radius of the electron may play the role of the universal constant in this new complementarity.[18]

16. See Pauli, "Zur Älteren und neueren Geschichte des Neutrinos," in *Wolfgang Pauli: Collected Scientific Papers* (New York, 1964), 2, 1313-1337, and the introduction by F. Rassetti to Fermi's papers on beta-decay, in *Enrico Fermi, Collected Papers* (Chicago, 1962), 1, 538-540.
17. Bohr to Dirac, 24 Nov. 1929; Dirac to Bohr, 26 Nov. 1929; Bohr to Dirac, 5 Dec. 1929; Dirac to Bohr, 9 Dec. 1929; Bohr to Dirac, 23 Dec. 1929 (BSC). These letters are in English.
18. "My view was that the difficulties in your theory might be said to reveal a contrast between the claims of conservation of energy and momentum on one

It should be noticed Bohr is drawing a quite different connection between the permanence of particles and energy conservation than he did in the manuscript. In the latter the possibility that the electron does not persist in the nucleus with mechanical individuality opens the possibility that conservation fails. Here, instead of suggesting mechanical individuality and the mechanical conservation laws becoming invalid together, Bohr poses the validity of the first as excluding the validity of the second; that is, if one can follow a given electron in its transition from a positive to a negative kinetic energy state, then one cannot at the same time have energy conservation. It is noteworthy that Bohr brings a fundamental length, the classical electron radius, into the theory, typically introducing it by means of a complementarity relation. Bohr did not pursue this idea for long, however. He gave it up in favor of the attempt to resolve the paradox by the analysis of the experimental conditions, which we shall find in his second letter to Dirac.

Dirac replied that he had already heard Gamow speak on Bohr's views at Kapitza's Club. "My own opinion of this question," he wrote, "is that I should prefer to keep rigorous conservation of energy at all costs and would rather abandon even the concept of matter consisting of separate atoms and electrons than the conservation of energy. There is a simple way of avoiding the difficulty of electrons having negative kinetic energy." Dirac proceeded to sketch the solution Bohr had inquired about, in which transitions to negative energy states are barred because these states are almost completely filled by an infinite "sea" of electrons, and in which holes in this sea exhibit themselves as positively charged protons.

Bohr answered with a fundamental objection against Dirac's suggestion. An infinite sea of electrons should produce an infinite electric charge density. He then explained his own position. It was not simply Dirac's interpretation of the transitions from positive

---

side and of the conservation of the individual particles on the other side. The possibility of fulfilling both these claims in the usual correspondence treatment would thus depend on the possibility of neglecting the problem of the constitution of the electron in non-relativistic classical mechanics. It appeared to me that the finite size ascribed to the electron on classical electrodynamics might be a hint as to the limit for the possibility of reconciling the claims mentioned." (Bohr to Dirac, 24 Nov. 1929 [BSC].)

to negative energy that he opposed, but the whole program of attempting to interpret them in terms of ordinary concepts. "In the difficulties of your old theory I still feel inclined to see a limit of the fundamental concepts on which atomic theory hitherto rests rather than a problem of interpreting the experimental evidence in a proper way by means of these concepts." As he pointed out later in the letter, Bohr's hope was that those parts of Dirac's theory which had been so successful in explaining experiments could be separated from those which involved the transitions between positive and negative energy states. Bohr then suggested that the key to the problem revealed by Klein might be a more critical inquiry into the experimental conditions under which the transitions show themselves. To achieve the field strength that is necessary a number of electrons would need to be massed together; the massing would be so great that one would already have a situation outside the reach of classical concepts.

Bohr then restated his ideas on beta-decay and added, "In the fact that the total charge of the nucleus can be measured before and after the β-ray disintegration and that the results are in conformity with conservation of electricity I see a support for upholding the conservation of the elementary charges even at the risk of abandoning the conservation of energy." In his contributions to the Rome meetings on nuclear physics in October 1931 and to the Solvay Congress of October 1933 similar comments appear. It was his willingness to allow charge conservation a preferred position over energy conservation that, among other things, separated Bohr from Pauli throughout their controversy over the interpretation of beta-decay.[19]

Of particular interest in Dirac's reply to Bohr is the first paragraph. "I am afraid," Dirac wrote, "I do not completely agree with your views. . . . I cannot see any reason for thinking that quantum mechanics has already reached the limit of its development. I think it will undergo a number of small changes . . . and by these means most of the difficulties now confronting the theory will be removed. If any of the concepts now used . . . are found to be incapable of

---

19. See *Structure et Propriétés des Noyaux Atomiques; Rapports et Discussions du Septième Conséil de Physique . . . Solvay* (Paris, 1934), p. 324 (Pauli), and Bohr, "Sur la méthode de correspondance dans la théorie de l'électron," pp. 227-228; see also Bohr, "Atomic Stability and Conservation Laws," *Convegno di Fisica Nucleare, Ottobre 1931* (Rome, 1932), p. 119.

having an exact meaning, one will have to replace them by something a little more general, rather than make some drastic alteration in the whole theory." Thus, in late 1929, Dirac did not agree with Bohr's appreciation of the severity of the crisis in physics. Clearly, the difference in their outlooks on the general situation turned in part on the difference in their evaluation of Dirac's concept of an "infinite sea of electrons." Dirac was optimistic. Bohr, in his final letter, indicated again that he found the infinite electron density a "fatal" objection. Dirac's optimism proved justified by the discovery of the positron. Nevertheless, six years later their positions were to be dramatically reversed. It was then to be Dirac—in despair over the difficulties of relativistic quantum theory—who was willing to countenance the idea of energy nonconservation in beta-decay. Bohr, on the contrary, encouraged by the agreement of experiment with the theory of beta-decay that Fermi had based on Pauli's hypothesis[20] was to consciously echo Dirac's letter of 1929: "I understand of course the weight of your arguments regarding the present difficulties of relativistic quantum mechanics," Bohr was to write to Dirac, "but I am inclined to think that the only way to progress is to trace the consequences of the present methods as far as possible in the same spirit as your positron theory."[21]

Certain points are sufficiently supported by these documents alone that they may be asserted here. First, there is Bohr's expectation that beta-emission and the behavior of nuclear electrons would not be explained by existing quantum physics. This failure is unambiguously stated in the manuscript fragment, where Bohr adduced the fact of the continuous beta-ray spectrum. Another important

20. Enrico Fermi, "Tentativo di una teoria dell' Emissione dei Raggi 'Beta'," *Ric. Scientifica, 2* (1933), 491-495; "Versuch einer Theorie der β-Strahlen. I," *Zeits. f. Phys., 88* (1934), 161-171.
21. Bohr to Dirac, 2 July 1936 (BSC). See Dirac, "Does Conservation of Energy Hold in Atomic Processes?" *Nature, 137* (1936), 298-299, and Bohr, "Conservation Laws in Quantum Theory," *Nature, 138* (1936), 25. Although Dirac's article was ostensibly based on Shankland's experiments on the Compton effect, a careful reading shows that it was the more general problems that troubled him. See also his letter of 9 June 1936 to Bohr (BSC): "The non-relativistic nature of the present quantum theory appeared to me most strongly when I was writing my book. In the first edition where I tried to build up everything from a relativistic definition of state and observable, I found many things which were extremely awkward to explain. But these difficulties all vanish when one makes free use of non-relativistic ideas. I think there must be something fundamental underlying this."

reason for the failure of quantum physics is that the dimensions of nuclei are of the same order as the classical size of the electron; in his later Rome address, Bohr argued that nonrelativistic quantum mechanics can only be applied to systems so large with respect to their constituent electrons that the latter can be treated as points. Second, there is Bohr's vision of these nuclear phenomena as only part of the group for which a new physics would have to be found; the group included the structure of elementary electric particles and the Klein paradox. That Bohr foresaw that this new physics would have some new and surprising features is evident throughout these documents. Not least, it is reflected in the last of Dirac's letters to Bohr. Finally, in seeking the new theory Bohr proceeded in a characteristic way. He pondered over the concepts and laws of the existing theory, trying to distinguish which were likely to remain valid in the coming theory and which were likely to fail. It was not only the mechanical conservation laws Bohr weighed; it was also the law of conservation of charge, and the concept, closely connected with the conservation principles, of the permanence of particles. As we have seen, Bohr opted to retain charge conservation and relinquish mechanical conservation.

The works in which Bohr presented his thoughts on the nucleus for publication were the Faraday and Rome Lectures, composed at roughly the same time. The published Faraday Lecture was based on the talk Bohr gave at the Chemical Society in London on 8 May 1930, and on three lectures given at Cambridge University during the same visit to England. Although Bohr did some preliminary writing on the article as early as the summer of 1930, the work only began in earnest in 1931. He mailed the section on the nucleus to England in the fall of 1931, and he gave the entire manuscript an extensive revision in December 1931. The lecture deals with all of quantum theory, and is organized historically. The problems of the nucleus form the subject of the final section.[22]

22. Bohr, "Faraday Lecture . . . Chemistry and the Quantum Theory of Atomic Constitution," *Journ. of the Chem. Soc.* (1932), Part 1, pp. 349-384. For the origins of the published paper, compare the publications with the manuscripts. For the dating of the writing, see Léon Rosenfeld, "Nogle Minder om Niels Bohr," *Niels Bohr: Et Mindeskrift* (Copenhagen, 1963), pp. 69-70, and also Bohr to S. E. Carr, 15 Sept. 1931; Carr to Bohr, 30 Oct. 1931; Bohr to Carr, 4 Jan. 1932 (Bohr Administrative Correspondence). I am indebted to Mrs. S. Hellmann for bringing the latter correspondence to my attention.

The Rome Lecture was based on the informal remarks Bohr gave at the meeting on nuclear physics in Rome in October 1931. He began writing on the train trip back from Rome, mailing the finished paper to Fermi for inclusion in the proceedings on 12 March 1932. The title of the article is "Atomic Stability and Conservation Laws," and Bohr's introduction explains that, "Serious doubt has recently arisen, whether the concept of energy can find an unambiguous application to radioactive disintegrations in which electrons are expelled from atomic nuclei. . . . The following remarks may serve as an introduction to a discussion of this problem."[23] It is the Rome Lecture, therefore, which was specifically directed to the question of the validity of energy conservation for beta-decay. Since, in addition, it gives Bohr's views at a moment just before he learned of the discovery of the neutron, it deserves outlining here. In doing so, I shall supplement it with the section on the nucleus from the Faraday Lecture; this is legitimate because it was written about the same time and is consistent with the Rome Lecture in content.

Bohr's Rome article discussed in succession the position of the mechanical conservation laws in ordinary quantum theory, in relativistic quantum theory, and in nuclear physics. The ordinary quantum mechanics, Bohr began, rests on two independent foundations: the classical theory of particles, which "is so constructed that it satisfies the laws of conservation of energy and momentum," and the quantum of action (p. 120). Ordinary quantum theory has certain limits. First, it incorporates a limitation belonging to the classical theory of particles. The classical theory contains a contradiction: on the one hand, it treats the electron as a point mass; on the other, it attributes a minimum diameter to the electron to avoid an infinite self-energy from the electron's interaction with its own field. This contradiction is circumvented by restricting the theory to the description of systems of diameter large with respect to the electron diameter. A second limitation of quantum mechanics is that it is not relativistically invariant (pp. 120 and 122). The reasons that quantum mechanics has nevertheless been successful in describing the behavior of atoms are twofold: first, the dimensions of atoms are very

---

23. Bohr, "Atomic Stability and Conservation Laws," p. 119. Hereafter, references to this paper will be made in the text by page number. For the dating of the writing, see Bohr to Fermi, 13 Nov. 1931, and 12 March 1932 (BSC).

large compared with the classical diameter of the electron; second, the relativistic effects are small (pp. 121-123).

Bohr next took up Dirac's relativistic theory of the electron and the contemporary theories of quantum electrodynamics. Both, Bohr explained, come up against fundamental difficulties. For the Dirac theory, it was still the existence of the transitions to states of negative mass that Bohr saw as critical; for quantum electrodynamics, it was the appearance of infinite energy. Bohr envisioned a final solution in a theory where, in contrast to quantum mechanics, "the elementary particles and the quantum of action appear as inseparable features" (p. 123). It is necessary to go beyond this general appraisal, Bohr continued, and pose the question of whether, to some extent, "the present theory offers a reliable guidance for the analysis of the phenomena" (p. 123). There follows a rich and complex analysis of the application of these two parts of theory to a series of cases (pp. 123-126). The result is that the theory is not sufficient to yield a conclusion about the conservation laws, and, therefore, we can reason only from experiment. Bohr implied that it is necessary to separate atomic from nuclear phenomena: "As far as the extra-nuclear electrons in atoms are concerned," Bohr concluded, "there is no experimental indication of a failure of the conservation laws" (p. 127).

Bohr treated the experimental evidence for the validity of the energy law in the nucleus in detail in the final section. He gave first place to the argument from nuclear statistics to energy nonconservation. The information on nuclear spins and statistics, which had increased considerably from 1929 to 1932,[24] posed the following problem. One can tell unequivocally how many protons and electrons each nucleus possesses from its charge and mass, and since protons and electrons each have spin $1/2$, one can tell the total number of spin $1/2$ particles each nucleus contains. Quantum mechanics gives the rule that a system composed of an odd number of spin $1/2$ particles obeys Fermi-Dirac statistics. This was in conflict with the evolving experimental information on statistics of nuclei. As Purcell pointed out, the reaction of physicists to this problem was to say that the electron's spin was suppressed when it entered the nucleus. In Bohr's formulation, "the idea of spin is found not to be applicable to intra-nuclear electrons."[25]

24. See Purcell, pp. 126-127, and Brink, pp. 5-6.
25. Bohr, "Faraday Lecture," p. 380.

Bohr continued: "This remarkable 'passivity' of the intra-nuclear electrons in the determination of the statistics is a very direct indication, indeed, of the essential limitation of the idea of separate dynamical entities when applied to electrons. Strictly speaking, we are not even justified in saying that a nucleus contains a definite number of electrons, but only that its negative electrification is equal to a whole number of elementary units, and, in this sense, the expulsion of a β-ray from a nucleus may be regarded as the creation of an electron as a mechanical entity."[26] "We cannot therefore be surprised," he added to the last thought, "if these processes should be found not to obey such principles as the conservation laws of energy and momentum, the formulation of which is essentially based on the idea of material particles" (p. 128).[27] This is an elaborated version of the argument already sketched in the 1929 fragment, and it is extremely important. Bohr is making use of a particular form of the concept of the creation of material particles in regarding them as created with respect to their mechanical properties. This provides a model of the nucleus which, in a partial sense, is free of electrons. Electrons are in the nucleus with respect to their existence as discrete units of charge, but not with respect to their existence as mechanical entities.

At the end of the section Bohr brings in the beta-ray spectrum. He poses two alternatives. Either energy conservation applies, and the individual nuclei of the product of a radioactive disintegration are left with differing amounts of energy and are hence different, or energy conservation does not apply, so that, although differing energies are released in their formation, the nuclei are identical. In the first alternative, it becomes difficult to account for experiments showing the existence of definite rates of decay for radioactive ele-

26. *Ibid.* Almost the same phrases appear in the Rome article, pp. 127-128.
27. Bohr pointed out that the laws of quantum mechanics—in particular the quantum mechanical rules for statistics—do hold for nuclear protons in sharp contrast to electrons (pp. 128-129). In seeking an understanding of the difference, he made the suggestion that the ratio m/M, the mass of the electron over the mass of the proton, "plays a fundamental part in the question of the stability of atomic nuclei. In this respect the problem of nuclear constitution exhibits a characteristic difference from that of the constitution of the extra-nuclear electron configuration, since the stability of this configuration is essentially independent of the mass-ratio." (Faraday Lecture, p. 379.) The significance of this suggestion is that it is one of a series of attempts Bohr made throughout the thirties to exhibit and fix the essential difference between atomic and nuclear systems.

ments as well as for facts pointing to the essential identity of nuclei with the same numbers of protons and electrons among nonradioactive elements; thus, evidence favors the second alternative, that of renouncing the energy law (pp. 129–130). There was a third alternative, however, and Pauli was at the Rome meeting to speak for it.[28]

In an article written for a memorial volume to Bohr, Heisenberg characterized Bohr's style of work in these words: "His insight into the structure of the theory was not a result of a mathematical analysis of the basic assumptions, but rather of an intense occupation with the actual phenomena." Bohr "feared that the formal mathematical structure would obscure the physical core of the problem, and in any case, he was convinced that a complete physical explanation should absolutely precede the mathematical formulation."[29] It was the priority Bohr gave to the phenomena that lay behind his reluctance to entertain Pauli's hypothesis of a neutrino. Bohr felt that experiment should decide whether energy and momentum were truly indefinite in beta-emission or whether some hitherto unknown particle was carrying the missing portions away. A theory could only be built after the facts; this is the sense of his remark in the 1929 fragment that we shall have no grounds in theory for rejecting nonconservation in case "experimental evidence should really corroborate the suggestion." It is also behind his remarks on the neutrino in the proceedings of the 1933 Solvay Congress. In his Solvay paper, completed more than a month after he had received one of Fermi's articles on beta-decay, Bohr still expressed the opinion that "before one has new experiments in this area, it seems to me . . . to be difficult to take a position on the subject of Pauli's interesting suggestion."[30]

In the event, theory preceded experiment. Most physicists came to accept Pauli's hypothesis long before the neutrino was detected in 1956. Bohr himself had come to favor it over his suggestion of nonconservation by the time of his controversy with Dirac in 1936. In justifying his new view, however, he again stressed the priority of

---

28. Rassetti, cited above, note 16, p. 538.
29. Heisenberg, "Quantum Theory and Its Interpretation," *Niels Bohr*, ed. S. Rozenthal (Amsterdam, 1967), pp. 95 and 98.
30. Bohr, "Correspondance dans la théorie de l'électron," *Septième Conséil de Physique . . . Solvay*, p. 228. For the dates on which Bohr received Fermi's paper and completed his Solvay manuscript, see Bohr to Fermi, 31 Jan. 1934, and Bohr to Pauli, 3 March 1934 (BSC).

phenomena, grounding his position in "the suggestive agreement between the rapidly increasing experimental evidence regarding β-ray phenomena and the consequences of the neutrino hypothesis of Pauli so remarkably developed in Fermi's theory."[31]

Despite the problematic character of nuclear physics in the years 1929–1932, Bohr had not held back from continuous reflection on it. There are a number of reasons for this. First, Bohr was always interested in obtaining a view of the whole of physics and of the relations of its parts. No clearer indication of this side of Bohr's thought is needed than the Rome Lecture itself. This preoccupation would have led him to wrestle with physics' more puzzling aspects whether or not he saw them as ripe for solution. Second, although he was certainly in the camp of the radicals, he was more sanguine than the others about the possibility of progress. "I do not share the pessimistic attitude [about the fundamental problems] that you and Pauli like to express so humorously," he wrote to Heisenberg in December 1930, explaining that he was preparing a fresh attack upon these problems.[32] Finally, this optimism is connected with Bohr's intellectual style. He confronted the phenomena directly with the concepts, without having to first possess a mathematical formulation. This admitted of an easier access to problems than Pauli's and Heisenberg's approaches. For, on the one hand, it was not necessary for Bohr to take a large step in the initial formulation of a theory, while, on the other, he could employ smaller, more subtle adjustments in altering the theory. As a consequence of his procedure, he was, by the end of this period, in possession of an elaborate analysis of the situation in nuclear theory within the framework of the totality of physics. As it happened, this analysis was given a completed form on the eve of the discovery of the neutron.

3. BEFORE THE NEUTRON: HEISENBERG

Heisenberg had been on a trip through the United States and to the Far East during most of 1929. His first letter to Bohr after his return to Leipzig was a Christmas greeting and comment on the

31. Bohr, "Conservation Laws in Quantum Theory," p. 25.
32. Bohr to Heisenberg, 8 Dec. 1930. This letter and all the others from Bohr to Heisenberg cited in this paper are in Danish. See also the letter of 18 Feb. 1931.

scientific problems of the day.³³ Heisenberg wrote that despite some improvements, his and Pauli's quantum electrodynamics³⁴ "remains a very grey theory" so long as the difficulty of transitions in Dirac's theory is unsolved. About Dirac's own solution, Heisenberg is "quite sceptical." Moreover, "a great, but interesting misfortune also seems to enter with nuclear spins." It appears as if the electrons do not contribute to the spin at all, and this, along with the continuous beta-ray spectrum, suggests that "there no longer really are electrons in the nucleus (es nicht eigentlich 'Electronen' mehr im Kern gibt)." Shortly after, Heisenberg launched a new and radical attack on the difficulties he had enumerated; this work is recorded in letters he wrote to Bohr on 26 February and 10 March 1930.³⁵ The most compelling motivation for Heisenberg's attack was the desire to solve the problem of the infinite self-energy of the electron. In analogy to classical theory, Heisenberg introduced the classical electron radius as a fundamental length. This length is of the same order of magnitude as nuclear diameters and as the quantity h/Mc, where M is the mass of the proton. Since the proton was the heaviest particle then known, h/Mc—the proton's Compton wavelength—was the smallest length within which the uncertainty relations allowed a particle to be localized. This indeterminacy may be seen as giving the theorist a freedom to alter physical laws for dimensions less than h/Mc. Accordingly, Heisenberg proposed the construction of a "lattice-world" ("Gitterwelt") of cells of volume $(h/Mc)^3$, in which new relations would hold within the cells. He characterized this as the "crudest method" by which a fundamental length could be introduced. A theory of this kind must be constructed so that systems that are large with respect to the cells obey quantum mechanics, while systems of the same order of magnitude obey new laws. In particular, the nucleus was a system which would be governed by any new

33. Heisenberg to Bohr, 20 Dec. 1929 (BSC). All letters from Heisenberg cited in this paper are in German.
34. Heisenberg and Pauli, "Quantentheorie der Wellenfelder," *Zeits. f. Phys.*, 56 (1929), 1-61, and 59 (1929), 168-190.
35. Following Heisenberg's trip, a period of close communication began between the two physicists. Approximately fifty letters between Bohr and Heisenberg from the start of 1930 to the end of April 1932 are preserved in the Bohr Scientific Correspondence. In addition, the two met at least eight times: in April, June, September, and October of 1930, in March of 1931, and in January, March, and April of 1932. The meeting dates are deduced from the letters.

results that might follow from the theory. One of the results Heisenberg hoped for was "a kind of 'reconnaissance in force' [gewaltsame Erkundung] . . ., which allows one to anticipate everything that can occur in nuclear physics."[36]

As a point of departure for a mathematical formulation, Heisenberg wrote down the Klein-Gordon differential equation, which gives a relativistic quantum-mechanical description of the electron. He confined himself to the one-dimensional case, introducing the cell length, $a = h/Mc$, by converting the differentials with respect to length into differences.[37] The resulting difference equation is similar to one in the theory of metal lattices.[38] Possibly with the aid of this analogy, Heisenberg sketched out the curve of the energy as a function of wave number. The crucial feature of the energy dependence derived from the difference equation, as opposed to the differential equation, is that it is periodic. Heisenberg interpreted the difference equation as describing a particle that behaves like an electron in the neighborhood of energy minima and like a proton near the maxima. A similar curve can be derived for light quanta; it represents a situation where the wave packet for light has a vanishing group velocity in the region of the energy maxima.

Heisenberg found that within systems of dimensions smaller than cell-dimensions, neither energy, momentum, nor charge was conserved. "That is, these laws all hold as approximations in ordinary atomic physics, but fail in nuclear physics. A further interesting result would be this: that atomic nuclei would consist only of protons and (slow) light quanta of mass M, not of electrons. For in order to build wave packets of nuclear dimensions, one can only use waves in the neighborhood of the maximum of the E-curve." He concluded: "I don't know whether you regard this radical attempt as completely crazy. But I have the feeling that nuclear physics is

---

36. Heisenberg to Bohr, 23 March 1930 (BSC).
37. The Klein-Gordon equation is $\Box \phi + m^2 c^2 \phi = 0$, where $\Box$ is the differential operator corresponding to $-E^2/c^2 + p^2$. Assuming that $\phi = T(t) u(x)$ for the one-dimensional case and treating time in the ordinary manner, Heisenberg gets $-(E/c)^2 u_n + (h/2\pi i a)^2 [u_{n+1} - 2u_n + u_{n-1}] + m^2 c^2 u_n = 0$. The energy approximates mc^2—the electron's rest energy—near minima and $hc/2\pi a \sim Mc^2$—the proton's rest energy—near maxima.
38. I am indebted to Dr. Franco Iachello for pointing this out, and for an elucidation of the physics in these letters.

not to be had much more cheaply."³⁹ Thus Heisenberg, like Bohr, was willing to abandon the mechanical conservation laws. He was even ready to jettison conservation of charge. It was here that Bohr objected on grounds of correspondence considerations. He explained that he felt Heisenberg's treatment of electric charge did not relate properly to the charge concept in classical electrodynamics. In addition, however severely the classical theory's area of application is limited, Bohr felt "doubtful that these boundaries can be marked out in as simple a manner as you seem to hope." "I have been occupied with precisely this question," Bohr informed Heisenberg, and he enclosed a copy of his beta-ray manuscript and one of his letters to Dirac to give Heisenberg "the direction my thoughts have taken."⁴⁰

At this point, with Bohr and Heisenberg in possession of each other's hypotheses, the correspondence breaks. Heisenberg was due in Copenhagen the second week in April, the occasion being one of the informal meetings on current problems in physics held at the Institute for Theoretical Physics in these years. Pauli was also there, and Gamow had come from Cambridge where he had been occupied with the problem of nuclear electrons. Among the others present were L. Landau, R. Peierls, W. Heitler, and, from Sweden, I. Waller. From the list of participants, as well as the correspondence preceding the meeting, it can be inferred that both Heisenberg's lattice-world and Bohr's ideas were discussed.⁴¹

After Heisenberg returned to Leipzig, he stopped work on the lattice-world model. Whether or not the Copenhagen conference discussions were decisive is unclear. Heisenberg had already raised weighty objections against his own model in his letter of 10 March, where, above all, he had pointed out that it did not satisfy the criterion of relativistic invariance. Whatever the cause, his first letter to Bohr on his return from the conference sketches an entirely new

39. Heisenberg to Bohr, 10 March 1930 (BSC).
40. Bohr to Heisenberg, 18 March 1930 (BSC).
41. For example Heisenberg wrote to Bohr on 23 March 1930 (BSC), "[Ich] möchte die Diskussion auf unser Beisammensein in Kopenhagen verschieben. Ich möchte auch gern ausführlich dann Deinen Standpunkt kennen lernen." See also Gamow to Bohr, 25 Feb. and 23 March 1930 (BSC). The names of participants are taken from the "Udenlandske Gæster paa Universitetets Institut for teoretisk Fysik." This register of foreign guests, preserved in the Niels Bohr Archive, contains arrival entries from 1919 to 1956.

approach to the problem of the electron's self-energy, one based on the study of very fast particles.[42] In an article embodying this program, completed in August 1930, Heisenberg included a passage which is probably a history of his shift. If one decides to introduce the classical electron radius, r_o, into quantum theory, Heisenberg wrote, "it would first of all seem plausible to introduce it in such a way that one divides space into cells of finite magnitude r_o^3, and replaces the previous differential equations with difference equations. . . . Although such a lattice-world has . . . interesting properties, one must also observe that it leads to deviations from the present theory which do not seem plausible from the point of view of experiment. In particular, the assumption that a minimum length exists is not relativistically invariant, and one can see no way to bring the demand for relativistic invariance into conformity with the introduction of a fundamental length. It would therefore seem more correct for the present *not* to introduce the length r_o in the foundations of the theory, but to hold fast to relativistic invariance. If one takes this second viewpoint, an essential simplification of the problem is achieved by considering only the motions of electrons and protons whose velocity approaches that of light."[43]

Heisenberg's changed program of investigation related to cosmic rays, rather than to nuclear physics, as his lattice-world model had done. There is nothing I have found to indicate a parallel attack on nuclear constitution in the succeeding eighteen months.[44] Years

42. Heisenberg to Bohr, 26 April 1930 (BSC).
43. Heisenberg, "Die Selbstenergie des Elektrons," *Zeits. f. Phys.*, 65 (1930), 4-5. The new approach proved capable of locating the source of the difficulties more accurately rather than of solving them.
44. I base this conclusion on Heisenberg's papers and his letters to Bohr. The ten papers Heisenberg published between the article on the self-energy of the electron and the first nuclear paper fall into four groups. Two are on field theory: one on the mathematics of his and Pauli's electrodynamics ("Bemerkungen zur Strahlungstheorie," *Ann. d. Phys.*, 9 [1931], 338-346); and the other on energy fluctuations ("Über Energieschwankungen in einem Strahlungsfeld," *Berichte Sächs. Akad. Wiss. Leipzig*, 83 [1931], 3-9). Three deal with applied problems: one on the quantum theory of ferromagnetism ("Zur Theorie der Magnetostriktion und der Magnetisierungskurve," *Zeits. f. Phys.*, 69 [1931], 287-297); one on the treatment of atomic shell structure ("Zum Paulischen Ausschliessungsprinzip," *Ann. d. Phys.*, 10 [1931], 888-904); and the third is a short calculation of the scattering of X-rays from atoms, carried out at the request of Debye ("Über die inkohärente Streuung von Röntgenstrahlen," *Phys. Zeits.*, 32 [1931], 737-740). Three are popular or philosophic articles: a lecture on uncertainty ("Die Rolle der Unbestimmtheitsrelationen in der modernen Physik,"

later, Heisenberg recalled that at this time no ground was felt to exist for doing nuclear physics. "There was one exception. That was the theory of Gamow for alpha-decay. But one had the feeling that this was a very lucky case where one could do something without really understanding the nucleus. There was just no basis, however, for coming to a real theory of the nucleus."[45]

There was thus a break of about a year and a half in Heisenberg's investigations into nuclear structure between the winter of 1930 and the months after the discovery of the neutron. This gap is related to his way of doing physics. In his article in the Bohr memorial volume, Heisenberg contrasted the role of mathematics in his and in Bohr's work in the years 1924–1927 when quantum mechanics was being created. It came more naturally to him than to Bohr to use "a formal mathematical standpoint." Unlike Bohr, moreover, he placed a certain confidence on purely logical deductions from the initial propositions as a means of arriving at an interpretation of them.[46] A letter Heisenberg wrote shortly after the "Self-Energy" paper implies the same work style. Here he described his work as somewhat "grey on grey." "I am trying to think about relativistic quantum theory, but up to now, I have found absolutely no formal point of attack (formale Angriffspunckt). Perhaps one must indeed first see the entire development of nuclear physics, before one can

---

Monatshefte f. Math. u. Phys., 38 [1931], 365-372); a short review article on ferromagnetism ("Fortschritte in der Theorie des Ferromagnetismus," *Metallwirtschaft, 9* [1930], 843-844); and an article on causality ("Kausalgesetz und Quantenmechanik," *Erkenntnis, 2* [1931], 172-182). Finally, two are on cosmic rays. The first is a detailed treatment of the interaction of fast particles and energetic radiation with matter; the nucleus is considered insofar as the contribution to the scattering of gamma rays from the nuclear alpha particles and "free" electrons respectively is taken up ("Theoretische Überlegungen zur Höhenstrahlung," *Ann. d. Phys.,* 13 [1932], 430-452). The other is a related note ("Über die durch Ultrastrahlung hervorgerufenen Zertrümmerungsprozesse," *Naturwiss.,* 20 [1932], 365-366).

45. Interview with Heisenberg on nuclear theory in the thirties, by the author, 16 June 1970. The interview is on deposit in the Niels Bohr Archive, at the Niels Bohr Institute in Copenhagen, and in the Center for the History and Philosophy of Physics at the American Institute of Physics. I am grateful to Professor Heisenberg for permission to reproduce parts of it here.

46. Heisenberg, "Quantum Theory and Its Interpretation," pp. 98 and 104.

get farther here."[47] The phrases "formal mathematical standpoint" and "formal point of attack" refer to a method in which the initial physical insights are embodied in a set of mathematical propositions as the first step in the construction of the theory. The next step, then, is to use mathematical deduction to reveal new physical results. Heisenberg's lattice-world is an illustration of this procedure. The difference equation, which embodies the idea of a fundamental length, serves as the theory's formal point of departure. By mathematical deduction, the equation reveals a new picture of the behavior of matter and photons and a new nuclear model. At the time, Heisenberg judged the lattice-world and, hence, the model of the nucleus which followed from it to be unsatisfactory. The discovery of the neutron was to provide him with the clue to a new mathematical starting point in the spring of 1932. In the intervening years, he could not see a way to proceed and did not occupy himself further with the problem.

4. THE RESPONSE TO THE NEUTRON

Bohr

It was in Heisenberg's ski hut in the mountains of Bavaria that Bohr finished the manuscript of his Rome address in March 1932. When he returned to Copenhagen, one of the tasks awaiting him was that of organizing the year's informal Institute Conference on current problems in physics. It was to take place at Easter, and Bohr would see Heisenberg again at that time. Also waiting for Bohr was a letter from James Chadwick, containing a page-proof of Chadwick's note in *Nature* announcing the neutron. "As you will see," Chadwick told Bohr, "I have put this forward rather cautiously, but I think the evidence is really rather strong."[48] Bohr's interest was

47. Heisenberg to Bohr, 18 Sept. 1930 (BSC). In a sense, Heisenberg followed the program he suggests here. In investigations beginning in 1936, he used Fermi's treatment of beta-decay as a starting point for an attempt to solve problems in relativistic quantum mechanics by the introduction of a fundamental length in a more sophisticated way. See, in particular, the review article, "Theorie der Elementarteilchen und universellen Länge," *Ann. d. Phys., 32* (1938), 20-33.
48. Chadwick to Bohr, 24 Feb. 1932 (BSC). Chadwick, "Possible Existence of a Neutron," *Nature, 129* (1932), 312.

strong and immediate, as was Heisenberg's.[49] Their responses, however, were sharply different. Bohr immediately grasped that the new neutral particle could illuminate the treatment of collision problems in quantum mechanics; he therefore sought an explanation for the surprisingly low frequency of neutron-electron collisions,[50] which was a reason the neutron had not been detected sooner. Heisenberg, in contrast, took up the implications of the neutron for the unsolved problems of nuclear structure.

Bohr's reactions to the neutron as a nuclear component appear in the introduction and concluding remarks of some manuscripts on collision problems. It is worth sketching them, chiefly as an aid to understanding Heisenberg. Let me recall first the well-known fact that when Rutherford first suggested the possibility of a neutron in his Bakerian lecture of 1920, he had in mind a compound system of proton plus electron. He believed, mistakenly, that he had just discovered helium-3, composed of three protons and one electron. By analogy and extrapolation, he suggested the existence of hydrogen-2 and of the neutron. "If we are correct in this assumption," he wrote, referring to the structure of helium-3, "it seems very likely that one electron can also bind two H nuclei and possibly also one H nucleus."[51] When Chadwick made his discovery in the winter of 1932, he and Rutherford thought the neutral particle was probably the collapsed hydrogen atom Rutherford had suggested.[52] Not unnaturally, it was this model of the neutron Bohr took as his point of departure.

The neutron is therefore an analogue of the alpha-particle, exactly as in Rutherford's original proposal. "From a formal point of view," Bohr wrote, "the neutron may be considered as a nucleus

---

49. Bohr to Heisenberg, 22 March 1932, and Heisenberg to Bohr, 24 March 1932 (BSC).
50. Bohr did not publish his solution. It is in manuscript: "Properties of the Neutron," April 1932; "Foredrag i Fysisk Forening," 9 May 1932; and "Atomic Collision Problems," probably 1932.
51. Rutherford, "Nuclear Constitution of Atoms: Bakerian Lecture," *The Collected Papers of Lord Rutherford of Nelson*, ed. Sir J. Chadwick (London, 1965), *3*, 34. Rutherford does not use the word "neutron."
52. Rutherford, "Origin of the Gamma Rays," *Nature, 129* (132), 458. Chadwick, "The Existence of a Neutron," *Foundations of Nuclear Physics*, ed. R. T. Beyer (New York, 1949), p. 19.

THE IMPACT OF THE NEUTRON: BOHR AND HEISENBERG

of an element of atomic number zero."[53] The significance of its discovery is "above all . . . that we have augmented [the range of] our knowledge of nuclei."[54] Bohr thus assimilated the neutron to the analysis of nuclei he had just completed. In his talk at the Easter conference, he reminded his listeners of his treatment of nuclei in the Faraday lecture, which had just been published. He indicated that the considerations on nuclear statistics may be carried over to the neutron. "We shall thus expect that it will obey the exclusion principle [that is, Fermi-Dirac statistics] just as the electron or the proton, and that if it could be broken up by some external agency, we may expect that the idea of energy conservation would find no simple application." Further, "as regards [the neutron's] constitution, it is at the present stage of atomic theory not possible to offer any detailed explanation. Of course, its mass and charge suggest that the neutron is formed by a combination of a proton and an electron, but we can not explain why these particles combine in this way, as little as we can explain why four protons and two electrons should combine to form a helium nucleus."[55]

Bohr's conclusion "with regard to the question of nuclear construction itself" was that "we stand without any aids to understand it. This is an area where one is collecting data. . . . These investigations . . . have the unusually great physical interest, that they relate to an accumulation of experience, and one may thereby hope to be able to find a point of departure for a further expansion of the theoretical methods."[56] Thus, Bohr's analysis left the difficulties of nuclear structure almost exactly where they were before.

Bohr's response to the neutron was surely conditioned by his life-long concern with collision problems. But it seems probable that it was also affected by the fact he had just thought his way through to a detailed and comprehensive analysis of the nucleus, reducing the chance that he would see the neutron as a clue to nuclear structure; as concerns its relation to the nucleus, he was ready with a concep-

53. Bohr, "Extract from an address delivered at the conference on . . . atomic problems . . . Copenhagen, 7-13 April 1932," p. 1, contained in "Properties of the Neutron," cited in note 50.
54. Bohr, "Foredrag," p. 5.
55. Bohr, "Extract," pp. 1-2.
56. Bohr, "Foredrag," p. 17.

tual pigeon hole for it. Of course, he saw very quickly that one need no longer have "free" electrons in the nucleus, but that one could pack all nuclear electrons into neutrons. But at the initial moment, nuclear structure remained a mystery, and the neutron became a new, special case of this mystery.

Heisenberg

Heisenberg's work on nuclear structure was carried out at the same time that Bohr expressed these views. Heisenberg published his theory in three parts, submitted in June, July, and December of 1932.[57] Of these, the third does not fall within the period of this paper, and will be considered only in so far as it elucidates the ideas of the first two.[58]

Undoubtedly Heisenberg saw very soon after hearing of the neutron that one could explain the charge and mass of nuclei in terms of numbers of neutrons and protons without having to use the hypothesis of nuclear electrons. He must also have realized from the start that one could also explain beta-decay without electrons, if one accepted the assumption that a neutron could disintegrate into a proton and an electron. He surely also saw right away that one could account for nuclear spins and statistics by assigning the neutron a spin of $1/2$ and the property of obeying Fermi-Dirac statistics. These ideas occurred to a number of physicists independently.[59] It seems certain that Bohr had also thought of them by the time he met Heisenberg at the Easter meeting; this is implied in comments he sent Heisenberg in acknowledgment of the June paper. Moreover, Bohr's Easter lecture is entirely consonant with these insights.

57. Heisenberg, "Über den Bau der Atomkerne," *Zeits. f. Phys., 77* (1932), 1-11; *78* (1932), 156-164; *80* (1933), 587-596. Hereafter, these will be cited in the text by volume and page number. The first paper and part of the third are translated on pp. 144-160 of Brink. The translation contains errors and omissions, however.

58. The third paper reflects the objections of the theory's critics. It also contains concepts that belong to a new stage of theory. Of particular interest is the new formulation Heisenberg gives to the problem of the nature of the neutron. See *80,* 595, penultimate paragraph, or my article, "Heisenberg's Papers on Nuclear Structure," *Proceedings of the XIIth International Congress on the History of Science, Paris, August, 1968* (in press).

59. They were published first by D. Ivanenko, "The Neutron Hypothesis," *Nature, 129* (1932), 798.

There was a difference between Bohr's and Heisenberg's view of the neutron, however. While Heisenberg concurred in Bohr's catalogue of the problems inhering in a compound neutron, he recognized an additional difficulty: the neutron seemed to be elementary as well as complex. I do not believe Bohr shared this view in the spring of 1932. It may have occurred to Heisenberg as a result of his concern with nuclear structure, and hence may have arisen after the Easter conference. In 1970, Heisenberg portrayed it as having been linked with the empirical facts on the numbers of neutrons and protons in nuclei. "At that time we had an unclear feeling that the neutron somehow can be considered as consisting of proton and electron, but also somehow not, because after all the neutron seemed to be very similar to the proton. I had from the very beginning the idea that the neutron was a kind of brother to the proton . . . from the fact that an approximately equal number of protons and neutrons were in the nucleus. . . . And, on account of this symmetry, it was not so nice to say that the neutron is a compound particle consisting of proton and electron, while the proton is an elementary particle."[60]

Heisenberg's June paper contains expressions of this state of uncertainty throughout. He wrote in his introduction that "if one wishes to picture the neutron as composed of proton and electron, one must ascribe Bose statistics and zero spin to the electron. It does not seem purposeful, however, to go into such a picture in more detail (ein solches Bild näher auszuführen). The neutron shall be regarded rather as an independent, fundamental constituent, which under proper circumstances can split into a proton and an electron, in a process where the conservation laws for energy and momentum are probably no longer applicable" (77, 1–2). In this statement, Heisenberg favors the view of the neutron as an elementary particle. In the concluding paragraph, however, he recognizes the compound neutron as indispensable for understanding phenomena like scattering from nuclei (77, 10–11).[61]

The double view of the neutron was incorporated in the mathe-

60. Interview with Heisenberg, 16 June 1970, p. 2.
61. See also 78, 160-163. This is connected with Heisenberg's researches in the months just before April. See Heisenberg, "Theoretische Überlegungen zur Höhenstrahlung," cited in note 44, pp. 440-442.

matical formulation Heisenberg's theory. He achieved a "formal point of departure" by making use of two analogies, the first of which is based on the idea of the neutron as elementary, the second on the idea of the neutron as compound. He added to the customary four coordinates which give the position and the spin of each of the heavy nuclear particles a fifth coordinate, p^ζ, which takes the value $+1$ for the neutron and -1 for the proton. This coordinate, which Heisenberg called the "ρ-spin," later came to be called the "isotopic spin." It represents the neutron and proton in a formal sense as two different states of the same system. It was now possible to make an analogy between the isotopic spin and the two spin states of the electron, thereby carrying over the mathematical apparatus that describes electron spin. It is clear that with respect to this aspect of the theory, the neutron and proton are entirely symmetrical.

At the same time, Heisenberg drew an analogy between a system composed of a proton and a neutron and a hydrogen molecule ion and an analogy between a system of two neutrons and a hydrogen molecule. This enabled him to use the physics and mathematics of the quantum theory of chemical binding. Between the neutron and proton he postulated a charge-exchange force similar to that which holds the H_2 ion together. Between the two neutrons he introduced a weaker interaction, corresponding to the force binding a neutral hydrogen molecule. Finally, using the same comparison, he assumed no force between the bare protons except the Coulomb repulsion. This part of the theory, therefore, rests on a conception of the neutron as compound and the proton as elementary. Moreover, the asymmetry in the models of neutron and proton leads to asymmetry in the forces themselves. The neutron-proton force is stronger than the neutron-neutron force, and the proton-proton nuclear force is nonexistent.

The mathematical expression Heisenberg arrived at in this way was the energy operator, or Hamiltonian, of the nucleus conceived as composed entirely of heavy neutrons and protons.[62] The most

---

62. The Hamiltonian is:
$$H = 1/2M \sum_k p_k^2 - 1/2 \sum_{k>l} J(r_{kl}) (\rho_k^\xi \rho_l^\xi + \rho_k^\eta \rho_l^\eta)$$
$$- 1/4 \sum_{k>l} K(r_{kl}) (1 + \rho_l^\zeta)(1 + \rho_k^\zeta)$$
$$+ 1/4 \sum_{k>l} e^2/r_{kl} (1 - \rho_l^\zeta)(1 - \rho_k^\zeta) - 1/2 D \sum_k (1 + \rho_k^\zeta).$$

extended use he made of this operator was to deduce criteria predicting which isotopes exist in nature and which members of the radioactive decay series emit beta- and which alpha-rays.[63] To determine whether a given nucleus will be stable or not, he used the Hamiltonian to estimate the binding energies of the lowest states of various nuclei. For example, at the start of the second paper, he takes up the question of light nuclei. He considers all nuclei with a fixed $n = n_1 + n_2$, where n_1 is the number of neutrons in a given nucleus and n_2 is the number of protons. The short-range character of the neutron-proton and neutron-neutron forces is expressed by the assumption that the binding of a given neutron or proton is dependent on the ratio n_1/n_2 (77, 7). Heisenberg uses the Hamiltonian to derive a curve of the binding energy as a function of n_1/n_2. The curve descends to a minimum for some $n_1/n_2 = a$ and then rises again. He concludes that nuclei for which $n_1/n_2 > a$ are unstable in beta-decay and hence nonexistent (78, 156–157).[64] Thus Heisenberg used stability criteria based on the principle of energy conservation even in those cases where he treated beta-decay. This is remarkable, since he conjectured at the start of the paper that the process of beta-decay is probably not governed by energy conservation. He explicitly called attention to this discrepancy at the place where he first introduced considerations like that described above. "Although the application of energy and momentum conservation laws to the decay of a neutron appears completely questionable (durchaus fraglich erschient) in light of the findings on the continuous beta-spectrum, nevertheless, . . . use will be made here of an energy-balance for beta-radiation" (77, 6). In his December paper, he returned to this point: in principle, "this attempt to retain certain consequences of energy conservation even outside the boundaries of its applicability is,

---

Here p_k is the momentum of the kth particle, so that the first term is the kinetic energy. ρ_k^ξ, ρ_k^η, and ρ_k^ζ are 2x2 matrices constructed in analogy to spin matrices. The second term in H, operating on a neutron-proton pair, exchanges their charges and gives an attractive force dependent on their separation. The third term gives an attractive neutron-neutron force. The fourth term is the Coulomb repulsion between protons, and the fifth the binding energy of the electrons in the neutrons. Here D is the mass defect of the neutron (77, 3).

63. Besides these applications, which are developed in 77, 6-11 and 78, 156-160, Heisenberg used his Hamiltonian to give brief explanations of the special stability of the alpha particle (77, 4-5) and of interatomic forces (77, 5).

64. Heisenberg makes no attempt to estimate a; he immediately proceeds to a more detailed treatment in which odd and even nuclei are separated.

logically, completely possible." Nevertheless, he continued, it left little hope that present theories would not sooner or later reveal inner difficulties. Therefore "these stability criteria represent the least secure part of the considerations carried out here" (*80*, 596).

It is worthwhile to outline the discussion of radioactive series which forms the major part of the first paper, for it acquaints us with the first results of the theory as his contemporaries received them. Heisenberg starts by considering a member of a decay series with even atomic number and which decays with alpha emission. The ratio n_1/n_2 will increase with successive alpha decays until it reaches a certain critical value. At this value, the loss in binding energy which would be occasioned by the loss of a neutron—the *n-n* energy plus the neutron mass defect—would be more than compensated by the gain occasioned by the gain of a proton—the *p-n* energy minus the Coulomb term. Beta-decay now occurs and, consequently, n_1/n_2 decreases. An odd atomic number nucleus results from this first beta-decay, which means that there is at least one proton which cannot be bound into an (exceptionally stable) alpha particle. Hence another beta-decay can now occur at a lesser value of the critical ratio n_1/n_2, after which alpha-decay resumes until the ratio again becomes sufficiently large so as to exceed the first critical value (*77*, 9). In this way Heisenberg managed to extract from his analysis a quasi-numerical prediction: each successive beta-decay must correspond to the ratio's exceeding one of two values. He then compared this prediction with n_1/n_2 calculated for the three natural radioactive series; it fit the thorium and actinium series, but not the radium series (*77*, 9–10).

Bohr's reaction to the first of Heisenberg's papers is recorded in a letter of 27 June. "In the hope that this letter can still reach you in Leipzig, I am hurrying to write how much we all have liked your wonderfully fine paper. As you may imagine, it was not completely strange to me that all difficulties can be shoved over onto the neutrons; but that such a simple systematics with regard to the connection between α- and β-radiation can be achieved by this was a great and happy surprise."[65]

The insights that the neutron is compound and also analogous to

---

65. Bohr to Heisenberg, 27 June 1932 (BSC).

the proton were not united into a consistent model in Heisenberg's 1932 papers. To reconcile them at that date would probably have required a heroic creative act. Heisenberg would have had to foresee both the existence of the positron, announced only in August 1932, and the decay of the proton by positron emission, announced by the Joliots only at the beginning of 1934. He would have had to guess, in addition, that the nuclear force between neutron and proton is approximately equal to the forces between neutron and neutron and between proton and proton; this phenomenon of charge independence was not discovered until 1936.[66]

An inconsistency that Heisenberg found even more troublesome at the time was that between the compound neutron and the experimental evidence on the neutron's mass. Insofar as Heisenberg pictured the neutron as a compound particle, he pictured it as a combination possessing exceptional stability. In this respect, it resembled the alpha particle. But the alpha particle has a binding energy of approximately 28 million electron volts, while the measurements of Chadwick pointed to a binding energy of only 1 to 2 million electron volts for the neutron.[67] That such a slightly bound system should show extraordinary stability presented a paradox (*78*, 163).

Heisenberg was aware of both these inconsistencies. It is perhaps the greatest strength of his papers that he did not allow himself to be stopped by them. He justified his theory on two grounds. First, he pointed out that the difficulties connected with his neutron model rested on the application to it of the laws of quantum mechanics. But "the very existence of the neutron contradicts the laws of quantum mechanics in their present form. Both the validity of Fermi statistics for neutrons—hypothetical, to be sure—and the negation of the energy conservation law in beta-decay show the inapplicability of the current quantum mechanics to the structure of the neutron" (*78*, 163). This is precisely the view of the neutron Bohr gave at the Easter conference, and it is based on the view of nuclei which Bohr and Heisenberg both adhered to before 1932.

66. C. D. Anderson, "The Positron," *Science, 76* (1932), 238. I. Curie and F. Joliot, "Un nouveau type de radioactivité," *Comptes Rendus, 198* (1934), 254-256. For the researches establishing charge independence, see Brink, p. 61.
67. Chadwick, "The Existence of a Neutron," p. 702.

Second, Heisenberg defended his theory on the ground of its consequences. If, because of the problem of explaining the neutron's particular stability, one chose to relinquish the idea that the neutron can disintegrate with the emission of an electron, one could picture it as an elementary, indestructible particle ("einen unzerstörbaren Elementarbaustein" [*80*, 594]). In that case, electrons must exist in radioactive beta emitters on the same footing as protons, neutrons, or alpha particles. This picture entails an overwhelming difficulty: if electrons exist in nuclei this way, they must be coupled to the heavy components, for example, to the alpha particles, by strong forces. But this led one back again to the mystery of why alpha particles are emitted with definite energies and electrons are not. More generally, it then is impossible to comprehend why heavy nuclear particles obey quantum mechanics, since they are bound to electrons "whose behavior in the nucleus lies entirely outside the domain of quantum mechanics." By contrast, if one postulates the stability of the neutron, one "makes possible a clean separation of the region accessible to quantum mechanics from that which is not accessible, since by virtue of this stability a purely quantum mechanical system of protons and neutrons can be built up in which the new features entering because of beta-decay do not give rise to difficulties. This possibility of a sharp division between the quantum mechanical features and those new features which are characteristic for the nucleus seems to be lost when the electrons are regarded as independent nuclear constituents" (*80*, 595). Thus, the acceptance of a seemingly inconsistent model of the neutron was not only defensible on the basis that physical theory was impotent to evaluate *any* neutron model, but it was a precondition for doing nuclear physics at all.

5. CONCLUSIONS

In the years before 1932, Bohr regarded nuclear physics as one part of a larger area of physics having no satisfactory theory. His reasons included the size of the nucleus, the continuous beta-ray spectrum, and the anomalous behavior of nuclear electrons. He became interested in the hypothesis of energy nonconservation first as an explanation for the beta-emission spectrum. Nevertheless, it was possible for Bohr to make use of such an explanation precisely because nuclear physics belonged to the domain which stood outside the known laws. Furthermore, his interest in the validity of the conservation laws for beta-decay was a special case of his interest in finding out which concepts of ordinary quantum mechanics could be applied to this domain and which could not. This way of proceeding—by probing the limits of validity of existing theory—was characteristic of Bohr. In applying it to the nucleus, he arrived at the idea that not only the mechanical conservation laws but the related concept of mechanical permanence failed for nuclear electrons. He proposed a nucleus that was free of electrons with respect to its mechanical properties, but which included electrons with respect to its electrical properties. This picture satisfied the data on nuclear charges and beta-decay, but it rested on the anticipation of a fundamentally new theory.

Heisenberg had a similar outlook. On the one hand, he felt that the existence of nuclear electrons made it impossible to solve the problems of the nucleus by ordinary quantum mechanics; on the other hand, he attempted to picture an electron-free nucleus that can yet emit betas. Heisenberg's model was constructed of protons and equally massive light quanta. No less than Bohr's, it was based on the assumption that the nucleus demanded radically new laws. Heisenberg sought a formal, mathematical expression as a starting point, and in this he differed markedly from Bohr. Perhaps the most striking illustration of their difference is in the use of the classical electron radius, r_o. Bohr sought to make it the keystone of a new complementarity relation. Heisenberg brought it into his theory as a dimension fixing the cell-size of a lattice-world. In Heisenberg's lattice-world, the boundary between the domain of ordinary quantum mechanics and the domain of the new physics is defined by the length r_o. It was a concomitant of his theory of nuclear structure that it was possible to redefine this boundary in a more subtle way. It ceased to be marked out in terms of a length and became instead a division between the nucleus conceived as a collection of heavy particles and the constituent particles themselves.

It was not a surprise to Bohr and Heisenberg that this boundary had to be characterized in some new way; both had felt from the start that Heisenberg's use of r_o was crude. It was a surprise to them, however, that the boundary could (metaphorically) be drawn down the middle of the nucleus, and this was the novelty that resulted from the experimental discovery of the neutron. While the anticipation had been that the solution to nuclear structure would be a part of the solution of relativistic quantum theory, the solution (partial, to be sure) was in fact a shift of nuclear physics out of the realm of relativistic quantum mechanics. Heisenberg put the new situation with particular clarity in his review of nuclear theory at the 1933 Solvay Congress. "Because the experimental facts concerning the structure of atomic nuclei have not carried us, up to now, to new physical notions going outside of quantum mechanics, it is necessary to examine at the start . . . to what extent quantum or wave mechanics can be used in this new domain. As precise as possible a limitation of the possibilities of the application of quantum mechanics is one of the first tasks of nuclear theory."[68]

The part of the nucleus that remained outside ordinary quantum mechanics was the neutron. Here both Heisenberg and Bohr inclined to the view they had held before; i.e., they expected the energy and momentum laws to be violated in the beta-disintegration of neutrons. Yet the discovery of the neutron together with Heisenberg's 1932 theory tended to make their view less tenable. The neutron increased the plausibility of Pauli's neutrino; furthermore, Heisenberg made fruitful use of energy conservation in his treatment of beta-decay. At the time, Heisenberg regarded his use of the energy law as an especially dubious part of his theory. In the event, the argument was reversed, and the success of the conservation law in treating beta emission became a strong reason for retaining it.[69]

It remains to discuss the factors that helped Heisenberg arrive at his theory. Although he had attempted to build the nucleus out of protons and equally massive neutral particles as early as the winter of 1930, I do not think that this attempt motivated his 1932 theory. There is no evidence that he continued to try to find ways of consructing a nucleus from a combination of positive and neutral particles. His 1930 attempt was based on a desire to find a way of

---

68. Heisenberg, "Considérations théoriques générales sur la structure du noyau," *Septième Conséil de Physique . . . Solvay*, p. 289.
69. Pauli, in the "Discussion du Rapport de M. Heisenberg," *Septième Conséil de Physique . . . Solvay*, pp. 324-325.

constructing an electron-free nucleus. That he held this program meant that he was particularly sensitive to new clues for structuring nuclei without electrons and, hence, that he was sensitive to the neutron. A year and a half fell between Heisenberg's efforts of 1930 and his neutron-proton nuclear model. The break may well have given him a freshness that facilitated the emergence of new ideas. I have connected the time lapse with his method of using mathematics in the creation of theories. By contrast, Bohr's opposite method of work helped place him in the position of confronting the neutron at a moment when he had just finished a detailed consideration of nuclear problems. This may have been one reason why Bohr's initial response to the neutron was so different from Heisenberg's.

Finally, I have argued that the view of the nucleus Heisenberg held before 1932 must also be counted among the preconditions for his theory. It was necessary that he should be able to sustain an inconsistent neutron model. The basis for doing so was the view that placed the nucleus outside quantum mechanics. He shared this view with Bohr, and indeed it had been Bohr who had given it public expression and who had extended it to the neutron. Heisenberg made use of its extension to the neutron precisely to show that the analysis was no longer necessary for the rest of the nucleus.

ACKNOWLEDGMENTS

Many people have extended themselves to help me with this paper. I should like first of all to thank Professor Werner Heisenberg for taking time to discuss the subject with me, and Professor Gerald Holton for his support and encouragement at the inception of my researches. Among the many at the Niels Bohr Institute who gave me assistance, I should like to single out Professor Léon Rosenfeld for his informative discussions, Professor Aage Bohr for access to the Niels Bohr Archive and permission to quote documents, and Mrs. Sophie Hellmann for moral as well as practical support. Drs. Franco Iachello and Philip Siemens discussed the physics with me. Finally, I thank Professor Brookes Spencer, Dr. Peter Heimann, and Professor Paul Forman for the important service of criticizing the paper as history.

The work was supported in part by a grant from the National Science Foundation.

The Origin of G. N. Lewis's Theory of the Shared Pair Bond

BY ROBERT E. KOHLER, JR.[*]

1. INTRODUCTION: THE IMPORTANCE OF G. N. LEWIS'S SHARED PAIR BOND

The half century following August Kekulé's conception of structure theory in 1859 was the golden age of structural organic chemistry. Deduction of structural formulas was the order of the day, and organic chemists made the most of the opportunities provided by art and a prodigal nature. By 1900 structural chemistry was a highly sophisticated science recently crowned by the elucidation of the sugars, purines, and proteins by Emil Fischer (1852–1919), the paragon of the classical tradition. Although structural chemistry continued to flourish in the twentieth century, by the 1920's a second flowering of organic chemistry had begun to attract the best young minds as the classical tradition had forty years before. This new development was the study of reaction mechanisms; as nineteenth-century chemistry was concerned with the statics of chemical structure, the new chemistry dealt with the dynamics of chemical change.

The key to the new chemistry was the chemical bond itself. For the classical structural chemists the chemical bond was a simple line, and where the aim was a structural formula, this abstract representation sufficed. But to understand how bonds are made and broken and rearranged, it is essential to have a detailed and concrete picture of the chemical bond in terms of atomic structure. The inadequacy of the many attempts to represent reaction mechanisms before 1910, such as those of Arthur Lapworth, for example, illustrate that any representation of the mechanism of chemical change could be only

[*] Burndy Library, Norwalk, Connecticut 06852.

as revealing as the symbols used. Without a proper theory of the chemical bond, the rich legacy of chemical knowledge and experience could never be used.

The first satisfactory picture of the chemical bond was proposed early in 1916 by Gilbert N. Lewis (1875–1946),[1] the American physical chemist better known to some for his work on thermodynamics. His book, *Valence and the Structure of Atoms and Molecules* (1923),[2] which elaborated the picture of the bond as a shared pair of electrons, was the textbook of the new generation of mechanistic chemists. Without Lewis's conception of the shared pair bond, the interpretation of reaction mechanisms already begun by the English school of A. Lapworth (1872–1941), T. M. Lowry (1874–1936), C. K. Ingold (b. 1893), and R. Robinson (b. 1886) would not have gotten very far. Likewise, without the idea of the shared pair bond, then being used with increasing confidence and success by organic chemists, the application of quantum mechanics to the chemical bond in the late 1920's by H. London, E. Schrödinger, and L. Pauling would have begun on far less certain ground. Lewis himself, however, did not participate in the harvest he helped to sow. The popularization of the shared pair bond was due mainly to the papers and lectures of Irving Langmuir (1881–1957) between 1919 and 1922. The few papers on bond theory that Lewis published after 1916 are now mainly of historical interest: in both organic and quantum chemistry, in which Lewis's theory proved most serviceable, Lewis was never really at home.

When it was first proposed, Lewis's theory was completely out of tune with established belief. For nearly twenty years it had been almost universally believed that all bonds were formed by the complete transfer of *one* electron from one atom to another. The paradigm was the ionic bond of Na^+ Cl^-, and even the bonds in compounds such as methane or hydrogen were believed to be polar, despite their lack of polar properties. From the standpoint of the polar theory the idea that two negative electrons could attract each other or that two atoms could share electrons was absurd. Even the

---

1. G. N. Lewis, "The Atom and the Molecule," *J. Am. Chem. Soc.*, 38 (1916), 762-785. Received 26 January 1916.
2. G. N. Lewis, *Valence and the Structure of Atoms and Molecules* (New York, 1923). Reprinted, New York, 1966.

idea of a special bond for nonpolar compounds was admitted at first by only a very few dissenters. Lewis was particularly anxious to make clear that his was not a dualistic view: the polar and nonpolar bonds differed not in kind but only in degree. The less equally a pair of electrons was shared, the more polar the bond was; in the extreme case ionization occurred, with the net transfer of one electron. The polar theory thus became a special case of Lewis's more general theory.

On one point the polar orthodoxy and Lewis's theory agreed: in a stable molecule each atom (except hydrogen) must be surrounded by eight electrons. This "rule of eight" reflected the familiar fact that each row of the periodic table contained eight elements and that the valence of each element corresponded to the number of electrons it had to gain or lose to complete (or empty) a stable "octet." But whereas in the polar theory an octet could be completed only by transfer of electrons, in Lewis's theory two octets could share pairs

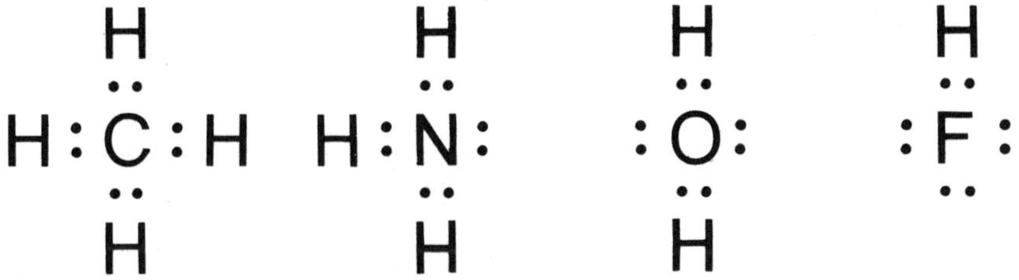

$$H:\overset{..}{\underset{..}{C}}:H \quad H:\overset{..}{\underset{..}{N}}:\quad :\overset{..}{\underset{..}{O}}:\quad :\overset{..}{\underset{..}{F}}:$$

(with H above C, N, O, F and H below C, N)

FIGURE 1. G. N. Lewis's double dot representation of shared electron pair bonds and free pairs.

of electrons. Lewis proposed that shared pair bonds and the *free pairs* making up the octet be represented by double dots, and by 1923 this contention had been widely adopted (Fig. 1). Since one shared pair counted in the octets of two atoms, Lewis was able to explain the existence of the diatomic molecules, O_2, N_2, Cl_2, etc. (Fig. 2). By transfer of electrons, only one octet could be filled.

$$:\overset{..}{\underset{..}{Cl}}:\overset{..}{\underset{..}{Cl}}:\quad :\overset{..}{\underset{..}{O}}::\overset{..}{\underset{..}{O}}:\quad :N:::N:$$

FIGURE 2. G. N. Lewis's representation of the diatomic molecules as shared and free electron pairs.

The most striking feature of Lewis's theory is the concrete form which the "rule of eight" took—the cubic atom. Lewis believed that the eight electrons of an octet formed the eight corners of a cube, and that all atoms consisted of a positive nucleus surrounded by electrons in fixed concentric cubic shells. Thus Lewis conceived of the single bond as two cubic atoms with a shared edge, and a double bond as two cubes with a common face (Fig. 3). The disadvantages

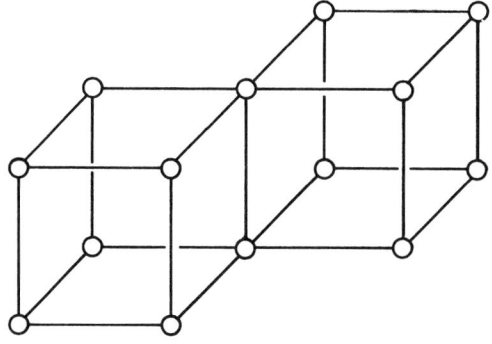

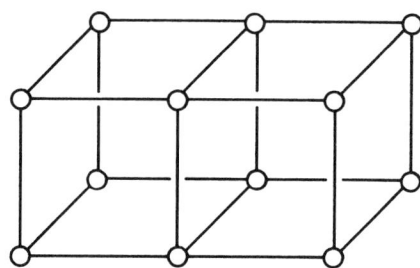

FIGURE 3. G. N. Lewis's representation of the single and double bonds as cubic atoms with a shared edge and face.

of the cubic atom are obvious: it cannot explain the tetrahedral stereochemistry of the carbon atom, or free rotation around a single bond, or the triple bond of acetylene. To fit these facts Lewis already in 1916 was in the process of modifying his cubic atom to a tetrahedral atom by drawing together the electrons at the corners of the cube into four pairs at the corners of a tetrahedron[3] (Fig. 4). But

3. G. N. Lewis, "The Atom and the Molecule," p. 780.

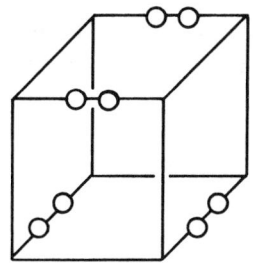

FIGURE 4. G. N. Lewis's tetrahedral atom with four electron pairs.

though the tetrahedral atom figures most prominently in *Valence*,[4] Lewis's theory of the shared pair originally came to him in terms of the cubic atom.

The cubic atom is the one main feature of Lewis's theory that did not survive. The work of Niels Bohr on atomic structure and the quantum chemical calculation of orbital shapes in the 1920's very soon made the cubic atom seem a naive and somewhat embarrassing relic. Since the cubic atom is not essential to the shared pair theory it was conveniently forgotten. But historically the cubic atom was perhaps the most important part of Lewis's theory. It was Lewis's first step into valence theory, which he took in 1902, and there is reason to think that it also provided him with the crucial insight in 1915 into the shared pair bond.

2. G. N. LEWIS AND THE CUBIC ATOM, 1902–1905

Lewis was born in 1875 in West Newton, a suburb of Boston. In 1884 his father, a lawyer, moved his family to a farm outside the rough Midwestern town of Lincoln, Nebraska.[5] Lewis's father was evidently a very independent man, and Gilbert was educated entirely at home. He read at the age of three and showed precocious intellectual abilities.[6] After two years at the University of Nebraska, he transferred in 1893 to Harvard where he obtained his B.S. in 1896. After a year as schoolmaster at Phillips Academy in Andover, he returned to Harvard to study with T. W. Richards (1868–1928), the

---

4. G. N. Lewis, *Valence*, pp. 82-83.
5. J. H. Hildebrand, "G. N. Lewis," *Biog. Mem. Nat. Acad. Sci., 31* (1958), 210-235.
6. W. F. Giauque, "G. N. Lewis," *Am. Phil. Soc. Yearbook 1946*, pp. 317-322.

brightest and most rapidly rising star in American physical chemistry. He obtained his Ph.D. with Richards in 1899 for his dissertation on electrochemical potentials.

Lewis's student years were boom years in the world of physical chemistry. Svante Arrhenius' ionic theory, proposed in 1887, was finally enjoying widespread use, and thermodynamics was thriving under the leadership of J. H. van't Hoff, Wilhelm Ostwald, and H. W. Nernst. Richards too was actively pursuing the opportunities in this growing science, with the cooperation of his able student. Lewis himself published elaborate theoretical papers in 1900[7] and in 1901.[8] After a year as assistant to Richards, Lewis made the customary year's pilgrimage to Europe to work with Ostwald at Leipzig and Nernst at Göttingen. He met van't Hoff and learned the pleasures of shocking the chemical establishment with his novel ideas, a taste he retained throughout his life.[9] While Ostwald did not fully accept Lewis's views, he was impressed by his young visitor: "All in all I think Dr. Lewis is a competent and learned man, who will turn in an excellent performance."[10] On returning from Europe in 1901 Lewis became an instructor at Harvard, teaching courses in electrochemistry and physical chemistry.[11]

The most exciting event of Lewis's student years was J. J. Thomson's discovery of the electron in 1896. This first glimpse into the inner structure of the atom fulfilled a long-standing expectation that chemical affinity would turn out to be electrical, the most notable prophecy having been Hermann Helmholtz' celebrated Faraday Lecture of 1881.[12] It became clear that inorganic compounds were formed simply by the transfer of single electrons. Not surprisingly, the electron transfer bond was also extended to organic compounds.

---

7. G. N. Lewis, "A New Conception of Thermal Pressure and a Theory of Solutions," *Proc. Am. Acad., 36* (1900), 145-168.
8. G. N. Lewis, "The Law of Physico-Chemical Change," *Proc. Am. Acad., 37* (1901), 49-69.
9. Letter from G. N. Lewis to T. W. Richards, Leipzig, 13 January 1901. T. W. Richards Archive, Harvard University.
10. Letter from W. Ostwald to T. W. Richards, 30 March 1901. T. W. Richards Archive, Harvard University.
11. *Harvard University Catalogue*, Division of Chemistry, 1901-1905.
12. H. Helmholtz, "Modern Developments of Faraday's Theory of Electricity," *J. Chem. Soc., 39* (1881), 277-304.

In 1901 William A. Noyes[13] (1857–1941), later professor at the University of Illinois, realized that an electron could be transferred either from atom A to atom B or from atom B to atom A, and that the resulting compounds would not be the same. He predicted that NCl_3, for example, should exist in two isomeric forms, with nitrogen carrying a charge of either +3 or −3. The same idea was taken up independently by Julius Stieglitz (1867–1937), a rising young chemist at the University of Chicago.[14] The search for the missing "electromere" of NCl_3 was pursued (fruitlessly) for over twenty years.

The discovery of the electron also provided a new rationale for the familiar polarity and periodicity of valences. The best known theory of electrovalence was proposed in 1904 by Richard Abegg (1869–1910), professor of inorganic chemistry at Breslau.[15] According to Abegg, each element had a definite maximum number of *principal valences,* either positive or negative, and also a certain number of latent *contravalences,* of opposite sign to the principal valences. The sum of these principal and latent valances was always eight (Fig. 5).

	Li	Be	B	C	N	O	F
Principal valence	+1	+2	+3	±4	−3	−2	−1
Contravalence	—	—	—	±4	+5	+6	+7

FIGURE 5. R. Abegg's table of valences and contravalences.

An electropositive atom and an electronegative atom formed a bond by the simultaneous action of one positive and one negative valence, i.e., by the transfer of one electron. If both atoms were electronegative or electropositive, a contravalence of the needed sign could be brought into play. The H-H bond was formed by one positive valence and one negative contravalence; while in $HClO_4$ chlorine used all seven positive contravalences to combine with electronegative oxygen.

13. W. A. Noyes and A. C. Lyon, "The Reaction Between Chlorine and Ammonia," *J. Am. Chem. Soc.,* 23 (1901), 460-463. On Noyes, see R. Adams, "W. A. Noyes," *Biog. Mem. Nat. Acad. Sci.,* 27 (1952), 179-208.
14. J. Stieglitz, "On Positive and Negative Halogen Ions," *J. Am. Chem. Soc.,* 23 (1901), 797-799.
15. R. Abegg, "Die Valenz und das periodische System," *Z. Anorg. Chem.,* 39 (1904), 330-380.

Abegg saw the possibility of a special nonpolar bond for compounds lacking saltlike properties. But since there was no sharp boundary between the polar and nonpolar types Abegg shrank from assuming two completely different kinds of bonds. Besides, the idea of a special "atom-affinity" unrelated to electrons struck Abegg as pure metaphysics.[16] For Abegg the only kind of chemical affinity was electrostatic attraction, and all bonds, even in nonpolar, symmetrical molecules, were electron transfer bonds. In the post-electron period, his was the majority view.

Abegg's theory also explained why the sum of principal valences and contravalences was always eight: eight was simply the number of attachment sites for electrons on the atom.[17] The positive valence number was the number of filled sites, and the negative the number of unfilled sites. Why every atom should attach just eight electrons and no more Abegg did not say. His theory was purely schematic; it was not a theory of the structure of the atom itself.

Two years before Abegg's paper appeared, a similar idea had occurred to Lewis in the more concrete form of a cubic atom. In *Valence* Lewis described the origin of the cubic atom and reproduced part of a memorandum of 28 March 1902, in which he first set his idea down. The cubic atom was first conceived as a graphic teaching device to illustrate the rule of eight:

> In the year 1902 (while I was attempting to explain to an elementary class in chemistry some of the ideas involved in the periodic law) becoming interested in the new theory of the electron, and combining this idea with those which are implied in the periodic classification, I formed an idea of the inner structure of the atom which, although it contained certain crudities, I have ever since regarded as representing essentially the arrangement of electrons in the atom.
>
> In accordance with the idea of Mendeleef, that hydrogen is the first member of a full period, I erroneously assumed helium to have a shell of eight electrons. Regarding the disposition in the neutral charge which balanced the electrons in the neutral atom, my ideas were very vague; I believe I inclined at that time toward the idea that the positive charge was also made up of discrete particles, the localization of which determined the localization of the electrons.[18]

16. For other arguments for polar bonds, see *ibid.*, pp. 345-348, 369.
17. *Ibid.*, p. 380.
18. G. N. Lewis, *Valence*, pp. 29-30.

G. N. LEWIS'S THEORY OF THE SHARED PAIR BOND

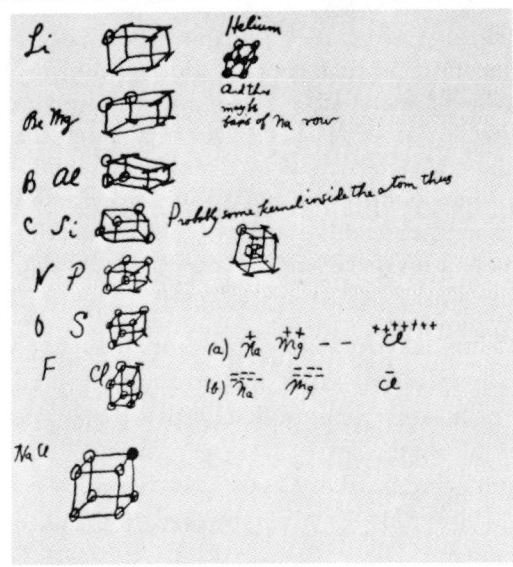

FIGURE 6. Memorandum of 1902.

The cubic atom had all the advantages of Abegg's scheme, explaining electropositive and electronegative bonds by the loss and gain of electrons, and the additional advantage of being much more concrete. The unique stability of eight electrons, which Abegg simply accepted, was for Lewis a necessary consequence of the cubic structure of the atom. Not surprisingly what began as a graphic embodiment of the empirical rule of eight soon became a serious theory of atomic structure.

For the time being, Lewis apparently did little more with his cubic atom. His first interest lay in thermodynamics, and there was little more that could be done with the idea, given the state of the art. Besides, as Lewis later confided to Robert Millikan, he had received little encouragement to pursue his speculations:

> I went from the Middle-west to study at Harvard, believing that at that time it represented the highest scientific ideals. But now I very much doubt whether either the physics or the chemistry department at that time furnished real incentive to research. In 1897 I wrote a paper on the thermodynamics of the hohlraum which was read by several members of the chemistry and physics departments. They agreed unanimously that the work was not worth doing, especially as I postulated

a pressure of light, of which they all denied the existence. They advised me strongly not to spend time on such fruitless investigations, all being entirely unaware of the similar and more successful work that Wien was then doing. A few years later I had very much the same ideas of atomic and molecular structure as I now hold, and I had a much greater desire to expound them, but I could not find a soul sufficiently interested to hear the theory. There was a great deal of research work being done at the university, but, as I see it now, the spirit of research was dead.[19]

It is not surprising that Lewis's cubic atom was not well received. American chemistry at the time was strongly empiricistic. Careful determination of atomic weights, Richards's forte, and precise measurement of physical properties were the order of the day, and an imaginative young man like Lewis was bound to feel unappreciated. During 1902-1904 Lewis published nothing, either speculative or experimental.

In 1904 Lewis asked for and was granted a leave of absence for the final two years of his appointment, and sailed for the Philippines to become superintendent of weights and measures in Manila. There he began his careful measurements of electrode potentials which occupied him for the next twenty years. In 1905 he returned to Boston to join the group of bright young physical chemists around A. A. Noyes (1866-1936) at the Massachusetts Institute of Technology.[20] The atmosphere at MIT was apparently more favorable for the "spirit of research" than it was at Harvard. It was probably this period that Lewis had in mind when he wrote that his cubic atom was "discussed freely with my colleagues and in my classes." But it was still "given no further publicity."[21]

Only a few scraps of evidence have survived bearing on Lewis's thoughts on valence and the cubic atom in 1905-1913.[22] The main problem for Lewis seems to have been the numerous elements that formed compounds with less than the maximum number of bonds,

19. Letter from G. N. Lewis to R. A. Millikan, 28 October 1919. G. N. Lewis Archive, University of California, Berkeley.
20. On A. A. Noyes, see L. Pauling, "A. A. Noyes," *Biog. Mem. Nat. Acad. Sci., 31* (1958), 322-346.
21. G. N. Lewis, *Valence,* p. 30.
22. Only three manuscripts survive in the G. N. Lewis Archive at Berkeley: one from Harvard, one from Manila, and one from MIT. A detailed analysis of this material will be published in the future.

i.e., without completion of the octet. In one note, for example, Lewis observed that "SO_2 not perfect [cube]"; yet it is a perfectly stable compound. He also noted the oxides of nitrogen, with their valence numbers from one to five. Compounds of two electronegative or two electropositive elements, such as Cl_2 or NCl_3, also gave him trouble, since with such compounds both octets could not be completed at the same time. And LiH left ". . . at least 2 electrons outside"—outside the octet of lithium, presumably. In retrospect, it was not so much the cubic atom *per se* that proved troublesome, but the basic assumption behind it—that octets could be filled only by transfer of electrons, i.e., that all bonds were polar bonds. The problems that troubled Lewis were precisely those which in 1913–1914 led him and others to adopt the dualistic view that in addition to the usual polar bond a special nonpolar bond must also exist.

3. THE POLAR ORTHODOXY AND LEWIS'S DUALISM, 1905–1913

Meanwhile, the polar theory of valence was winning greater and greater popularity, especially after the publication of J. J. Thomson's *The Corpuscular Theory of Matter* in 1907.[23] In this enormously popular book Thomson set forth a rigorous theory of atomic structure and chemical bonding based on his electron theory. His atom, the celebrated "plum-pudding" atom, consisted of a sphere of uniform positive electricity in which the negative "corpuscles" (electrons) were embedded. By classical electrodynamics Thomson calculated that such atoms were stable only if the corpuscles were arranged in concentric rings, each with a strictly defined number of corpuscles. The outer, incomplete rings were responsible for chemical bonding, since they would tend to gain or lose corpuscles to attain a more stable state.[24] Chemical bonds were formed by transfer of electrons,[25] as Abegg had simply assumed. Thomson, however, was able to back this assumption with rigorous mathematics. For example, the calculated properties for atoms having fifty-nine to sixty-seven corpuscles corresponded exactly to the properties of the first row of eight elements, from lithium to fluorine.[26]

23. J. J. Thomson, *The Corpuscular Theory of Matter* (London, 1907).
24. *Ibid.*, p. 119.
25. *Ibid.*, p. 122.
26. *Ibid.*, pp. 114-119.

Thomson recognized that a nonpolar bond could exist in theory: two equal positive spheres could overlap with a symmetrical distribution of corpuscles (Fig. 7). But only in theory: in fact, he

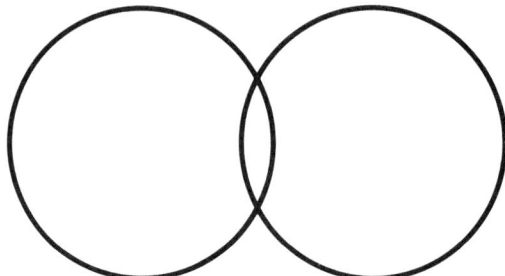

FIGURE 7. J. J. Thomson's model of a nonpolar bond as overlapping positive spheres.

argued, jostling of the molecule would upset the delicate equilibrium and result in a transfer of electrons.[27] Thomson appealed to the physical analogy of a flexible siphon between two bottles of water suspended from springs. Equilibrium was possible with an equal distribution of water, but if by jiggling, one bottle got the slightest excess, the system would rapidly reach a new, permanently stable state with one bottle full and the other dry. Likewise, though a nonpolar bond of overlapping spheres was conceivable, in reality all bonds were polar bonds:[28]

> For each valency bond established between two atoms the transference of one corpuscle from the one atom to the other has taken place. . . . This electrical process may be represented by the production of a unit tube of electric force between the two atoms. . . . In this way we can give a physical interpretation to the lines by which in graphical formulae the chemists represent the valency bonds, these lines representing the tubes of force which stretch between the atoms connected by the bond.[29]

Thomson proposed that these tubes of force be represented by vectors rather than by the nondirectional lines of structural formulas, since vectors revealed things about chemical structure that

27. *Ibid.*, pp. 121, 127-129.
28. *Ibid.*, p. 131.
29. *Ibid.*, p. 138.

were hidden in the usual symbols.[30] For example, as Noyes and Stieglitz had already pointed out, vector bonds revealed the possibility of an electronic isomerism in many compounds not yet known to have more than one form.[31]

Thomson's prestige and his display of mathematical physics were powerful arguments for the polar theory. Around 1910 a vigorous and rather self-conscious movement began, especially in America, to exploit Thomson's theory. K. G. Falk of MIT and J. M. Nelson of Columbia initiated a long series of systematic studies of the "electronic conception of valence."[32] H. S. Fry at the University of Cincinnati likewise delved into the mysteries of the vector bond.[33] By 1912 W. A. Noyes and Julius Stieglitz had become two of the most successful American chemists of the day, and their long-standing interest in the polar theory gave the new movement prestige and respectability. By 1913 or so the polar theory completely dominated chemistry, and it did until it was replaced by Lewis's theory in the early 1920's.

In 1913 the first dissenting voices were raised against the polar orthodoxy in short papers by W. C. Bray and G. Branch,[34] and by Lewis.[35] In 1912 Lewis had accepted an invitation to direct the college of chemistry of the University of California at Berkeley, which was badly in need of revitalization. Lewis was given generous financial backing and a free hand in recruiting new faculty. With him he brought a group of bright young men, eager to put new life into American chemistry. Among them were William Bray (1879–1946)[36] from MIT, Wendell Latimer (1893–1955)[37] from Kansas, and Joel Hildebrand (b. 1881) from Pennsylvania. These men all became

30. *Ibid.*, pp. 138-139.
31. *Ibid.*, pp. 132-133.
32. K. G. Falk and J. M. Nelson, "The Electron Conception of Valence," *J. Am. Chem. Soc., 32* (1910), 1637-1654, and many subsequent papers.
33. H. S. Fry, "A Critical Survey . . . ," *J. Am. Chem. Soc., 34* (1912), 664-675, and subsequent papers.
34. W. C. Bray and G. Branch, "Valence and Tautomerism," *J. Am. Chem. Soc., 35* (1913), 1440-1447.
35. G. N. Lewis, "Valence and Tautomerism," *J. Am. Chem. Soc., 35* (1913), 1448-1455.
36. See J. H. Hildebrand, "William C. Bray," *Biog. Mem. Nat. Acad. Sci., 24* (1951), 13-24.
37. See J. H. Hildebrand, "Wendell M. Latimer," *Biog. Mem. Nat. Acad. Sci., 32* (1958), 221-237.

distinguished teachers, and through their teaching and textbooks they brought American chemistry for the first time to a level with European chemistry.

The open, iconoclastic spirit of Lewis's reign has been vividly described by Hildebrand. Breadth of learning and interest were encouraged; and no one was allowed to become an entrenched and narrow specialist. At weekly department research conferences the whole range of chemistry was discussed with the utmost freedom by everyone, including the beginning students:

> The members of the department became like the Athenians who, according to the Apostle Paul, "spent their time in nothing else, but either to tell or to hear some new thing." Any one who thought he had a bright idea rushed out to try it on a colleague. Groups of two or more could be seen every day in offices, before blackboards or even in the corridors, arguing vehemently about these "brainstorms." It is doubtful whether any paper ever emerged for publication that had not run the gauntlet of such criticism. The whole department thus became far greater than the sum of its individual members.[38]

This was the "spirit of research" that Lewis missed during his years at Harvard, and in this spirit, as Bray and Branch stated,[39] their own and Lewis's papers were made public.

Bray and Branch objected that the polar theory had been extended far beyond its proper limits. Since the low dielectric constant, non-ionizability, and unreactivity of many organic compounds simply did not suggest the presence of polar bonds, Bray and Branch proposed a dualistic scheme. They distinguished two distinctly different aspects of the concept of valence—*polar* valence and *total* valence. Instead of one valence number for each atom in a molecule, they proposed two, one for the number of electrons transferred, and one for the total number of bonds, both polar and nonpolar. The polar number could still be used for inorganic compounds, while the total valence number, expressed graphically in structural formulas, would serve for most organic compounds. The vector bond could still be used to signify polar bonds, but Bray and Branch insisted that its use be strictly limited to those bonds shown experimentally to be polar.[40]

38. J. H. Hildebrand, "G. N. Lewis," p. 212.
39. W. C. Bray and G. Branch, "Valence and Tautomerism," p. 1440.
40. *Ibid.*, p. 1443.

This dualistic terminology was no mere artifice: it reflected a real difference in chemical bonding: "we have suggested that there are two distinct types of union between atoms: polar, in which an electron has passed from one atom to the other, and nonpolar, in which there is no motion of an electron."[41]

Lewis enthusiastically endorsed Bray and Branch's dualistic view and officially adopted their new terminology.[42] His only complaint was that they had not gone far enough:

> The independence of these two conceptions, valence number and polar number, is, I believe, even more complete than Bray and Branch have considered it to be. Apparently we must recognize the existence of two types of chemical combination which differ not merely in degree but in kind.[43]

In one important particular Lewis felt that Bray and Branch had not been true to the dualistic view: namely, in their adoption of the vector bond. Since electrostatic forces are radially symmetrical, the forces holding polar structures together could not be directed in space, as nonpolar bonds were:

> The nonpolar molecule, subjected to changing conditions, maintains essentially a constant arrangement of the atoms; but in the polar molecule the atoms must be regarded as moving freely from one position to another, falling now into one place, now into another, like the bits of glass in a kaleidoscope.[44]

Lewis argued that the vector bond was an illegitimate attempt to represent the nondirected polar bond as having the rigid geometry characteristic of nonpolar bonds:

> The arrow purports to show between which atoms an electron has passed, but since all electrons are alike, and presumably leave no trail behind them, we cannot say that atom A loses an electron to atom B and atom C to atom D, but only that atoms A and C have each lost an electron and atoms B and D have each gained one.[45]

41. *Ibid.*, p. 1443.
42. G. N. Lewis, "Valence and Tautomerism," p. 1448.
43. *Ibid.*, p. 1448.
44. *Ibid.*, p. 1449.
45. *Ibid.*, p. 1452.

In *Valence* he made the same point in a more striking way: "We do not think of an electron leaving a trail behind it, as a spider weaves its web, but if not, what is the significance of an arrow?"[46] Lewis's repugnance toward the vector bond thus outlived the dualistic theory from which his feeling was derived in 1913.

On one crucial point, however, both Bray and Lewis were silent: the physical nature of the nonpolar bond. Lewis defined valence number in strictly noncommittal terms as "the number of positions, or regions, or points (bond termini) on the atom at which attachment to corresponding points on other atoms occurs."[47] He could hardly have been more specific, since his model of atomic structure—the cubic atom—gave no hint how a nonpolar bond might be depicted in terms of electrons. The cubic atom was, after all, a product of the polar tradition, designed to explain the formation of positive and negative valences by the transfer of electrons. It was strictly a chemist's atom, providing no insight into the physical nature of a nonpolar bond.

Bray and Lewis did not long remain the sole advocates of the dualistic view. In 1914 and early 1915 three dualistic theories were brought to Lewis's attention. All were associated with new theories of atomic structure, and all included detailed physical pictures of the nonpolar bond. Each one was also flawed in ways that would have jarred with Lewis's dualistic viewpoint of 1913. But from each, when put into terms of the cubic atom, something remarkably like the shared pair emerges almost automatically. In this way, I believe, the shared pair bond was conceived.

4. J. J. THOMSON'S TWO ELECTRON BOND: 1914

While the apostles of polar bonds were proselytizing for Thomson's theory, Thomson himself turned apostate and decided that all bonds were not polar bonds after all. One reason for his change of mind was some unexpected results of experiments with "positive rays"—the streams of ionized fragments of molecules broken down by an electric discharge in a Crookes tube.[48] For example, if the carbon atom in CO really had a positive charge and

46. G. N. Lewis, *Valence*, p. 73.
47. G. N. Lewis, "Valence and Tautomerism," p. 1448.
48. J. J. Thomson, *Rays of Positive Electricity* (London, 1913), pp. 63-68.

the oxygen atom a negative charge, then the positive rays from CO should be made up largely of carbon ions. In fact, roughly equal numbers of carbon and oxygen ions were observed. By 1912 the evidence was strongly against the existence of "intramolecular ionization" in a variety of small molecules.

About the same time, Thomson also abandoned his plum-pudding atom. Rutherford's discovery of the nucleus in 1911 ruled out the positive sphere, and since any system of electrons circulating around a point positive charge was unstable by classical electrodynamics, Thomson was forced to conclude that the electric fields within the atom were not uniformly spread through space. By March 1913 Thomson had drawn up a new picture of the atom:[49] if the electrical forces originating in subatomic charges were confined to narrow tubes connecting the positive nucleus with the outer electrons, a stable atom could be formed.

In May 1914, Thomson published a long paper applying his new theory to the chemical bond.[50] His main point was that all chemical compounds could be divided according to their properties into two distinct groups—polar and nonpolar—and that these groups reflected the existence of two distinct types of bonds. Polar molecules of course contained electron transfer bonds;[51] but in nonpolar molecules there was no transfer of electrons, and the atoms themselves were only weakly dipolar owing to the shorter length of the atomic dipole. Thomson conceived the nonpolar bond as a tube of force stretching from an electron in one atom to the nucleus of a second.[52] An electron anchored by a tube of force to its own nucleus, as in a free atom or ion, was perfectly mobile, and was thus highly reactive. But when the tube was anchored to the nucleus of another fixed atom, the electron was fixed and immobile, i.e., not reactive. To satisfy the empirical rule of eight, Thomson assumed ad hoc that a group of eight electrons "saturated itself." For atoms with less than a complete group of eight, the condition of saturation was that every electron be anchored by a tube to an adjacent atom.[53]

49. J. J. Thomson, "On the Structure of the Atom," *Phil. Mag.*, 26 (1913), 792-799.
50. J. J. Thomson, "The Forces between Atoms and Chemical Affinity," *Phil. Mag.*, 27 (1914), 757-789.
51. *Ibid.*, p. 760.
52. *Ibid.*, pp. 780-789.
53. *Ibid.*, pp. 780-781. See also p. 765.

But for nonpolar bonds to be strictly nondipolar, one further condition had to be met:

> When the atoms are electrically neutral . . . for each tube of force which passes out of an atom, another must come in; and thus each atom containing n corpuscles must be the origin of n tubes going to other atoms and also the termination of n tubes coming from other atoms.[54]

This modest condition introduced a remarkable novelty. For every nonpolar bond between two atoms, *two* physical bonds, i.e., two tubes of force, were postulated to exist; *two* electrons, one from each atom, were involved in each single bond. For this reason Thomson proposed that the number of bonds in structural formulas of nonpolar compounds be doubled[55] (Fig. 8). Polar bonds could continue to be represented by the single vector bond of Thomson's earlier

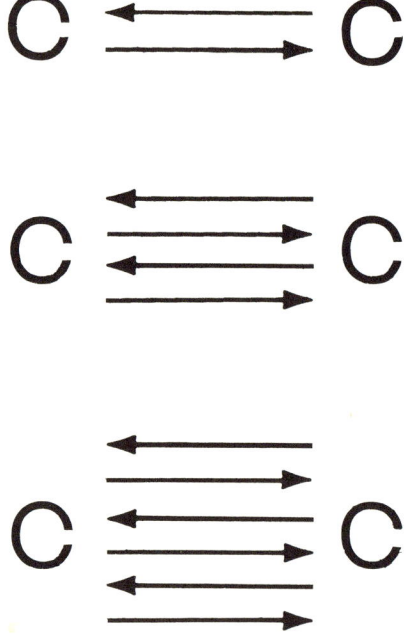

FIGURE 8. J. J. Thomson's two tube nonpolar bond in ethane, ethylene, and acetylene.

54. *Ibid.*, p. 782.
55. *Ibid.*, pp. 783-784.

theory, which represented the single molecular dipole.[56] The difference between the one and two line bonds was no mere formalism: for Thomson the polar and nonpolar bonds differed not in degree but in kind.

Thomson also distinguished between two kinds of nonpolar bond, corresponding to the electropositive and electronegative bonds of the polar theory. The normal bond between an outer electron in one atom and the nucleus in another Thomson called a *positive* bond. But a nearly filled octet—say with seven electrons—could also be formally regarded as a full octet plus one positive charge in the outer octet. The fictive octet saturated itself, while the fictive positive position charge was saturated by a tube of force to an electron in a second atom. This *negative* bond, being shorter, must be stronger than the positive bond[57] (Fig. 9). An atom with n electrons could thus exer-

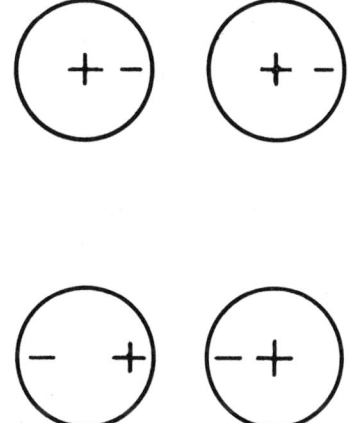

FIGURE 9. J. J. Thomson's positive and negative nonpolar bonds (top and bottom, respectively).

cise n positive valences or $8-n$ negative ones, as Abegg's theory required. Thomson thus exercised considerable ingenuity to include this traditional duality of the polar theory in his own theory of the nonpolar bond: this was one tie with his past he was unable to break.

56. *Ibid.*, pp. 787-788.
57. *Ibid.*, pp. 786-787.

5. W. C. ARSEM'S SHARED ELECTRON BOND: 1914

In August 1914 a paper appeared by William C. Arsem, a physical chemist at the General Electric Company in Schenectady,[58] which suggested that the nonpolar bond consisted of *one* electron *shared* between two atoms. Arsem's single contribution to valence theory, like Lewis's, was the product of a youthful inspiration. Arsem had been a student of Bray's at MIT, and it was then, about ten years earlier, that the idea of a shared electron bond came to him. He had been diffident at the time about publishing such a novel idea, but the appearance of similar ideas in print led him to make his views public ". . . with a feeling that they are in harmony with the present trend of scientific speculation."[59] A trend may not have begun, but a spirit of change was evidently in the air.

Like Bray and Lewis, Arsem could not believe that all molecules contained only polar bonds; symmetrical molecules such as H_2 must have a bond that did not involve actual electron transfer. In Arsem's view the nonpolar bond consisted of a single electron, as in the polar theory, but this one electron oscillated between the two atoms. The nonpolar bond was not a dipolar force between atoms, but a coextension of atoms: the oscillating electron was an intrinsic part of both atoms at the same time[60] (Fig. 10). From Arsem's deceptively simple modification of the one electron bond some very remarkable conclusions followed. Obviously the H_2 molecule could not dissociate into

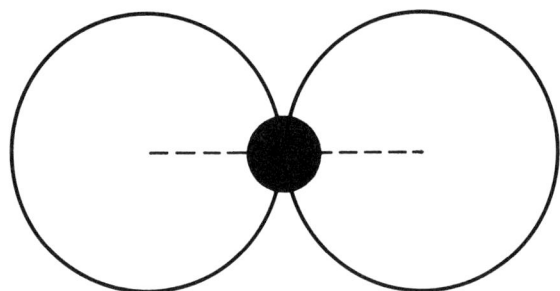

FIGURE 10. W. C. Arsem's oscillating single electron nonpolar bond.

58. W. C. Arsem, "A Theory of Valency and Molecular Structure," *J. Am. Chem. Soc., 36* (1914), 1655-1675.
59. *Ibid.,* p. 1656.
60. *Ibid.,* p. 1659.

free atoms, as it had always been assumed to do, but only into ions. The "hydrogen atom," indeed atoms in general, were fictitious entities, which had no existence outside of molecules—the real units of matter. Moreover, since a neutral H_2 molecule consisted of two H^+ ions and one electron, the electronic charge must be twice the value calculated from ionic theory. These bold ideas, which challenged the foundations of chemical theory, unfortunately obscured the sound idea that electrons could be shared. In retrospect it is clear that Arsem erred only in not breaking completely away from the one electron polar paradigm.

Lewis never referred to Arsem, and we do not know if he was influenced by Arsem's shared electron bond. But thanks to Dr. Ludwig Rosenstein, we know that Lewis was made acquainted with Arsem's paper, and in a most interesting way.[61] At the time Rosenstein was a young instructor in Lewis's department. He had studied with Lewis at MIT and had joined the migration to Berkeley, where he received his Ph.D. with Bray in 1914. It was also Bray who drew his attention to Arsem's paper. Arsem's "half-electron bond" made a profound impression on Rosenstein at the time, but he also recognized that Arsem's idea was seriously flawed. Assuming that the hydrogen atom had one electron, which almost everyone believed it did, then the H_2 molecule must have a *two* electron bond. Rosenstein first discussed this exciting idea with Thomas Hine, an advanced graduate student, who agreed that Arsem was wrong in assuming that free neutral atoms could not exist. The fact that sodium metal vapor was nonconducting left no doubt that atoms, not ions, were present. A two electron bond seemed shocking at first, since two negative charges must strongly repel each other, but it seemed inescapable.

Rosenstein and Hine then went to Lewis. When they arrived at his office, Lewis was dictating a paper on thermodynamics to Merle Randall and was too engrossed to pay any attention to half electron or two electron bonds. Rosenstein and Hine continued to discuss these ideas, but there was no further communication with Lewis.

61. I am grateful to Dr. L. Rosenstein for calling my attention to Arsem's paper and for the recollections recounted here. It is only fair to him to repeat his judicious warning to me: even when his memories were most clear he did not fully trust them.

However, Rosenstein recalls that some months later at a departmental research conference, Lewis presented his theory of the cubic atom and drew upon some "old notes," undoubtedly his memorandum of March 1902. Rosenstein does not recall that Lewis mentioned an electron pair bond. Whether or not Lewis's revival of the cubic atom was connected with Arsem's paper or his abortive discussions with Rosenstein is not known. But this event was another portent of coming changes in valence theory; the indications were all pointing toward a two electron nonpolar bond.

6. ALFRED PARSON AND THE CUBIC OCTET: 1914–1915

The most likely cause of Lewis's renewed interest in the cubic atom in the fall of 1914 was an unpublished paper by a visiting graduate student, Alfred Parson, in which appeared both a two electron bond and a *cubic octet*.

Alfred L. Parson was born in Tucknawij, India in 1889, the son of Methodist missionary parents.[62] While studying chemistry as an undergraduate at Oxford, where he obtained his B.S. in 1911, Parson conceived the idea that the force responsible for chemical bonding was not electrical but magnetic. In this he followed the French physicist P. Langevin (1872–1946), who in 1905 had proposed that the electron was a zero resistance electric circuit, i.e., an electromagnet.[63] Parson's enthusiasm for a magnetic bond was shared by his tutor, Herbert B. Baker (1862–1935).[64] Parson read an early version of his "magneton theory" to the Alembic Club in 1912. He elaborated his theory at Harvard, where he spent the academic year 1913–1914 as an Austin Fellow in chemistry. There he had the enthusiastic help of David Webster (b. 1888), a young instructor of physics who provided him with the background in physics he himself lacked.[65]

62. A. L. Parson, *The Realities of these Emergent Times, Including the Theomony and Cosmonomy* (Oxford, privately printed, 1964), p. 27.
63. P. Langevin, "Magnétisme et Théorie des Electrons," *Annales de Chimie*, 5 (1905), 70-122. Langevin did not apply his idea to chemical bonding.
64. A. L. Parson, letter to the author, 8 February 1969. On Baker see J. C. Philip, "Herbert B. Baker," *J. Chem. Soc., 138* (1935), 1893-1896. It is a pleasure to thank Mr. Parson for his patient and helpful replies to the author's inquiries.
65. A. L. Parson, letter to the author, 8 February 1969.

In 1914 it was arranged for Parson to spend a year with Lewis at Berkeley. From there, Parson sent his manuscript to the *Journal of the American Chemical Society*. Owing to its length and its purely theoretical nature, it was rejected. But the editor, W. A. Noyes, who was still keenly interested in the theory of valence, saw its value. He wrote to the Smithsonian Institution, which published longer monographs, that he was "strongly of the opinion that the paper is fully worthy of publication."[66] Parson mailed the manuscript from New York on 26 May, on his way back to England and the war.[67] It was enthusiastically refereed by David Webster[68] and appeared in print on 29 November 1915.[69]

Parson intended to exploit his theory experimentally[70] and built an electrometer in the hope of detecting molecular dipoles.[71] But the war intervened; Parson enlisted, suffered shell-shock in the trenches,[72] and never returned to experimental science.

Parson's atom consisted of a Thomsonian positive sphere in which the magnetons were arranged not in concentric rings but at the corners of cubic octets. The magneton itself was a rapidly revolving circular band of electricity, and the magnetic moment it generated was the source of chemical affinity and of the stability of the octet. Parson constructed a model of the cubic octet (Fig. 11) in which the eight electromagnets did indeed arrange themselves in a number of stable configurations.[73] Parson tried to justify the unique stability of the cubic octet mathematically, but as he admitted he had ultimately to appeal to its symmetry and its serviceability in explaining the rule of eight.[74] Parson's atom was derived from the pre-Rutherford physics of his undergraduate days. Parson argued at

66. Letter from W. A. Noyes to C. D. Wolcott, 20 May 1915. Smithsonian Institution Archives.
67. Letter from A. L. Parson to C. D. Wolcott, 26 May 1915. Smithsonian Institution Archives.
68. Referees' reports, Smithsonian Institution Archives.
69. A. L. Parson, "A Magneton Theory of the Structure of the Atom," *Smithsonian Miscellaneous Collections, 65* (1915), 1-80.
70. *Ibid.*, p. 80.
71. A. L. Parson, "A Highly Sensitive Electrometer," *Proc. Nat. Acad. Sci., 1* (1915), 400-401. Letter from A. L. Parson to the author, 27 May 1969.
72. A. L. Parson, *Realities*, p. 34.
73. A. L. Parson, "A Magneton Theory," plates following p. 18.
74. *Ibid.*, pp. 18, 19, 22.

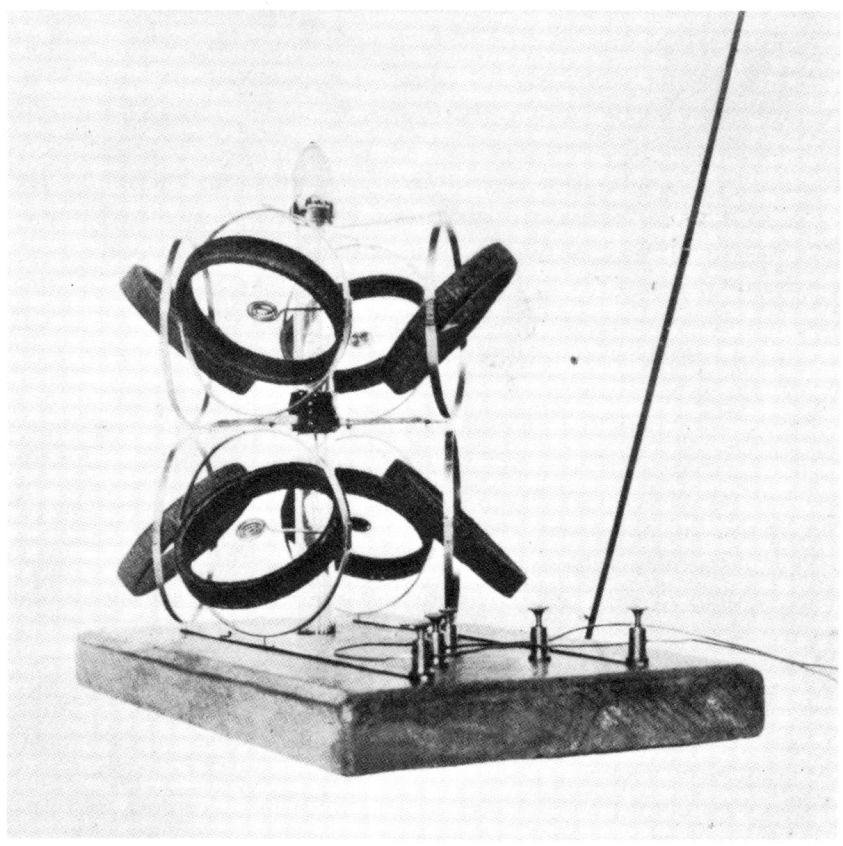

FIGURE 11. A. L. Parson's model of the magneton and the cubic octet.

length against Rutherford's nuclear theory, which by 1914 was widely accepted. He also believed, as Lewis had in 1902, that helium was the last of a row of eight elements, and he disputed Moseley's highly acclaimed work of 1914 that showed helium to have two not eight electrons.[75] In short, Parson's atom and his uneasy physics retained too much of the plum-pudding atom, which Thomson himself had discarded some years earlier.

But the application of Parson's atom to chemical bonding was highly novel and suggestive.[76] Like Bray and Lewis, Parson saw the necessity of a dualistic view: the polar theory failed to explain the

75. *Ibid.*, pp. 2-3, 6-7, 25-28.
76. *Ibid.*, pp. 28-34.

simplest molecule, H_2, and in more complex organic molecules it failed to assign polarities to each bond without gross inconsistencies. However, magneton pairing provided Parson with a mechanism of bonding that did not involve transfer of electrons and which was ". . . as simple and ready to hand as the stroke that is used in a structural formula to represent its action."[77] He thus postulated two distinct kinds of "chemical action." Atoms of electronegative elements—such as chlorine, whose nearly filled octet is easily completed by transfer of magnetons—formed bonds by magneton transfer. This tendency Parson called *negative action*. Atoms of electropositive elements—such as hydrogen, which has only a few magnetons in its octet—could not be filled without generating large electrostatic charges. Bonds were formed instead by pairing of individual magnetons. This bonding Parson called *positive action*; it was this action that was responsible for the formation of nonpolar bonds.

It is important to note that Parson's dualism is based less on the distinction between polar and nonpolar bonds, as Lewis's dualism was, than on the mechanistic distinction between completion of the octet and pairing of individual magnetons. Parson's ideas were in this respect much closer to Abegg's traditional electrochemical duality of positive and negative valences. Thus while Lewis was limited to two kinds of bond, Parson was led to the logical conclusion that from two types of chemical action *three* distinctly different kinds of bond could be formed.[78]

First, two atoms both acting positively formed a *positive bond*, as in H-H. But since each hydrogen atom had one magneton, the H-H bond had to consist of a pair of magnetons (Fig. 12). The objection that two electrons would repel each other was no obstacle, for two

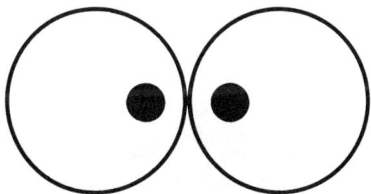

FIGURE 12. A. L. Parson's two electron positive bond in H_2.

77. *Ibid.*, p. 14.
78. *Ibid.*, p. 29.

magnetons would readily form a stable pair. This positive bond was thus a nonpolar bond.

Second, from one atom with positive and one with negative action a *neutral bond* was formed, as in H-Cl. Despite its name, this bond was a polar bond, formed by the transfer of one electron (Fig. 13).

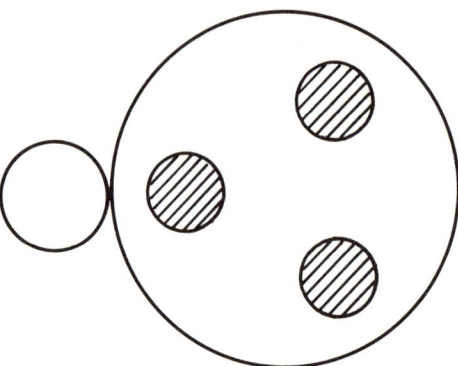

FIGURE 13. A. L. Parson's neutral bond in HCl.

Finally, two atoms both acting negatively formed a *negative bond,* as in Cl_2. The negative bond required further explanation, since it is logically impossible for two atoms with incomplete octets to complete them at the same time. Parson proposed that the magnetons from both atoms formed a "mobile group," which oscillated between the atoms, forming a full octet now in one atom and now in the

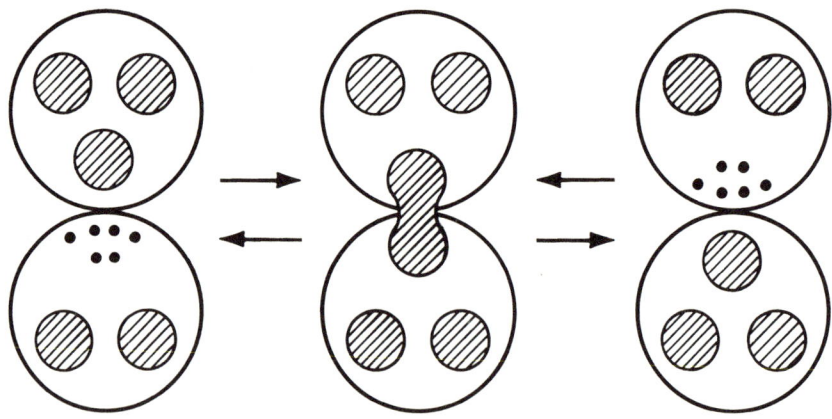

FIGURE 14. A. L. Parson's oscillating negative bond in Cl_2.

other[79] (Fig. 14). Since free magnetons would certainly disrupt the completed octets, Parson represented the leftover magnetons in negative bonds as the more stable *free pairs,* which he denoted by a carat.[80] Thus O_2 with its twelve mobile magnetons had two free pairs; N_2 with its group of ten had only one.[81] Parson recognized that positive and negative chemical actions were extreme types and that in a row of elements the balance between these opposing tendencies must shift gradually:

> This gradation in the tendency to form the group of eight leads us to the conclusion that there must be . . . a kind of tautomerism or dynamical equilibrium between the two possible modes of union, as follows:

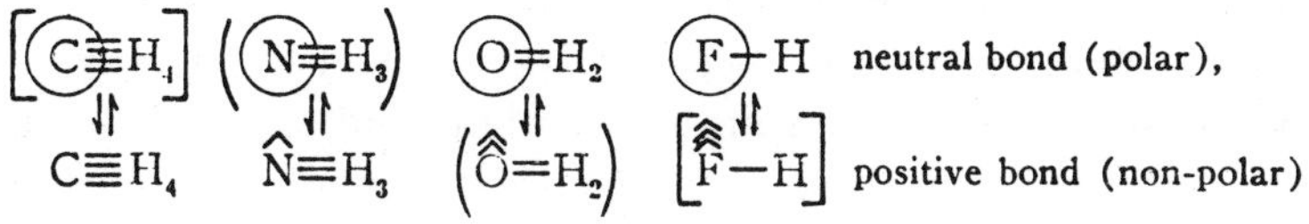

$$\begin{array}{cccc} [C\equiv H_4] & (N\equiv H_3) & (O\equiv H_2) & (F-H) \quad \text{neutral bond (polar)}, \\ \Updownarrow & \Updownarrow & \Updownarrow & \Updownarrow \\ C\equiv H_4 & \hat{N}\equiv H_3 & (\hat{O}=H_2) & [\hat{F}-H] \quad \text{positive bond (non-polar)} \end{array}$$

FIGURE 15. A. L. Parson's tautomeric pairs of polar and nonpolar forms. Parentheses and brackets indicate minor and trace components, respectively.

> the proportion of polarized molecules increasing regularly from CH_4, where it is very small, to HF, in which it greatly predominates. In view of the incessant vibrations of all molecules, this is mechanically a more likely condition than the statical one in which the carbon atom just does not and the nitrogen atom just does succeed in forming the group of eight.[82]

Bray and Branch had also considered compounds such as H_2SO_4 and Cl_2 as mixtures of polar and nonpolar forms, and they noted that Lewis had made similar suggestions himself.[83] But Bray and Lewis saw nonpolar bonds as mere lines; they could not count electrons. Parson's more detailed picture of the chemical bond enabled him to see the hidden implications in the dualistic view: he could account for all the electrons in a molecule in octets, in two electron bonds, and, most important, in free pairs.

In his 1916 paper Lewis noted that though Parson's monograph had just been published, he had "had an opportunity of looking it

---

79. *Ibid.,* pp. 29-30.
80. *Ibid.,* pp. 30, 42.
81. *Ibid.,* pp. 29-30.
82. *Ibid.,* pp. 32-33.
83. W. C. Bray and G. Branch, "Valence and Tautomerism," p. 1447.

over with the author over a year ago,"[84] that is, about September 1915. Parson recalls having had only two interviews with Lewis about his theory, one at the beginning of his stay (probably the one Lewis referred to) and one shortly before he left in May 1915. He does not recall that there was any detailed discussion of his theory, except that Lewis drew a cubic atom and remarked that "I once had the idea of a cube corresponding to the octave law."[85] Since Parson's manuscript amounted to eighty printed pages, it is quite conceivable that Lewis never read it closely. There must have been some discussion, however, since Lewis later confessed that he was partly responsible for the idea that the magneton constituted a unit magnetic force.[86] But besides these two meetings there was apparently little communication between the two. Parson was a shy, solitary person, who worked mostly at night and avoided discussing his ideas with his colleagues.[87] He declined to work on the research problem Lewis proposed, and Lewis allowed him to pursue his own interests.[88] Lewis could be tolerant of eccentric talent, perhaps because of his own experiences during his student years. In 1902 Lewis was the obscure young man who could find no interest among established men for his novel views. In 1914 he was an established figure, and to see his own neglected cubic atom being elaborately readied for print must have seemed like an echo from his own past. It was probably Parson's cubic octet that caused Lewis to exhume his eleven year old notes on the cubic atom to display at the department seminar.

7. THE ORIGIN OF THE SHARED PAIR: 1915

There has been remarkably little interest among chemists in the origin of the shared pair. The few who have thought about it seem to have adopted the view that the shared pair was induced from two empirical facts: first, Moseley's demonstration that helium, an inert gas, has two electrons, and second, the fact that with a few exceptions

84. G. N. Lewis, "The Atom and the Molecule," p. 773.
85. A. L. Parson, letter to the author, 8 February 1969.
86. G. N. Lewis, *Valence,* p. 32.
87. L. Rosenstein, E. Q. Adams, and J. H. Hildebrand, personal communications.
88. A. L. Parson, letter to the author, 27 May 1969.

the number of electrons in all compounds is even. The exceptions, which Lewis called "odd molecules," proved the rule, for they are unstable, reactive, and tend to form compounds by pairing.[89] Thus in 1928 W. H. Rodebush casually assumed that Lewis derived the shared pair from these empirical facts.[90] Helium and the odd molecules were indeed Lewis's most important pieces of evidence for the reality of the shared pair. But there is nothing in the bare facts to suggest the rich meaning of the shared pair. Rodebush's empiricistic view reveals only what chemists at the time thought their science ought to be like.

It seems fairly clear that Lewis derived the shared pair in some way from the rich and suggestive speculations of Thomson, perhaps Arsem and Rosenstein, and Parson on the nonpolar bond. The similarity between Lewis's shared pair and Thomson's two tube bond or Parson's paired magneton bond is indeed striking. Lewis avoided explicitly mentioning Thomson's two tube bond,[91] and in the historical section of *Valence* he mentioned only Thomson's overlapping spheres.[92] There he also noted Parson's H-H bond and his oscillating bond, but not the free pair. Very likely Lewis feared that these ideas would be taken as a closer precedent to his own than they probably were in fact. Indeed, the issue of priority might well have arisen but for the general indifference of most chemists to the shared pair, the intervention of the war for Lewis and for Parson, and the fact that the postwar interests of Lewis and Thomson did not overlap. But in retrospect it is clear that the origin of the shared pair is an event far too complex to be dealt with by the question of simple priority.

What Lewis found in Thomson's and Parson's papers on the nonpolar bond was a tantalizing mixture of suggestive ideas and obvious flaws. The question is not what answers he found but what opportunities. We too must look at Thomson's and Parson's models with the eyes of the dualistic Lewis of 1913, keeping in mind how they might be improved by use of the cubic atom. Lewis certainly

---

89. G. N. Lewis, "The Atom and the Molecule," pp. 770-771, 774.
90. W. H. Rodebush, "The Electron Theory of Valence," *Chem. Rev., 5* (1928), 513.
91. G. N. Lewis, "The Atom and the Molecule," pp. 771-772.
92. G. N. Lewis, *Valence,* p. 74.

approved of Thomson's and Parson's dualistic views. As he noted in 1916, Thomson's "extremely interesting" paper presented the same dualistic scheme he himself had proposed in 1913. Lewis must have been all the more keenly interested because both Thomson's two tube bond and Parson's oscillating octet were strikingly more detailed and sophisticated than Lewis's own picture of the nonpolar bond, which in 1913 was still a simple line.

In other ways, however, Lewis probably did not go along with Thomson's or Parson's theories. To the strict dualist, both theories retained aspects of the purely polar orthodoxy. For example, Thomson represented his tube of force and his polar bonds as vectors—Lewis's *bête noire*. Thomson's conception of positive and negative nonpolar bonds likewise smacks of the Falk and Nelson scheme. It may well have seemed to Lewis that Thomson, even more than Bray and Branch, had failed to keep the two types of bond completely separate and distinct. Likewise Parson's positive and negative actions were strikingly reminiscent of Abegg's electrochemical duality, and like Thomson's theory Parson's too contained positive and negative nonpolar bonds. It is clear in retrospect that both Thomson and Parson had stayed too close to the polar orthodoxy.

Lewis was thus presented with several suggestive but only partly successful pictures of the nonpolar bond. Given Parson's cubic octet and Lewis's renewed interest in his own cubic atom, it would be astonishing if Lewis had not applied his cubic atom to Thomson's

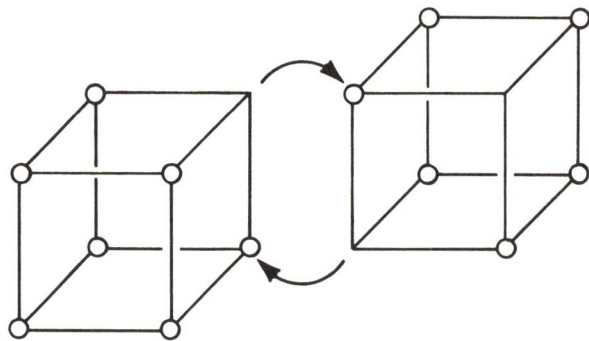

FIGURE 16. Possible representation of Thomson's two tube bond in terms of cubic atoms.

two tube bond and Parson's oscillating octet. Moreover, reconstructing the obvious ways he might have done so leads almost automatically to a picture remarkably similar to the shared pair bond. This striking coincidence, as well as circumstantial evidence, has led me to believe that Lewis did in fact conceive the shared pair while trying to fit the imperfect models of Thomson or Parson to his cubic atom model. Thus when Thomson's two tube bond is represented in cubic atoms it becomes quite clear that there cannot be two kinds of nonpolar bond (Fig. 16). In the abstract, Thomson's fictive octet and fictive positive charge seem logical, but in terms of concrete cubic atoms they are revealed as physical absurdities. Thomson's positive and negative bonds are revealed to be identical. It also seems a simple step to have two cubic atoms share an edge, thus eliminating altogether the need for vectors and tubes of force.

Thomson's dipolar atom was a physicist's atom, designed for mathematical treatment of the chemical bond in terms of dipoles and tubes of force. But it could not provide a full and concrete picture of an atom in a molecule. In contrast, Lewis's cubic atom was a chemist's atom; it said nothing about the physics of the nonpolar bond, but it had the advantage of providing a highly concrete picture of the location in space of all the bonds and electrons in an atom. Both atom models are rich in unexpected implications, and the advantages of the one make up for what the other lacks.

Parson's atom was also fundamentally a chemist's atom. But whereas Lewis's octets are arranged concentrically around the nucleus, Parson's cubic octets drift freely in the positive sphere. Parson's atom is not a cubic atom; it does not locate the electrons at definite points in space. Thus Parson's picture of the negative bond gives no sense of what an oscillating bond might look like in terms of three-dimensional atoms. (Parson did represent Cl_2 as a cross section of the cubic octet, but the result is not at all revealing.[93]) In contrast, the negative bond lends itself most remarkably to representation in terms of the cubic atoms; the same picture emerges as the one from the two tube bond (Fig. 16). It is obvious from the picture that the "oscillation" of the octet need not involve the movement of the whole octet, but only the motion of two electrons. The

---

93. A. L. Parson, "A Magneton Theory," p. 30.

breakup of the octet is a logical fiction. Moreover, it is no great leap to represent the shared "mobile group" of fourteen electrons as two cubes with a shared edge. But then the "oscillation" vanishes; clearly this nonpolar bond need not be a transitory state, but a nonpolar bond in which both octets are simultaneously filled—a feat Parson stated to be logically impossible. Parson's "tautomeric" forms of NH_3 etc. are likewise dramatically transformed by representation in cubic atoms. The octet is not dispersed in the nonpolar form; the free pairs and two electron bonds constitute the four pairs of the octet just as in the polar form (Fig. 17). The polar and nonpolar

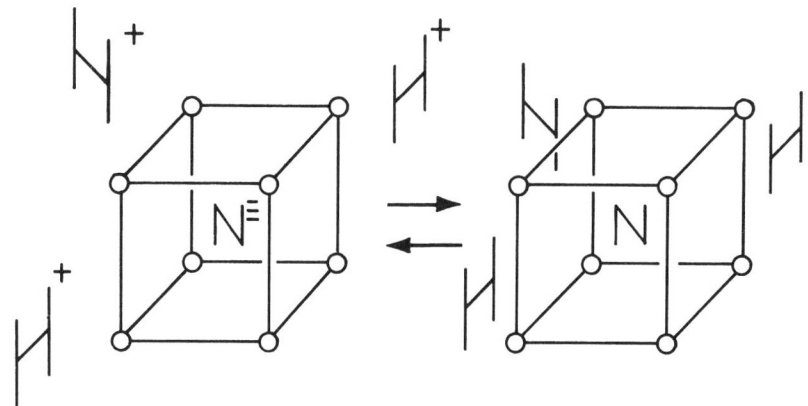

FIGURE 17. Possible representation of Parson's tautomeric pairs in terms of cubic atoms.

forms thus begin to seem different not in kind but only in degree. Here perhaps is the seed of Lewis's repugnance toward his polar-nonpolar dualism.

Thus when Thomson's and Parson's nonpolar bonds are drawn out in concrete cubic atoms the complexities resulting from their lingering association with the polar orthodoxy fall away, and a simpler, more unified view emerges almost automatically. The pictures themselves suggest the theory implicit in them.

Illuminated by these thought experiments, several quasi-historical statements that Lewis later made denying his former dualism begin to make sense:

> All of these excursions into the theory of valence seemed to lead but to an impasse. Thus the . . . polar theory, while offering an adequate explanation of the nature of extremely polar substances . . . proved

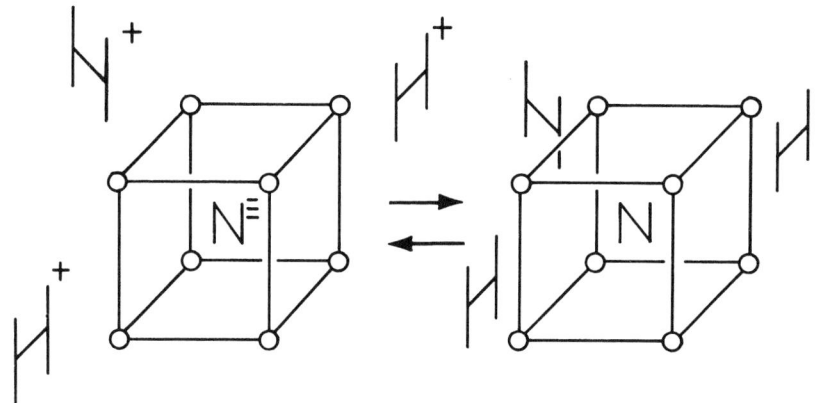

incompetent to explain the behaviour of the relatively non-polar compounds, especially those of organic chemistry. On the other hand, the pure structural theory gave complete satisfaction in the interpretation of the chief facts of organic chemistry, but seemed little qualified to account for those phenomena of a highly polar type. . . . Finally the suggestion of two entirely distinct kinds of chemical union, one for polar and the other for non-polar compounds, was repugnant to that chemical instinct which leads so irresistibly to the belief that all types of chemical union are essentially one and the same. Already, however, there were some hints of the way out of this perplexing quandary. If the properties of substances might not be explained by the mere assumption of charged atoms, might they not be explicable if we should no longer regard the atom as a unit, but rather if we might ascertain where the charge or charges resided within the atom itself.[94]

Since the only atom model that did locate the electrons in space was Lewis's cubic atom, it appears that Lewis was hinting that the cubic atom provided the key to the conception of the shared pair.

Or again, in a passage explicitly concerning his 1902 memorandum on the cubic atom Lewis wrote:

These hypotheses regarding the arrangement of electrons in the atom, while they were discussed freely with my colleagues and in my classes, were given no further publicity. Indeed, while this theory of structure seemed to offer a remarkably simple and satisfactory explanation of the process which occurs when sodium combines with chlorine to form sodium chloride, it did not seem to explain chemical combinations of a less polar type, such as occur in the hydrocarbons.

Yet I could not bring myself to believe in two distinct kinds of chemical union. It seemed rather that the union of sodium and chlorine and the union of hydrogen and carbon must represent extreme types of a method of combination which ultimately would be found to be common to all kinds of compounds. However, it was many years before I found it possible to reconcile this idea entirely with the idea of the cubical atom.[95]

Lewis certainly did not show any repugnance toward the dualistic view in 1913, as he claims here. He was obviously embarrassed about his radical change of tune; he alluded to it only once, and did his best to smooth it over.[96] But though there might be two different

94. Lewis, *Valence*, pp. 73-74.
95. *Ibid.*, p. 30.
96. *Ibid.*, p. 71.

kinds of bond, there was surely only one kind of atom, and Lewis probably did believe in 1913 that the nonpolar as well as the polar bond would be explained by the cubic atom. In the new context of 1923, when the dualistic hypothesis was the most prominent difference between his theory and the theories of Parson and Thomson, it might well have seemed to Lewis that his belief in the cubic atom had been a disbelief in his former dualism. But in fact the cubic atom is the one continuous thread from his electrochemical theory of 1902 through his dualistic theory of 1913 to the shared pair of 1915.

Thus Lewis's shared pair bond and his abandonment of dualism were probably the somewhat unexpected results of applying the cubic atom to Thomson's or Parson's pictures of the nonpolar bond. Lewis probably had hoped merely to refine his dualistic view. But the picture that emerged, almost automatically and without other theoretical groundwork, was so dramatic an improvement that it would not be surprising if he saw with repugnance the dualism with which he had begun. This is as far as the circumstantial evidence can take us in our historical reconstruction. Was Thomson's two tube bond or Parson's two electron bond the key? Or was it Thomson's overlapping spheres or Arsem's shared electron bond? There are too many good possibilities and no grounds for choice. But the very richness of opportunities open to Lewis makes it all the more probable that from one or more of these sources the shared pair bond emerged.

ACKNOWLEDGMENTS

This paper was prepared while the author was a Macy Fellow in the Department of History of Science at Harvard University. It is a pleasure to acknowledge the generous and timely support of the Macy Foundation.

It is also a pleasure to thank Dr. R. N. Lewis and Dr. E. S. Lewis for their interest in my study of their father's work, Mrs. Mary Lewis for permitting me to see her husband's unpublished papers, and Dean Harold S. Johnston for his invitation to visit the Lewis Archive.

Finally I wish to thank Mr. Alfred Parson and Dr. E. Q. Adams for their reminiscences and, especially, Dr. Ludwig Rosenstein for several rewarding afternoons discussing Lewis and the shared pair.

LIBRARY OF DAVIDSON COLLEGE

Books on re